Students:
Looking to improve your grades?

SAGE edge™

SAGE
Premium
Video

BOOST COMPREHENSION. BOLSTER ANALYSIS.

- SAGE Premium Video **EXCLUSIVELY CURATED FOR THIS TEXT**
- **BRIDGES BOOK CONTENT** with application & critical thinking
- Includes short, auto-graded quizzes that **DIRECTLY FEED TO YOUR LMS GRADEBOOK**
- Premium content is **ADA COMPLIANT WITH TRANSCRIPTS**
- Comprehensive media guide to help you **QUICKLY SELECT MEANINGFUL VIDEO** tied to your course objectives

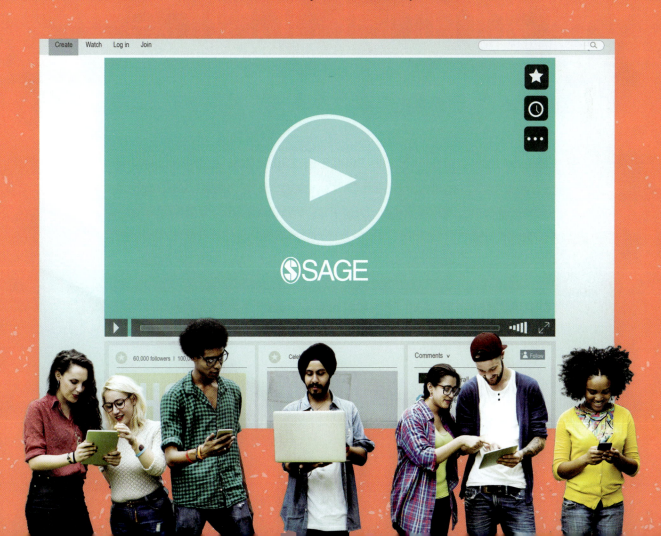

SAGE Criminology & Criminal Justice:
Our Story

Believing passionately in the **POWER OF EDUCATION** to transform the criminal justice system, **SAGE Criminology & Criminal Justice** offers arresting print and digital content that **UNLOCKS THE POTENTIAL** of students and instructors. With an extensive list written by renowned scholars and practitioners, we are a **RELIABLE PARTNER** in helping you bring an innovative approach to the classroom. Our focus on **CRITICAL THINKING AND APPLICATION** across the curriculum will help you prepare the next generation of criminal justice professionals.

A BRIEF INTRODUCTION TO CORRECTIONS

I dedicate this book to my wife, Gina, and our children, Tiffany, Ronnie, and Danny. I appreciate their support and understanding.

I dedicate this text to the Eastham UC. Through thick and thin, I could always count on you.

I dedicate this text to my brother, Guy Hanser, with the Texas Department of Criminal Justice—Institutional Division. I am proud of you and the work that you do.

Lastly but most importantly, I also dedicate this text to all the men and women who work, have worked, and will eventually work in the field of corrections, whether institutional or community-based. Your dedication to public safety and fair-minded actions under stressful circumstances are appreciated. All of us are depending on you.

A BRIEF INTRODUCTION TO CORRECTIONS

Robert D. Hanser
University of Louisiana at Monroe

Los Angeles | London | New Delhi
Singapore | Washington DC | Melbourne

FOR INFORMATION:

SAGE Publications, Inc.
2455 Teller Road
Thousand Oaks, California 91320
E-mail: order@sagepub.com

SAGE Publications Ltd.
1 Oliver's Yard
55 City Road
London EC1Y 1SP
United Kingdom

SAGE Publications India Pvt. Ltd.
B 1/I 1 Mohan Cooperative Industrial Area
Mathura Road, New Delhi 110 044
India

SAGE Publications Asia-Pacific Pte. Ltd
18 Cross Street #10-10/11/12
China Square Central
Singapore 048423

Acquisitions Editor: Jessica Miller
Content Development Editor: Laura Kearns
Editorial Assistant: Sarah Manheim
Production Editor: Laura Barrett
Copy Editor: Diane DiMura
Typesetter: C&M Digitals (P) Ltd.
Proofreader: Sarah J. Duffy
Indexer: Amy Murphy
Cover Designer: Scott Van Atta
Marketing Manager: Jillian Ragusa

Printed in Canada

Library of Congress Cataloging-in-Publication Data

Names: Hanser, Robert D., author.

Title: A brief introduction to corrections / Robert D. Hanser.
Other titles: Introduction to corrections

Description: Third Edition. | Thousand Oaks : SAGE Publishing, 2020. | Revised edition of the author's Introduction to corrections, [2017] | Includes bibliographical references and index. |

Identifiers: LCCN 2019035013 | ISBN 9781544382401 (paperback) | ISBN 9781071801819 (loose-leaf) | ISBN 9781544398112 (epub) | ISBN 9781544398129 (epub) | ISBN 9781544398105 (pdf)

Subjects: LCSH: Community-based corrections.

Classification: LCC HV8665 .H36 2020 | DDC 364.6—dc23

LC record available at https://lccn.loc.gov/2019035013

This book is printed on acid-free paper.

20 21 22 23 24 10 9 8 7 6 5 4 3 2 1

BRIEF CONTENTS

DETAILED CONTENTS

©iStockphoto.com/f8grapher

©iStockphoto.com/TriggerPhoto

©iStockphoto.com/utah778

SAUL LOEB/AFP/Getty Images

Arne Dedert/dpa/picture-alliance/Newscom

CHAPTER 5: Probation and Intermediate Sanctions   112

©iStockphoto.com/ Gatsi

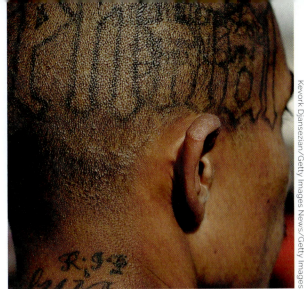

Kevork Djansezian/Getty Images News/Getty Images

Andrew Lichtenstein/Corbis Premium Historical/Getty Images

Stockbyte/Thinkstock

REUTERS/Lucy Nicholson

Jim West imageBROKER/Newscom

AP Photo/Houston Chronicle, Mayra Beltran

AP Photo/Paul Sancya

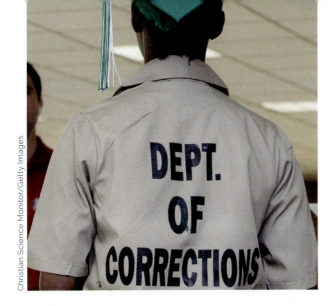

Christian Science Monitor/Getty Images

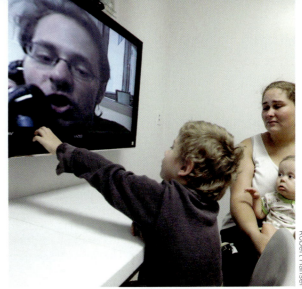

Robert Hanser

PREFACE

This text is intended to provide the reader with an overview of corrections in the United States in a manner that is streamlined and concise in presentation. As with our more comprehensive version of this text, we still provide the reader with a view of corrections that is both practitioner driven and grounded in modern research and theoretical origins. This text possesses practicality and realism in describing and explaining today's world of corrections. This single aspect of the book, along with its insightful portrayal of prison logic, exploration of subcultural issues in prison, and emphasis on persons who work within the field, both institutional and community based, is what sets this text apart from others in the correctional textbook market. Additionally, vignettes have now been included that provide a view of correctional issues and challenges from the vantage of correctional workers and offenders, which helps to further portray the day-to-day reality of the correctional experience.

A Brief Introduction to Corrections does illustrate how the typical practitioner conducts business in the field of corrections, including both institutional and community settings. At the same time, theoretical applications are made explicit to demonstrate to the student that contemporary punishment, incarceration, and supervision schemes are grounded in theories that are often overlooked. Indeed, this text shows that theory and the practical world do not have to be disjointed and disconnected from one another. Rather, each can serve to augment the other, and, in this book, each aspect provides the student with additional facets of *how* correctional practice is implemented (reflecting the world of the practitioner) and *why* it is implemented in that manner (rooted in theoretical perspectives).

This text is intended to serve as a stand-alone text for undergraduate students in introductory courses on corrections, correctional systems, or correctional practices. A special effort is made to tie the readings to practical uses that the majority of our students will encounter in the world of work. This includes discussions on qualifications of specific types of officers, stressors confronted in daily correctional work, examples of tools and instruments that are used in the field, and so forth. The organization of the book follows a logical flow through the correctional system, in terms of both historical evolution and operational developments in the field. In short, this text covers the full array of topics related to nearly any aspect of corrections—on an introductory level, of course.

Finally, this text is also unique in one other critical aspect. The data, figures, tables, and various programs showcased here are predominantly drawn from federal government documents and briefings. Thus, the data and programs selected are solid and tend to be of better quality than one might typically use. Federal research by the National Institute of Corrections abounds, and the right to public domain of much of this material has allowed the author to integrate it within the pages of this text. This provides for rich data and examples that are guaranteed to aid in student learning. Further, the sources have been subjected to rigorous scrutiny and consideration, ensuring that all information is valid and up to date.

NEW TO THIS EDITION

- Throughout the text, information has been updated to include new legislation, statistics, examples, and topics, such as sentencing practices, technological innovations, and offenders with special needs.

- Learning objectives continue to utilize Bloom's taxonomy and have a closer connection to the key concepts in each chapter.

- Some chapters have been combined and streamlined so that the result is a 12-chapter text that is more succinct, being workable for a variety of course schedules beyond the typical 16-week semester.

- New topics on the effects of realignment in California, medical care in jails, probation, compassionate prison design, female drug offenders, the use of reentry councils, and updated court decisions and controversies around the death penalty have been added.

- Recent statistics have been added to the figures and tables to provide students a contemporary snapshot of the status of corrections today.

- Engaging chapter-opening vignettes continue to be included to highlight important issues in corrections and allow students to understand the challenges corrections practitioners face each day.

APPROACH AND STRUCTURE OF THE TEXT

Significantly and perhaps uniquely, this text not only connects the practical world of corrections to the theoretical but also connects treatment and security aspects in the field of corrections to show the dichotomous relationship between these two types of approaches in offender management. Further, the practical aspects of this book are reinforced with specific exercises in which students themselves apply and synthesize the various concepts found throughout the chapters. In providing this content, this text consists of 12 chapters that cover all the basic aspects of correctional systems and practices. These chapters are summarized as follows:

Chapter 1: Early History of Punishment and the Development of Prisons in the United States

This chapter serves as an introduction to and overview of the historical development of corrections in Europe and the United States. Included in this chapter is a history of the development of sanctions as well as an overview of many classic figures in the history of corrections, including Charles Montesquieu, Cesare Beccaria, William Penn, and John Howard, among others. This chapter also discusses prisons in the United States, from the earliest prison used in the original 13 colonies to modern-day maximum-security facilities; the development of prisons and prison systems at both the state and federal levels is also discussed. The Pennsylvania system, the Auburn system, southern penology, the reformatory era, and the use of the Big Houses are all covered. Different models of correctional operation are provided, as is a brief overview of modern-day prison facilities.

Chapter 2: Ideological and Theoretical Underpinnings to Sentencing and Correctional Policy

This chapter revisits the purpose of corrections as a process whereby practitioners from a variety of agencies and programs use tools, techniques, and facilities to engage in organized security and treatment functions intended to correct criminal tendencies among the offender population. It is with this purpose in mind that a variety of philosophical underpinnings are presented, including retribution, incapacitation, deterrence, rehabilitation, restorative justice, and reintegration. Discussion regarding the use of incarceration as a primary tool of punishment is provided, and community-based sanctions are given extensive coverage. The death penalty is presented as the most serious sanction available, along with a discussion of the underlying philosophies associated with its use and implementation. Lastly, types of sentencing models, as well as disparities in both prison and death penalty sentences, are highlighted. In discussing the issue of disparity, the distinction between disparity and discrimination is made clear.

Chapter 3: Correctional Law and Legal Liabilities

This chapter demonstrates how there has been a constant interplay between state-level correctional systems and the federal courts recently in America. Amid this evolution of correctional operations, the interpretation of constitutional standards has been a central feature, as has the Supreme Court's interpretation of its own role in ensuring that those standards are met. Legal issues involving the death penalty have been included in this chapter, as well. The distinctions between federal suits and state suits are clarified. Lastly, a brief overview of injunctions and other forms of court-oriented remediation is presented. These actions are what ultimately led to the sweeping changes that we have seen in the field of corrections.

Chapter 4: Jail and Detention Facilities

Jail facilities are presented as complicated facilities that are not usually appreciated for the vital role that they play within the criminal justice system. The different types of tasks, such as the holding of persons prior to their court date, providing a series of unique sentencing variations, and the incarceration of persons who are technically part of the larger prison system, are all considered. The problems and challenges for jail facilities can be quite varied, and this creates a demanding situation for jail staff and administrators. Overall, jails have been given short shrift in the world of corrections, but they will be given much more attention in times to come. This chapter also provides a concise section addressing the use of jail facilities as immigration detention facilities. Lastly, data for this chapter are up to date.

Chapter 5: Probation and Intermediate Sanctions

The evolution of probation is presented, from the early days of recognizance and suspended sentences through modern-day uses. This chapter also includes a variety of different types of probation administrative models. Also discussed are the qualifications of officers, supervisory strategies, and responsibilities of offenders. Presentence investigation reports and revocation and legal procedures are also included. In addition, various intermediate sanctions, such as community service, the payment of fines, intensive supervision, GPS monitoring, home detention, and day reporting centers, are discussed. Together with community involvement, agency collaboration, and solid case management processes, intermediate sanctions are shown to be a key, interlocking supervision mechanism that improves the overall goal of public safety.

Chapter 6: Facility Design and Classification in Jails and Prisons

This chapter demonstrates that the physical features of a prison require forethought before ground is even broken at the construction site. Issues related to the location of the prison facility, the types of custody levels and security, the function of the facility, logistical support for the facility, and institutional services (such as laundry, kitchen, and religious services) are all important considerations. Technological developments and improvements in security, including cell block and electric fence construction, are presented. Challenges associated with technological innovations in security are also highlighted in this chapter. In addition, effective classification is presented as an essential aspect of both security needs and the needs of the inmate. It shows as well that classification processes are important for both security and treatment purposes.

Chapter 7: Prison Subculture and Prison Gang Influence

This chapter provides a glimpse of the "behind the scenes" aspects of the prison environment. The notion of a prison subculture, complete with norms and standards that are counter to those of the outside world, is presented. The effects of professionalism within the correctional officer ranks, the diversity of correctional staff, and the difference in this generation of inmates all have led to changes in the inmate subculture in modern times. Gangs have emerged as a major force in state prison systems. From this chapter, it is clear that prison gangs have networks that extend beyond the prison walls. Additional information is provided that covers the Prison Rape Elimination Act of 2003 (PREA) and how this has impacted prison subculture in today's contemporary correctional environment.

Chapter 8: Female Offenders in Correctional Systems

Female offenders, though a small proportion of the correctional population, are rapidly growing in number. The need for improved services and programming for female offenders is discussed in this chapter. Mother–child programming is presented as critical to female offender reformation. Legal issues specific to female offenders are discussed, and guiding principles to improve female offender reentry are provided. As with Chapter 7, this chapter also provides information related to the PREA and its impact on security and programming services for female offenders.

Chapter 9: Specialized Inmate Populations and Juvenile Correctional Systems

In this section, we include a discussion on the supervisory strategies used for a special offender population, which includes sex offenders, substance abusers, mentally ill offenders, and mentally disabled offenders. This growing population in the community presents special concerns for community safety and supervisory strategies. This section addresses some of these concerns and provides suggestions for effective supervisory strategies. This chapter contains numerous updates in data related to offenders with special needs. In addition, it provides a brief overview of strategies used for juvenile offenders including classification and assessment, treatment, detention, and incarceration when youth are tried as adults. Issues related to disparity and disproportionate minority contact (DMC) are provided. This chapter also includes a discussion of how the PREA has impacted the maintenance and operation of juvenile facilities in the United States. Updates on Supreme Court rulings related to juvenile corrections have also been included.

Chapter 10: Correctional Administration and Prison Programming

This chapter provides an overview of the organizational structure of both the federal and state prison systems in the United States. Styles of management and the delegation of responsibility are discussed. Private prison management is included in this chapter, along with the conclusion that such programs can be quite successful. This chapter also provides an overview of many of the typical programs offered to inmates within the prison environment. Educational, vocational, drug treatment, medical, recreational, and religious programs are all presented, but a more streamlined approach is utilized to give improved focus and clarity on the overall notion of offender programming inside institutions.

Chapter 11: Parole and Reintegration

This chapter provides an overview of the evolution of parole to its modern-day usage. Additional and up-to-date data and figures are included in the discussion of parole in the United States. Early historical figures who contributed to the development of parole, such as Sir Walter Crofton and Alexander Maconochie, are noted. This chapter also includes a variety of different types of parole administrative models that include not only the qualifications of officers but also the supervisory strategies and responsibilities of offenders. The use of prerelease planning and mechanisms, along with the parole board and parole revocation, are also included. The controversial nature of parole and other early release mechanisms is discussed.

Chapter 12: Program Evaluation, Evidence-Based Practices, and Future Trends in Corrections

This chapter illustrates the importance of evaluative research and distinguishes between process and outcome measures. The use of the assessment-evaluative cycle in corrections is discussed. Information on evidence-based practices is presented but is showcased in a more succinct manner to better illustrate how these practices aid agencies to excel in service delivery. A variety of future trends in the correctional field are also presented during the last few pages of this chapter.

PEDAGOGICAL AIDS

A number of pedagogical aids have been retained in each chapter of this briefer text on corrections. Their primary goal is to facilitate student learning and to aid the student in synthesizing the learning goal and applying it to the modern world of corrections. Through these added features, specific theories are identified and linked to a particular point in the correctional setting. Also, cross-national perspectives are provided within each chapter to acquaint the student with applications that exist in other nations around the globe. In addition, this text has a number of ancillaries that accompany it, all as a means of further improving student learning. The pedagogical features and ancillaries associated with this text are listed below.

- *Opening Vignettes:* At the very beginning of each chapter, a short story is provided that is related to the chapter's topic. Each story provides a high sense of realism in portraying issues that are encountered within the correctional environment.

- *Improved Chapter Learning Objectives:* At the beginning of each chapter is a set of learning objectives. These objectives serve as cues for the student and also provide for easy assessment of learning for the instructor. These points are germane to the chapter and prompt the student as to the information that will be covered. They also let the student know what is critical to the text readings. Each of these learning objectives is clearly linked to headings and subheadings throughout the text.

- *Focus Topics Boxes:* Many chapters include Focus Topic boxes that provide additional insight regarding specific points in the chapter. The topics typically help to add depth and detail to a particular subject that is considered important or interesting from a learning perspective. The inclusion of these boxes has been made with care and consideration to ensure that the material does indeed reinforce the learning objectives at the beginning of each chapter.

- *Key Terms and Key Cases:* At the end of each chapter is a list of key terms and most chapters list key cases that help to augment information relevant to the chapter learning objectives. The terms and cases are in bold throughout the text and are included in the glossary.

- *Discussion Questions:* At the end of each chapter is a list of five to seven discussion questions. These questions usually ask students about chapter content that is relevant to the learning objectives found at the beginning of each chapter. In this way, they serve the function of reinforcing specific knowledge that is applicable to the learning objectives and further clarify for the student the main points and concepts included therein.

- *"What Would You Do?" Exercises:* At the close of each chapter, these exercises present some sort of modern-day correctional scenario that the student must address. In each case, a problem is presented to students, and they must explain what they would do to resolve the issue or solve the problem. This feature provides an opportunity for students to apply and synthesize the material from the chapter and ensures that higher-order learning of the material takes place.

- ***Applied Exercise Features:*** These assignments require the student to perform some type of activity that integrates the material in the text with the hands-on world of the practitioner. In some cases, these assignments require that the student interview practitioners in the field, while in other cases students may need to utilize specific tools or instruments when addressing an issue in corrections. In each case, the student is required to demonstrate understanding of a particular aspect of the chapter readings and must also demonstrate competence in using the information, techniques, or processes that he or she has learned from the chapter. These exercises often also require the student to incorporate information from prior chapters or other exercises in the text, thereby building upon the prior base of knowledge that the student has accumulated.

- ***Text Glossary:*** A glossary of key terms is included at the end of the text. These key terms are necessary to ensure that students understand the basics of corrections. Definitions are provided in simple but thorough language.

INTERACTIVE eBOOK

Learn more at **edge.sagepub.com/hanserbrief/access**

Prison Tour Videos: In the Interactive eBook, these original videos feature inmates and correctional workers from Louisiana State Penitentiary in Angola, Louisiana, and Richwood Correctional Facility in Monroe, Louisiana.

Career Videos: In the Interactive eBook, interviews are available with criminal justice professionals discussing their day-to-day work and current issues related to technology, diversity, and cutting-edge developments in their field.

Criminal Justice in Action: In the Interactive eBook, original animations are available that give students the opportunity to apply the concepts they are learning and to check for a deeper understanding of how these concepts play out in real-world scenarios.

DIGITAL RESOURCES

To enhance this text and to assist in the use of this book, a variety of different digital resources have been created. This text is one of several by SAGE that strives to give the text its own sense of life through the use of various media and online resources that enhance the reading of the text, particularly online.

SAGE edge offers a robust online environment featuring an impressive array of tools and resources for review, study, and further exploration, keeping both instructors and students on the cutting edge of teaching and learning. SAGE edge content is open access and available on demand. Learning and teaching has never been easier!

http://edge.sagepub.com/hanserbrief

STUDENT RESOURCES

SAGE edge for students provides a personalized approach to help students accomplish their coursework goals in an easy-to-use learning environment.

- Mobile-friendly **eFlashcards** strengthen understanding of key terms and concepts.
- Mobile-friendly practice **quizzes** allow for independent assessment by students of their mastery of course material.
- Carefully selected chapter-by-chapter **video and multimedia links** enhance classroom-based explorations of key topics.
- **Learning objectives** reinforce the most important material.
- Lively and stimulating **chapter activities** can be used in class to reinforce active learning. The activities apply to individual or group projects.

INSTRUCTOR RESOURCES

SAGE COURSEPACKS FOR INSTRUCTORS makes it easy to import our quality content into your school's LMS. Intuitive and simple to use, it allows you to

Say NO to . . .

required access codes

learning a new system

Say YES to . . .

using only the content you want and need

high-quality assessment and multimedia exercises

For use in: Blackboard, Canvas, Brightspace by Desire2Learn (D2L), and Moodle

Don't use an LMS platform? No problem, you can still access many of the online resources for your text via SAGE edge.

SAGE coursepacks and SAGE edge include the following features:

- Our content delivered **directly into your LMS**

- An **intuitive, simple format** that makes it easy to integrate the material into your course with minimal effort

- Pedagogically robust **assessment tools** that foster review, practice, and critical thinking, and offer a more complete way to measure student engagement, including: Diagnostic chapter **coursepack quizzes** that identify opportunities for improvement, track student progress, and ensure mastery of key learning objectives

- **Test banks** built on Bloom's taxonomy that provide a diverse range of test items with ExamView test generation

- **Instructions** on how to use and integrate the comprehensive assessments and resources provided

- A comprehensive, downloadable, easy-to-use ***Media Guide in the Coursepack for every video resource***, listing the chapter to which the video content is tied, matching learning objective(s), a helpful description of the video content, and assessment questions

- **Sample course syllabi** for semester and quarter courses providing suggested models for structuring your courses

- Editable, chapter-specific **PowerPoint® slides** that offer complete flexibility for creating a multimedia presentation for your course

- **Lecture notes** that summarize key concepts by chapter to help you prepare for lectures and class discussions

- **Video resources** that bring concepts to life, are tied to learning objectives, and make learning easier

- **Chapter-specific discussion questions** to help launch engaging classroom interaction while reinforcing important content

ACKNOWLEDGMENTS

At this time, I would like to thank Jessica Miller, the acquisitions editor of this project, who provided me with the opportunity to further expand my presentation of the correctional system with *A Brief Introduction to Corrections*. Jessica has been extremely supportive and has even served as a valuable sounding board during challenging times when writing texts for SAGE. She continues to be instrumental in several projects, even though she holds her current role in acquisitions. I would like to also thank the content development editor, Laura Kearns, who has been very easy to work with and very supportive throughout the organization and compilation of the text information. I have found Laura to be both detail oriented but also practical and pleasant when providing feedback on needed updates or revisions.

I would like to extend special gratitude to all of the correctional practitioners who carry out the daily tasks of our correctional system, whether institutional or community based. These individuals deserve the highest praise as they work in a field that is demanding and undervalued—I thank you all for the contributions that you make to our society.

I would also like to thank Secretary LeBlanc of the Louisiana Department of Public Safety and Corrections; Ms. Pam Laborde, communications director for the Louisiana DPS&C; and all those who allowed me to interview and showcase various elements of that state's correctional system. In addition, Warden Burl Cain and Assistant Warden Cathy Fontenot deserve thanks and gratitude for allowing the filming and photo shoots at Louisiana State Penitentiary Angola.

In addition, I would like to thank Mr. Billy McConnell, Mr. Clay McConnell, and Warden Ray Hanson, as well as other personnel and staff of LaSalle Southwest Corrections. Their support for this text and willingness to be interviewed and allow filming and photo shoots at Richwood Correctional Center added a unique, useful, and educational element to the text.

I am also grateful to the many reviewers (see below) who spent time reading the document and making a considerable number of recommendations that helped to shape the final product. Every effort was made to incorporate those ideas. Their suggestions and insights helped to improve the final product that you see here, a product that, in truth, is a reflection of all those who were involved throughout its development.

Reviewers

LeAnn Cabage, Kennesaw State University

Joel George, Delgado Community College

Suman Kakar, Florida International University

Wesley Maier, Walla Walla Community college

William Mixon, Columbus State University

Paul Odems, Kingsborough Community College

Rebecca Pastrana, University of Texas at El Paso, El Paso Community College

Marina Saad, Rutgers University

Ronald Smith, Cuyahoga Community College

Carol Trent, Saint Francis University

Lauren Wright, Northeastern State University

1

Early History of Punishment and the Development of Prisons in the United States

Learning Objectives

1. Define *corrections* and the role it has in the criminal justice system.
2. Identify early historical developments and justifications in the use of punishment and corrections.
3. Discuss the influence of the Enlightenment and key persons on correctional reform.
4. Discuss the development of punishment in early American history.
5. Describe the changes to prison systems brought about by the Age of the Reformatory in America.
6. Identify the various prison systems, eras, and models that developed in the early and mid-1900s in America.
7. Explain how state and federal prisons differ and identify the Top Three in American corrections.

Prisoner Number One at Eastern Penitentiary

In 1830, Charles Williams, prisoner number one at Eastern State Penitentiary, contemplated his situation with a sense of somber and solemn reflection. He did this undisturbed due to the excruciating silence that seemed to permeate most of his incarceration. On occasion, he could hear keys jingling, and he might hear the sound of footsteps as guards brought his food or other necessities. Sometimes he could hear the noise of construction, as the facility was not yet finished and would not be fully functional for years to come. Otherwise, there was no other sound or connection to the outside world, and silence was the most common experience throughout most of the daylight hours and the entire night.

To be sure, Charles had all of his basic needs met at Eastern. He had his own private cell that was centrally heated and had running water. He had a flushing toilet, a skylight, and a small, walled recreation yard for his own private use. In his high-pitched cell, Charles had only natural light, the Bible, and his assigned work (he was involved in basic weaving) to keep him busy throughout the day. He was not allowed interaction with the guards or other inmates, and his food was delivered to him via a slot in the door. In addition, he was to not leave his cell for anything other than recreation in his own walled yard, and even then he was required to wear a special mask that prevented communication with other guards or inmates while he entered the yard.

Charles was a farmer by trade. He had been caught and convicted of burglary after stealing a $20 watch, a $3 gold seal, and a gold key. He was sentenced to 2 years of confinement with hard labor and entered Eastern on October 23, 1829. He had served 7 months of his sentence and already he felt as if he had been incarcerated for an eternity. He reflected daily (and quite constantly) on his crime. Before his arrival, he had had no idea what Eastern State Penitentiary would be like. As it turned out, it was quite numbing to Charles's sense of mental development, and he sometimes felt as if he did not even exist. Charles remembered his first glimpse of the tall, foreboding exterior of the unfinished prison as his locked carriage approached. It was an intimidating sight, and Charles, who was only 18 at the time of his sentencing, felt remorseful. He remembered when Warden Samuel R. Wood received him and explained that he would be overseeing Charles's stay at Eastern. The warden was very direct and matter-of-fact and exhibited a mean-spirited temperament. Charles found the warden to be reflective of his entire experience while serving in prison cell number one at Eastern. He thus had determined that he did not want to spend any more of his life in such confinement.

Charles considered the fact that he still had 18 months on his sentence—an eternity for most 18-year-olds. He knew that other inmates would soon follow his stay in the expanding prison. However, he was not the least bit curious about the future of Eastern. He was indeed repentant, but not necessarily for the reasons that early Quaker advocates might have hoped when they advocated for the penitentiary. Rather than looking to divine inspiration as a source of redemption from future solitary incarceration, he simply decided that he would never again be in a position where he could be accused of, guilty of, or caught in the commission of a crime. He just wanted to go back to simple farming and leave Eastern State Penitentiary out of both sight and mind for the remainder of his years.

DEFINING CORRECTIONS: A VARIETY OF POSSIBILITIES

Corrections: A process whereby practitioners engage in organized security and treatment functions to correct criminal tendencies among the offender population.

In this text, **corrections** will be defined as a process whereby practitioners from a variety of agencies and programs use tools, techniques, and facilities to engage in organized security and treatment functions intended to correct criminal tendencies among the offender population. This definition underscores the fact that corrections is a process that includes the day-to-day activities of the practitioners who are involved in that process. Corrections is not a collection of agencies, organizations, facilities, or physical structures; rather, the agencies and organizations consist of the practitioners under their employ and/or in their service, and the facilities or physical structures are the tools of the practitioner. The common denominator between the disparate components of the correctional system is the purpose behind the system. We now turn our attention to ancient developments in law and punishment, which, grounded in the desire to modify criminal behavior, served as the precursor to correctional systems and practices as we know them today.

The Role of Corrections in the Criminal Justice System

Generally speaking, the criminal justice system consists of five segments, three of which are more common to students and two of which are newer components, historically speaking. These segments are law enforcement, the courts, corrections, the juvenile justice system, and victim services. Of these, it is perhaps the correctional system that is least understood, least visible, and least respected among much of society. The reasons for this have to do with the functions of each of these segments of the whole system.

Unlike the police, who are tasked with apprehending offenders and preventing crime, correctional personnel often work to change (or at least keep contained) the offender population. This is often a less popular function to many in society, and when correctional staff are tasked with providing constitutional standards of care for the offender population, many in society may attribute this to "coddling" the inmate or offender.

On the other hand, the judicial or court segment is held in much more lofty regard. The work of courtroom personnel is considered more sophisticated, and jobs within this sector are more often coveted. Further, there tends to be a degree of mystique to the study and practice of law, undoubtedly enhanced by portrayals in modern-day television and the media. In this segment of the system, legal battles are played out, oral arguments are heard, evidence is presented, and deliberations are made. At the end, a sentence is given and the story concludes that all parties involved have had their day in court.

The juvenile justice system is unique from these other systems because much of it is not even criminal court but is instead civil in nature. This is because our system intends to avoid stigmatizing youthful offenders, hopes to integrate family involvement and supervision, and views youth as being more amenable to positive change. The juvenile justice system is designed to help youth and is, therefore, less punitive in theory and practice than the adult system. Again, the entire idea is that youth are at an early stage in life where their trajectory is not too far off the path; with the right implementation, we can change their life course in the future.

Victim services is, naturally, the easiest segment to sympathize with because it is tasked with aiding those who have been harmed by crime. The merits of these services should be intuitively obvious, but such programs are often underfunded in many states and struggle to help those in need. In addition to state programs, many nonprofit organizations are also dedicated to assisting victims.

After this very brief overview of each segment of the criminal justice system, we come back to the correctional system. The correctional system, despite its lesser appeal, is integral to the ability of the other systems to maintain their functions. As we will see later in this chapter, it is simply not prudent, realistic, or civilized to either banish or put to death every person who commits an offense. Indeed, such reactions would be extreme and quite problematic in today's world. Thus, we are stuck with the reality that we must do something else with those individuals who have offended. Naturally, some have committed serious crimes while others have not. Discerning what must be done with each offender based on the crime, the criminal, and the risk that might be incurred to society is the role of the correctional system. Further, it is the responsibility of this system to keep these persons from committing future crimes against society, a task that the other segments of the system seem unable to do.

PRACTITIONER'S PERSPECTIVE

"I have been a jail administrator for about seven and a half years . . . absolutely [the] best job I ever had."

Visit the IEB to watch Mitch Lucas's video on his career as a jail administrator.

Mitch Lucas
Jail Administrator

The correctional system is impacted by all of the other systems and, largely speaking, is at their mercy in many respects. Indeed, as police effect more arrests, more people are locked up and jails and prisons must contend with housing more inmates. When courts sentence more offenders, the same happens. A court has the luxury of engaging in plea agreements to modify the contours of a sentence, but the correctional system has few similar forms of latitude, other than letting offenders out early for good behavior—an option that many in society bemoan as the cause for high crime rates. Likewise, the juvenile system has a correctional segment that gets sufficient sympathy from the public, but state correctional facilities find themselves being given the "worst of the worst" of youthful offenders, making notions of rehabilitation more challenging than is desirable. And of course there is the victim services segment, through which the correctional system often attempts to redeem itself by ensuring that offenders are made accountable for their crimes and by generating revenue through fines, restoration programs, and compensation funds for victims. Amid this, correctional systems engage in victim notification programs and many include victim services bureaus for those who have questions or requests of the correctional system.

This complicated system of sanctioning offenders while operating within the broader context of the criminal justice system is the result of a long and winding set of historical circumstances and social developments. In this chapter, we will explore how this story has unfolded, starting with the reality that initially the role of corrections was simply to *punish* the offender. This punishment, it was thought, would be instrumental in *changing* the behavior of the offender. These notions are just as relevant in today's world of corrections, though the means of implementation have become much more complicated. Because these early debates, ideologies, and perspectives on corrections laid the groundwork to our current system, it is the role of this chapter to give the reader an understanding of how and why they developed as they did.

THE NOTION OF PUNISHMENT AND CORRECTIONS THROUGHOUT HISTORY

As might be determined by the title of this section, there has been a long-standing connection between the concepts of punishment and correction. It is as if our criminal justice system considers these two concepts as being one in the same. However, as we will find, these two terms are not always synonymous with one another. Rather, the purpose that underlies each is probably a better guide in distinguishing one from the other, not identifying their similarities. It is the application of penalties that has the longest history, and it is with this in mind that punishment is first discussed, with additional clarification provided in defining the more modern term of *corrections*. As we will see later in this chapter and in other chapters, the distinction between *corrections* and *punishment* may be quite blurred.

When applying punishments, it was hoped that the consequence would prevent the offender from committing future unwanted acts. Though one would consider it a good outcome if offenders are prevented from committing further crimes, this is not necessarily an act of *correction* regarding the offender's behavior. This is a very important point because it sets the very groundwork for what we consider to be corrections. Essentially, the common logic rests upon the notion that if we punish someone effectively, he or she will not do the crime again and is therefore corrected. Naturally, this is not always the final outcome of the punishment process. In fact, research has found cases where exposure to prison actually increases the likelihood of future criminal behavior (Fletcher, 1999; Golub, 1990). Likewise, some research has demonstrated higher rates of violent crime when the death penalty is applied, seemingly in reaction to or correlated with the use of the death penalty (Bowers & Pierce, 1980). This observation is referred to as the **brutalization hypothesis**, the contention of which is that the use of harsh punishments sensitizes people to violence and essentially *teaches* them to use violence rather than acting as a deterrent (Bowers & Pierce, 1980).

Brutalization hypothesis: The contention that the use of harsh punishments sensitizes people to violence and *teaches* them to use it.

Code of Hammurabi: The earliest known written code of punishment.

Lex talionis: Refers to the Babylonian law of equal retaliation.

Early Codes of Law

Early codes of law were designed to guide human behavior and to distinguish that which was legal from that which was not. These laws often also stated the forms of punishment that would occur should a person run errant of a given edict. Because laws reflected the cultural and social norms of a given people and tended to include punishments, it could be said that the types of punishment used by a society might give an outside observer a glimpse of that society's true understanding of criminal behavior as well as its sense of compassion, or lack thereof.

PHOTO 1.1 The Code of Hammurabi is one of the most ancient attempts to codify criminal acts and their corresponding punishments.

©iStockphoto.com/jsp

Babylonian and Sumerian Codes

The earliest known written code of punishment was the **Code of Hammurabi**. Hammurabi (1728–1686 BCE) was the ruler of Babylon sometime around 1700 BCE, which dates back nearly 3,800 years before our time (Roth, 2011). This code used the term ***lex talionis***, which referred to the Babylonian law of equal retaliation (Roth, 2011). This legal basis reflected the instinctive desire for humans who have been harmed to seek revenge. While Hammurabi's Code included a number of very harsh corporal punishments, it also provided a sense of uniformity in punishments, thereby organizing the justice process in Babylon (Stohr, Walsh, & Hemmens, 2013).

Roman Law and Punishment and Their Impact on Early English Punishment

Punishments in the Roman Empire were severe and tended to be terminal. Imprisonment was simply a means of holding the accused until those in power had decided the offender's fate. From what is known, it would appear that most places of confinement were simply cages. There are also recorded accounts of quarries (deep holes used for mining/excavating stone) used to hold offenders (Gramsci, 1996). One place of confinement in Rome that was well known was the Mamertine Prison, which was actually a sprawling system of underground tunnels and dungeons built under the sewer system of Rome sometime around 64 BCE. This was where the Christian apostles Paul and Peter were incarcerated (Gramsci, 1996).

Rome and other societies during this period considered convicted offenders to have the legal status of a slave, and they were treated as if they were essentially dead to society. In this "civil death," the offender's property would be excised by the government and the marriage (if any) between the offender and his or her spouse was declared void, providing the status of widow to the spouse.

Early Historical Role of Religion, Punishment, and Corrections

Perhaps the most well-known premodern historical period of punishment is the Middle Ages of Western Europe. The Middle Ages was a time of chaos in Europe during which plague, pestilence, fear, ignorance, and superstition prevailed. Throughout these dark times, the common citizenry, which consisted largely of peasants who could neither read nor write, placed their faith in religious leaders who were comparatively better educated and more literate.

While one might stand at trial for charges brought by the state, it was the **trial by ordeal** that emerged as the Church's equivalent to a legal proceeding (Johnson, Wolfe, & Jones, 2008). The trial by ordeal consisted of very dangerous and/or impossible tests used to prove the guilt or innocence of the accused. For instance, the ordeal of hot water required that the accused thrust a hand or an arm into a kettle of boiling water (Johnson et al., 2008). If after 3 days of binding the arm, the offender emerged unscathed, he or she was considered innocent. Of note was the general reason provided by the Church for its use of punishments. It would seem that the Church response to aberrant (or sinful) behavior was, at least in ideology, based on the desire to save the soul of the wayward offender. Indeed, even when persons were burned at the stake, the prevailing belief was that such burning would free their souls for redemption and ascension to Heaven. The goal, in essence, was to purify the soul as it was released from the body. This was especially true of persons who were convicted of witchcraft and who were believed to have consorted with spirits and/or were believed to be possessed by evil spirits.

Trial by ordeal: Very dangerous and/or impossible tests to prove the guilt or innocence of the accused.

Sanctuary

While the Church may have had a role in the application of punishments throughout history, it also provided some unique avenues by which the accused might avoid unwarranted punishment. One example would be the granting of sanctuary to accused offenders.

During ancient times, many nations had a city or a designated building, such as a temple or a church, where accused offenders could stay, free from attack, until such time that their innocence could be established (presuming that they were, in fact, innocent). In Europe, the use of sanctuary began during the fourth century and consisted of a place—usually a church—that the king's soldiers were forbidden to enter for purposes of taking an accused criminal into custody (Cromwell, del Carmen, & Alarid, 2002). In some cases, such as in England, **sanctuary** was provided until some form of negotiation could be arranged or until the accused was ultimately smuggled out of the area. If accused offenders confessed to their crimes while in sanctuary, they were typically allowed to leave the country with the understanding that return to England would lead to immediate punishment (Cromwell et al., 2002).

Sanctuary: A place of refuge or asylum.

This form of leniency lasted for well over a thousand years in European history and was apparently quite common in England. Eventually, sanctuary lost its appeal, and from roughly 1750 onward, countries throughout Europe began to abolish sanctuary provisions as secular courts gained power over ecclesiastical courts.

Early Secular History of Punishment and Corrections

The origin of law was one of debate during medieval times. Over time, secular rulers (often royalty and nobility) became less subservient to the Church and gained sufficient power to resist some of the controls placed upon them by the ecclesiastical courts. As such, much of the royalty, nobility, merchant class, and scholarly community advocated separation between government rule (at this time the king or queen) and the Church. Though this was an ultimately successful process, many did die as a result of their views.

It was at this time that criminal behavior became widely recognized as an offense against the state. Indeed, by 1350 CE, the royalty (consisting of kings, queens, and the like) had established themselves as the absolute power, and they became less tolerant of external factors that undermined their own rule; this meant that the Church continued to lose authority throughout Europe. Ultimately, all forms of revenue obtained from fines went to the state (or the Crown), and the state administered all punishments. This also led to the development of crime being perceived as an act in violation of a king or queen's authority.

Public and Private Wrongs

Public wrongs: Crimes against society or a social group.

Public wrongs are crimes against society or a social group and historically tended to include sacrilege as well as other crimes against religion, treason, witchcraft, incest, sex offenses of any sort, and even violations of hunting rules (Johnson et al., 2008). Among early societies, religious offenses were considered the most dangerous since these crimes exposed both the offender and the rest of the group to the potential anger and wrath of that culture's deity or set of deities. Witchcraft was commonly thought to entail genuine magical powers that would be used by the witch for personal revenge or personal gain; the use of such magic was considered bad for a social group because it drew evil spirits in the direction of the community.

The fear of witchcraft persisted for several hundred years, reaching its height of hysteria in the 1500s. Suspicion of witchcraft and the mass execution of suspected practitioners became commonplace during this time. Indeed, during the years between 1273 and 1660, Europe executed thousands of suspected witches, the majority of them women. The total number of persons executed due to witchcraft charges may have exceeded 100,000 (Linder, 2005).

Private wrongs: Crimes against an individual that could include physical injury, damage to a person's property, or theft.

In ancient times, resorting to private revenge was the only avenue of redress for victims who suffered a **private wrong**. These types of wrongs might have included physical injury, damage to a person's property, or theft. In such cases and in many areas of Europe, there was no official authority present; the victim was on his or her own to gain any justice that could be obtained. There was also additional incentive to retaliate against perpetrators, for if the victim was able to gain revenge this was likely to deter the perpetrator from committing future crimes against the victim. However, it is not surprising that in these cases the original perpetrator sometimes fought back against the retaliatory strike from the victim, regardless of who was wrong or right. This would then lead to a continual tit-for-tat situation that might ultimately develop into a perpetual conflict. Once social groups become more advanced, the responsibility for determining punishment shifted from the individual and/or family to society as a whole.

Retaliation Through Humiliation

During early parts of European history, retaliation also occurred through the use of humiliation. A number of punishments were utilized, some of which might even be considered corporal in nature (such as the ducking stool and the stocks and pillories), but they are included in this section because their distinctive factor lies more in their intended outcome: to humiliate and embarrass the offender (Johnson et al., 2008).

One early punishment was the gag, which was a device that constrained persons who were known to constantly scold others (usually their spouse) or were guilty of habitually and abusively finding fault with others, being unjustly critical, or lying about other persons (Silverman, 2001). An even more serious form of retaliatory punishment was the use of the bridle. The bridle was an iron cage that fit over the head and included a metal plate in the front. The plate usually had spikes, which were constructed so as to fit into the mouth of the offender; this made movement with the tongue painful and thereby reduced the likelihood that the offender would talk (Silverman, 2001).

The ducking stool was a punishment that used a chair suspended over a body of water. In most cases, the chair hung from the end of a free-moving arm. The offender was strapped into the chair, which was located near a riverbank. The chair would be swung over the river by the use of the free-moving arm and would be plunged into the water while the offender was restrained therein. In most cases, this punishment would be administered during the winter months when the water was extremely cold; this alone was a miserable experience. This was a punishment typically reserved for women—in particular, women who were known to nag others or use profane or abusive language. Women who gossiped were also given this punishment (Johnson et al., 2008).

Another common punishment in the Middle Ages was the stocks and pillories. Stocks consisted of wooden frames that were built outdoors, usually in a village or town square. A set of stocks consisted of a thick piece of lumber that had two or more holes bored into it. The holes were round and wide enough so that an offender's wrists would fit through. The board was cut into halves, and a hinge was used so that the halves could be opened and then closed. The boards would be opened, the offender would be forced to rest his or her wrists into the half-circle of the bottom half of the wooden board, and then the top half would be closed over the wrists. A lock on the side opposite the hinge kept the offender trapped, hands and wrists restrained by the board. The stock was usually constructed atop a beam or post set into the ground so that the offender would have to stand (rather than sit), sometimes for days or, in extreme cases, perhaps weeks.

The pillory was similar to the stock except the pillory consisted of a single large bored hole where the offender's neck would rest. When the pillory was shut and locked, the offender was restrained with his or her head immobilized and body stooped over. The device was specifically set atop a post at a height where most adult offenders could not fully stand up straight, adding to the discomfort of the experience. As with a set of stocks, the offender would be required to stand for several days and nights. In many cases, the offender was constrained by a combination of these devices, known as a stocks and pillory, where both the offender's head and hands were immobilized.

It was at this point that the use of branding became more commonplace. **Branding** was used to make criminal offenders, slaves, and prisoners of war easily identifiable. Offenders were usually branded on their thumb with a letter denoting their offense—for instance, the letter M for murder or T for theft. Harkening back to the connection between crime and sin, consider that even as late as the 1700s, the use of branding for humiliation occurred with the crime of adultery. In New Hampshire, a specific statute (1701) held that offenders guilty of adultery would be made to wear a discernible letter A on their upper-garment clothing, usually in red, but always in some color that contrasted with the color of the clothing. Students should go to Table 1.1 for a more succinct presentation of the various types of punishment that have just been discussed.

Branding: Usually on thumb with a letter denoting the offense.

Corporal Punishment

Up until the 1700s, corporal punishment tended to be the most frequently used punishment. This punishment was often administered in a public forum to add to the deterrent effect, thereby setting an example to others of what might happen if they were caught in the commission of a similar crime. Naturally, these types of punishment also included purposes of retribution. The most widely used form of corporal punishment was whipping, which dates back to the Romans, the Greeks, and even the Egyptians as a sanction for both judicial and educational discipline. Whippings could range in the number of lashes. A sentence of 100 lashes was, for most offenders, a virtual death sentence as the whipping was quite brutal; the lashes would fall across the back and shoulders, usually drawing blood and removing pieces of flesh.

TABLE 1.1

Types of Punishment in Early Correctional History

NAME OF PUNISHMENT	PURPOSE	DESCRIPTION
Trial by ordeal	Determine guilt or innocence	Very dangerous and/or impossible tests to prove the guilt or innocence of the accused
Gag	Humiliation	A device that constrained persons who were known to constantly scold others
Ducking stool	Humiliation and deterrence	Punishment that used a chair suspended over a body of water
Stocks	Humiliation	Wooden frames that were built outdoors, usually in a village or town square
Pillory	Humiliation	Similar to the stock except the pillory consisted of a single large bored hole where the offender's neck would rest
Branding	Humiliation and warn public	Usually on thumb with a letter denoting the offense
Whipping	Deterrence	Lashing the body of a criminal offender in front of a public audience
Capital punishment	Deterrence	Putting the offender to death in front of a public audience
Banishment and transportation	Deterrence	Exile from society
Hulk imprisonment	Retribution and incapacitation	Offenders kept in unsanitary decommissioned naval vessels
Indentured servitude	Retribution and incapacitation	Offender subjected to virtual slavery

Capital Punishment

This section will be brief due to more extensive coverage of the death penalty in Chapter 2. Historically speaking, the types of death penalties imposed are many and varied. Some examples include being buried alive (used in Western civilization as well as ancient China), being boiled in oil, being thrown to wild beasts (particularly used by the Romans), being impaled by a wooden stake, being drowned, being shot to death, being beheaded (especially with the guillotine), and being hanged. More contemporary methods include the use of lethal gas or lethal injection. By far, the most frequently used form of execution is hanging, which has been used throughout numerous points in history.

Banishment

In England between 1100 and 1700, there was an overreliance on the death penalty, and during this time the criminal code was nicknamed the "Bloody Code." Though the rich and powerful may have been supportive of the harsh penalties, there was an undercurrent of discontent among numerous scholars, religious groups, and the peasant population over the capricious and continuous use of the death penalty. Thus, **banishment** proved a very useful alternative that became used with increasing regularity in lieu of the death penalty.

Banishment: Exile from society.

The 1600s and 1700s saw the implementation of banishment on a widespread scale. Over time, banishment came in two versions, depending on the country in question and the time period involved. First, banishment could be permanent or temporary. Second, banishment could mean simple exile from the country or exile to and/or enslavement in a penal colony. The development of English colonies in the Americas opened up new opportunities for banishment that could rid England of her criminal problems on a more permanent basis. This form of mercy was generally only implemented to solve a labor shortage that existed within the American colonies, with most offenders shipped to work as indentured servants under hard labor.

Transporting Offenders

Transportation became a nearly ideal solution to the punishment of criminal offenders because it resolved all of the drawbacks associated with other types of punishment. The costs were minimal, it was difficult (if not impossible) for offenders to return to England, and offenders could become sources of labor for the new colonies. Johnson and his coauthors (2008) note that of those offenders who were subjected to transportation, the majority were male, unskilled, from the lower classes, and had probably resorted to crime due to adverse economic conditions.

PHOTO 1.2 The hulk prison ship was usually a vessel that was old and squalid inside. Little if any lighting was provided, and women, children, and men would be imprisoned together. The conditions were filthy, and rodents commonly lived among the offenders trapped therein.

Photos.com/Thinkstock Images

Indentured Servitude

Indentured servants in the American colonies included both free persons and offenders. Generally speaking, free persons who indentured themselves received better treatment due to the fact that they had some say in their initial agreement to working requirements prior to being transported to the colonies. Such persons came of their own accord in hope of making a better life in the New World. Most of these persons were poor and had few options in England. Though this meant that their lot was one of desperation, they were still not typically subjected to some of the more harsh treatment that offenders were subjected to when indentured into servitude.

Indentured status was essentially a form of slavery, albeit one that had a fixed term of service. During the time that persons were indentured, they were owned by their employer and could be subjected to nearly any penalty except death. It is estimated that nearly half of all persons who came to the Americas during the 1600s and 1700s were indentured servants (Johnson et al., 2008).

Hulks and Floating Prisons

When the American Revolution began in 1776, there was an abrupt halt to the transporting of convicts to those colonies. Thus, England began to look for new ideas regarding the housing of prisoners. One solution was to house offenders in hulks, which were broken-down, decommissioned war vessels of the British Royal Navy. These vessels were anchored in the River Thames. This practice started with the expectation that England would ultimately defeat the American colonies, thereby making the colonies available again for transportation. When it became clear that the colonies would maintain their independence, hulks were used as prisons for a more extended period. During the time when hulks were most widely used (1800s), there were over 10 such vessels that held over 5,000 offenders (Branch-Johnson, 1957).

Conditions aboard these decommissioned ships were deplorable. The smell of urine and feces, human bodies, and vermin filled the air. Overcrowding, poor ventilation, and a diet lacking appropriate nourishment left offenders in a constant state of ill health. Punishments for infractions were severe, and, as one might expect, there were no medical services. Further, all types of offenders were kept together aboard these vessels, including men, women, and vagrant youth. In many cases, there was no proactive effort to separate these offenders from one another. This then allowed for victimization of women and youth by other stronger and predatory offenders.

THE ENLIGHTENMENT AND CORRECTIONAL REFORM

As demonstrated earlier in this chapter, the roots of punishment tend to be ingrained in a desire for revenge. From this intent emerged a number of ghastly tortures and punishments. But beginning in the 1700s, a new mindset began to develop throughout Europe.

FIGURE 1.1

Major Correctional Thinkers in Early History

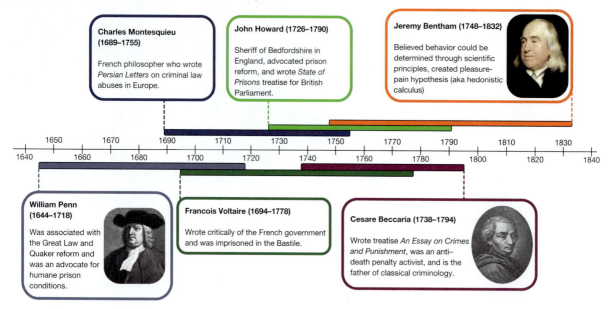

Photos.com; Wikimedia Commons; National Portrait Gallery

It was during this period, referred to as the Age of Enlightenment, that many of the most famous philosophers of modern Western history found their place and left their mark (Carlson, Roth, & Travisono, 2008). This is when thinkers and reformers such as William Penn, Charles Montesquieu, Francois Voltaire, Cesare Beccaria, John Howard, and Jeremy Bentham became known as leading thinkers on punishment as well as advocates of humane treatment for prisoners (see Figure 1.1).

William Penn, the Quakers, and the Great Law

William Penn (1644–1718) was the founder of the state of Pennsylvania and a leader of the religious Quakers. He was an advocate of religious freedom and individual rights (Carlson et al., 2008). He was also instrumental in spreading the notion that criminal offenders were worthy of humane treatment. The Quaker movement in penal reform did not exist just in America; it also took hold in Italy and England. In the process, it influenced other great thinkers, such as Cesare Beccaria, John Howard, and Jeremy Bentham, all of whom would achieve prominence after the death of William Penn.

Great Law: Correctional thinking and reform in Pennsylvania that occurred due to the work of William Penn and the Quakers.

The Quakers followed a body of laws called the **Great Law**, which was more humane in approach than the typical English response to crime. According to the Great Law, hard labor was a more effective punishment than the death penalty. This became a new trend in American corrections, where hard labor was viewed as part of the actual punishment for serious crimes rather than simply being something that was done prior to the actual punishment given to the offender (Johnston, 2009). This was also the first time that offenders received a loss of liberty (albeit while completing hard labor) as a punishment in and of itself. This same concept would later be adopted by a future scholar held in high regard: Cesare Beccaria.

Charles Montesquieu, Francois Voltaire, and Cesare Beccaria

Montesquieu and Voltaire were French philosophers who were very influential during the Age of Enlightenment, and they were particularly concerned with what would be considered human rights in today's society. Charles Montesquieu (1689–1755) wrote an essay titled *Persian Letters*, which was instrumental in illustrating the abuses of the criminal law in both France and Europe. *Persian Letters* is a collection of fictional letters from two Persian noblemen who visited Paris for the first time, and it reflects the thoughts of these two characters on European laws and customs as compared to those in Persia.

At about the same time, Francois Voltaire (1694–1778) became involved with a number of trials that challenged traditional ideas of legalized torture, criminal responsibility, and justice. Voltaire was intrigued with inequities in government and among the wealthy. Like his friend Montesquieu, Voltaire wrote critically of the French government. In fact, he was imprisoned in the Bastille (a fortified prison) for 11 months for writing a scathing satire of the French government. In 1726, Voltaire's wit, public behavior, and critical writing offended much of the nobility in France, and he was essentially given two options: He could be imprisoned or agree to exile. Voltaire chose exile and lived in England from 1726 to 1729. While in England, Voltaire became acquainted with John Locke, another great thinker on crime, punishment, and reform.

These two philosophers helped pave the way for one of the most influential criminal law reformers of Western Europe. Cesare Beccaria (1738–1794) was very famous for his thoughts and writings on criminal laws, punishments, and corrections. Beccaria was an Italian philosopher who wrote a brief treatise titled *An Essay on Crimes and Punishments* (1764). This treatise was the first argument among scholars and philosophers made in public writing against the death penalty. The text was considered a seminal work and was eventually translated into French, English, and a number of other languages.

Beccaria condemned the death penalty on two grounds. First, he claimed that the state does not actually possess any kind of spiritual or legal right to take lives. Second, he said the death penalty was neither useful nor necessary as a form of punishment. Beccaria also contended that punishment should be viewed as having a preventive rather than a retributive function. He believed that it was the certainty of punishment (not the severity) that achieved a preventative effect, and that in order to be effective, punishment should be prompt. Many of these tenets comport with classical criminological views on crime and punishment.

Due to Beccaria's beliefs and contentions, he became viewed as the Father of Classical Criminology, which was instrumental in shifting views on crime and punishment toward a more humanistic means of response. Among other things, Beccaria advocated for proportionality between the crime that was committed by an offender and the specific sanction that was given. Since not all crimes are equal, the use of progressively greater sanctions became an instrumental component in achieving this proportionality. **Classical criminology**, in addition to advocating proportionality, emphasized that punishments must be useful, purposeful, and reasonable. Beccaria contended that humans were hedonistic—seeking pleasure while wishing to avoid pain—and that this required an appropriate amount of punishment to counterbalance the rewards derived from criminal behavior. Further, Beccaria called for the more routine use of prisons as a means of incapacitating offenders and denying them their liberty. This was perhaps the first time that the notion of denying offenders their liberty from free movement was seen as a valid punishment in its own right.

Classical criminology: Emphasized that punishments must be useful, purposeful, and reasonable.

John Howard: The Making of the Penitentiary

John Howard (1726–1790) was a man of means who inherited a sizable estate at Cardington, near Bedford (in England). He ran the Cardington estate in a progressive manner and with careful attention to the conditions of the homes and education of the citizens who were under his stead. In 1773, the public position of sheriff of Bedfordshire became vacant, and Howard was given the appointment. One of his duties as sheriff was that of prison inspector. While conducting his inspections, Howard was appalled by the unsanitary conditions that he found. Further, he was dismayed and shocked by the lack of justice in a system where offenders paid their gaolers (an Old English spelling for jailers) and were kept jailed for nonpayment even if they were found to be innocent of their alleged crime.

Howard traveled throughout Europe, examining prison conditions in a wide variety of settings. He was particularly moved by the conditions that he found on the English hulks and was an advocate for improvements in the conditions of these and other facilities. Howard was impressed with many of the institutions in France and Italy. In 1777, he used those institutions as examples from which he drafted his *State of Prisons* treatise, which was presented to Parliament.

Jeremy Bentham: Hedonistic Calculus

Jeremy Bentham (1748–1832) was the leading reformer of the criminal law in England during the late 1700s and early 1800s, and his work reflected the vast changes in criminological and penological thinking that were taking place at that time. Born roughly a decade after Beccaria, Bentham was strongly influenced by Beccaria's work. In particular, Bentham was a leading advocate for the use of graduated penalties that connected the punishment with the crime. Naturally, this was consistent with Beccaria's ideas that punishments should be proportional to the crimes committed.

Bentham believed that a person's behavior could be determined through scientific principles. He believed that behavior could be shaped by the outcomes that it produced. Bentham contended that the primary motivation for intelligent and rational people was to optimize the likelihood of obtaining pleasurable experiences while minimizing the likelihood of obtaining painful or unpleasant experiences. This is sometimes called the pleasure-pain principle and is referred to as **hedonistic calculus**. Bentham's views are reflected in his reforms of the criminal law in England. Bentham, like Beccaria, believed that punishment could act as a deterrent and that punishment's main purpose, therefore, should be to deter future criminal behavior.

Hedonistic calculus:
A term describing how humans seem to weigh pleasure and pain outcomes when deciding to engage in criminal behavior.

PUNISHMENT DURING EARLY AMERICAN HISTORY: 1700s–1800s

With the exception of William Penn, the penal reformists all came from Europe and did the majority of their work on that continent. Indeed, none of these persons (Montesquieu, Voltaire, Beccaria, Howard, and Bentham) were influential until after Penn's death in 1718. In fact, Beccaria, Howard, and Bentham were not born until after William Penn had passed away, while Montesquieu and Voltaire were in their mid-to-late 20s at this time. The reason that this is important is twofold. First, it is important for students to understand the historical chronological development of correctional thought. Second, this demonstrates that while the American colonies experienced reform in the early 1700s, this reform was lost when the Great Law in Pennsylvania was overturned upon Penn's death in 1718. From the time of Penn's demise until about 1787, penal reform and new thought on corrections largely occurred in Europe, leaving America in a social and philosophical vacuum (Johnson et al., 2008).

Old Newgate Prison:
First prison structure in America.

PHOTO 1.3 Connecticut's Old Newgate Prison (pictured here) was the first official prison in the United States.

Wikimedia Commons

This digression in correctional thought continued throughout the 1700s and culminated with what is today a little-known detail in American penological history. The **Old Newgate Prison**, located in Connecticut, was the first official prison in the United States. The structure of this prison reflects the lack of concern for reforming offenders that was common during this era. Old Newgate Prison was crude in design and, in actuality, served two purposes: It was a chartered copper mine, and from 1773 to 1827, it was used as a colonial prison. This prison housed inmates underground and was designed to punish the offenders while they were under hard labor. Due to the desire to strengthen security of the facility (successful escape attempts had been made), a brick-and-mortar structure was built around the entry to the mine that consisted of an exterior walled compound and observation/guard towers. Thus, this facility truly was a prison, albeit a crude one. However, it was not built for correctional purposes; *its purpose was solely punishment.*

Students are encouraged to read Focus Topic 1.1: Escape From Old Newgate Prison for a very interesting tale and historical account of the development and use of this prison. This prison is hardly mentioned in most texts on American corrections; this should not be the case since this was a very significant development in American penological history. Further, Old Newgate Prison demonstrates how the development of prison construction and correctional thought occurred over the span of years with many lessons that were hard learned. The history of this prison is a critical beginning juncture in American penology and also demonstrates how modifications to prison structure became increasingly important when administering a system designed to keep offenders in custody. As we will see in future chapters, the concern with secure custody plagued correctional professionals throughout subsequent eras of prison development, with custody of the offender being the primary mandate of secure facilities.

The Walnut Street Jail

While the Old Newgate Prison was in full operation in Connecticut, advocates of prison reform in Pennsylvania were gaining momentum after several decades of apparent dormancy. A little over 60 years had elapsed after William Penn's death when, in the late 1780s, an American medical doctor and political activist by the name of Benjamin Rush became influential in the push for prison reform (Carlson et al., 2008). In 1787, Rush, the Quakers, and other reformers met together in what was then the first official prison reform group, the Philadelphia Society for Alleviating the Miseries of Public Prisons (which was later named the Pennsylvania Prison Society), to consider potential changes in penal codes among the colonies (Carlson et al., 2008). This group

Walnut Street Jail: America's first attempt to incarcerate inmates with the purpose of reforming them.

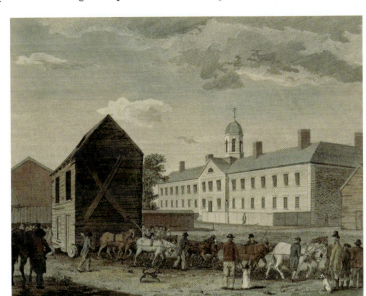

was active in the ultimate development of the penitentiary wing within the **Walnut Street Jail**, which was established in 1790 (Carlson et al., 2008). This development was America's first attempt to actually incarcerate inmates with the purpose of reforming them. A wing of the jail was designated an official penitentiary where convicted felons were provided educational opportunities, religious services, basic medical attention, and access to productive work activity. Thus, it is perhaps accurate to say that the Walnut Street Jail was also the first attempt at correction in the United States (Carlson et al., 2008). Eventually, counties throughout Pennsylvania were encouraged to transport inmates with long sentences to the Walnut Street Jail. This is thought to be the first move toward the centralization of the prison system under the authority of the state rather than of individual counties, as jails had until this time been organized.

PHOTO 1.4 The Walnut Street Jail, pictured here, was America's first attempt to actually incarcerate inmates with the purpose of reforming them.

The Miriam and Ira D. Wallach Division of Art, Prints and Photographs: Print Collection. The New York Public Library. (1800). Goal [i.e., jail] in Walnut Street Philadelphia. Retrieved from http://digitalcollections.nypl.org/items/510d47d9-7f13-a3d9-e040-e00a18064a99

While the Walnut Street Jail marked a clear victory for prison reformers, the jail (and its corresponding penitentiary wing) eventually encountered serious problems with overcrowding, time management, and organization as well as challenges with the maintenance of the physical facilities. Over time, frequent inmate disturbances and violence led to high staff turnover, and by 1835, the Walnut Street Jail was closed. This icon of reform stayed in operation only 8 years longer than the Old Newgate Prison.

However, it is extremely important that students read the following sentence very carefully: *The Walnut Street Jail was not the first prison in America; rather, it was the first penitentiary.* The difference is that a penitentiary, by definition, is intended to have the offender seek penitence and reform, whereas a prison simply holds an offender in custody for a prolonged period of time.

The Pennsylvania System

During the 1820s, two models of prison operation emerged: the Pennsylvania and Auburn systems (Carlson et al., 2008). These two systems came into vogue as the Old Newgate Prison was closed and once it became fairly clear that the Walnut Street Jail was not a panacea for prison and/or correctional concerns. With the approved allocation of **Western State Penitentiary** and **Eastern State Penitentiary**, the beginning of the Pennsylvania system was set into motion.

In 1826, the doors of Western State Penitentiary were open for the reception of inmates. The penitentiary opened with solitary cells for 200 inmates, following the original ideal to have solitary confinement without labor (Stanko, Gillespie, & Crews, 2004). However, doubts arose as to whether this would truly have reformative benefits among offenders and if it would be economical. Advocates of Western State Penitentiary contended that solitary confinement would be economical because offenders would repent more quickly, resulting in a reduced need for facilities (Sellin, 1970). While construction of Eastern State Penitentiary continued, planners were careful to learn from the mistakes of Western State Penitentiary. It is because of this that Eastern State Penitentiary has drawn most of the attention when historians and prison buffs talk about the Pennsylvania system of corrections.

Western State Penitentiary: Part of the Pennsylvania system located outside of Pittsburgh.

Eastern State Penitentiary: Part of the Pennsylvania system located near Philadelphia.

FOCUS TOPIC 1.1

Escape From Old Newgate Prison

Just a couple of years before the first shots of the American Revolution were fired, the Connecticut General Assembly decided that what the colony needed most was a good, heavy-duty gaol. In the legislators' wisdom, any new prison would have to meet certain specifications. It would have to be fairly close to Hartford; absolutely escape proof; self-supporting (i.e., inmates would have to be "profitably employed"); and—most important of all, then as now—cheap to build and maintain.

Near "Turkey Hills," in the region of northern Simsbury (now East Granby), there were some abandoned copper mines that had been sporadically dug with disappointing results since early in the century. The legislature immediately appointed a three-member study commission to "view and explore the copper mines at Simsbury."

The study group was mighty impressed with the prison potential of a many-shafted mine that ran deep under a mountain. Only 18 miles from Hartford, the mine boasted at least one cavern, 20 feet below ground, large enough to accommodate a "lodging room" that was 16 feet square. There were also lots of connecting tunnels where prisoners could be gainfully employed by being made to pick away at the veins of copper ore located there.

Better yet, according to the report, the only access to the mine from outside came from two air shafts: one 25 feet deep and the other 70 feet deep, the latter leading to "a fine spring of water." Still better was the low cost of mine-to-gaol conversion. By October 1773, the government had obtained a lease, carpenters had built the lodging room, and workmen had fitted a heavy iron door into the 25-foot air shaft, 6 feet beneath the surface. In the same month, the Connecticut General Assembly designated the place as "a public gaol or workhouse, for the use of this Colony"; named it Newgate Prison, after London's dismal house of detention; and appointed a "master" (or "keeper") and three "overseers" to administer the gaol.

Only men (never women) who had been convicted of the most dastardly crimes known to the colony—burglary, robbery, counterfeiting or passing funny money, and horse thieving—were eligible for a one-way trip into the state's dank, dark prison without walls. Chosen for the dubious honor of being Newgate's first prisoner was one John Hinson, a 20-year-old man about whom—considering his historic, "groundbreaking" status—surprisingly little is known. Convicted for some unrecorded crime and remanded to Newgate by the Superior Court on December 22, 1773, Hinson spent exactly 18 days in the "escape-proof" gaol before departing quietly for parts unknown. Although no one saw him leave, obviously, there was some evidence that he had used the 70-foot well shaft to climb out of the mine.

As a consequence of the successful escape of Hinson and, 3 months later, three more Newgate prisoners, it was ordered that modifications be undertaken that included, in 1802, the erection of a high stone wall around the prison.

Finally, in September 1827, after almost 54 years of operation, during which well over 800 prisoners were committed to its clammy, subterranean dungeons, Newgate Prison was abandoned, and the remaining inmates were transferred to the new state prison at Wethersfield. Significantly, the last escape attempt occurred on the night before the move to Wethersfield, when a prisoner fell back into the well—and drowned—as he tried to emulate old John Hinson of sainted memory. Coming when it did, at the bitter end of the facility's long, dark history, the death was a tragic, but somehow fitting, reminder of Newgate's most enduring legend. ●

Source: Philips, D. E. (1992). *Legendary Connecticut: Traditional tales from the nutmeg state*. Willimantic, CT: Curbstone Press. Copyright © 1992 by Joseph L. Steinberg. Reprinted by permission of Northwestern University Press.

In 1829, Eastern State Penitentiary opened. It was designed on a separate confinement system of housing inmates, similar to Western State Penitentiary. This system allowed inmates to reside in their cells indefinitely. Aside from unforeseen emergencies, special circumstances, or medical issues, inmates spent 24 hours a day in their cells. They had interactions with only a few human beings, most of them prison staff.

Eastern State Penitentiary was sometimes referred to as the Cherry Hill facility because it had been built on the grounds of a cherry tree orchard. The original structure had 252 cells, and each was much more spacious than those of Western State Penitentiary. Cells at Eastern were 12 feet long, 7 feet wide, and 16 feet high. The conditions within Eastern were quite humane and well ahead of their time. Indeed, as Johnston (2009) notes,

> Each prisoner was to be provided with a cell from which they would rarely leave and each cell had to be large enough to be a workplace and have attached a small individual exercise yard. Cutting edge technology of the 1820s and 1830s was used to install conveniences unmatched in other public buildings: central heating (before the U.S. Capitol); a flush toilet in each cell (long before the White House was provided with such conveniences); shower baths (apparently the first in the country). (p. 1)

It is clear that the physical conditions of this facility were sanitary even by today's standards. Further, the conditions of day-to-day treatment were also similar to what one might find in some prisons today.

Ultimately, the Pennsylvania system of separate confinement drew substantial controversy. The long periods of solitary confinement resulted in many inmates having emotional breakdowns, and various forms of mental illness emerged due to the extreme isolation. Prison suicide attempts became commonplace within the facility, which, by religious Quaker standards, meant that those inmates would not have their souls redeemed—an obvious failure at reform, both in the material world and in the spiritual world in which the Quakers believed. Eventually, the start of the Civil War made funds less available, and the practice of individual confinement was largely abandoned. Such was the demise of the Pennsylvania system of penitentiary management.

PHOTO 1.5 Western State Penitentiary, located outside of Pittsburgh, Pennsylvania, first opened with approximately 200 solitary cells for inmates in 1826.

© AP Photo/Keith Srakocic

The Auburn System

In 1816, 11 years before Old Newgate Prison closed in 1827, 19 years before the Walnut Street Jail closed in 1835, 10 years prior to the opening of Western State Penitentiary in 1826,

TABLE 1.2

Timeline for the Opening and Closure of Early American Prisons

PRISON	YEAR OPENED	YEAR CLOSED
Old Newgate Prison	1773	1827
Walnut Street Jail	1790	1835
Auburn Prison	1816	Still open. Renamed Auburn Correctional Facility.
Western State Penitentiary	1826	Closed in 2005 and reopened in 2007. Renamed State Correctional Institution at Pittsburgh.
Eastern State Penitentiary	1829	1971

Auburn system:
An alternative prison system located in New York.

and 13 years prior to the opening of Eastern State Penitentiary in 1829, the state of New York opened the Auburn Prison (see Table 1.2). The means that New York used to operate its prisons were different than the modes of operation in Pennsylvania. This alternative system was termed the **Auburn system** or congregate system, and under its provisions, inmates were kept in solitary confinement during the evening but were permitted to work together during the day. Throughout all of their activities, inmates were expected to stay silent and were not allowed to communicate with one another by any means whatsoever. Initially, this type of operation was implemented in Auburn Prison and the prison located in Ossining, New York. (Ossining would later be known as Sing Sing Prison.) The Auburn system was a significant turning point in American penology since it redefined much of the point and purpose of a prison facility.

Auburn designs tended to have much smaller cells than the Pennsylvania system, due to the fact that inmates were allowed out of their cells on a daily basis so that they could go to work. Auburn facilities were designed as industry facilities that had some type of factory within them. The economic emphasis throughout the Auburn system was one that became popular among other states and spread throughout the nation. In 1821, Elam Lynds was made warden at Auburn, and he was the primary organizer behind the development of the Auburn system. Warden Lynds contended that all inmates should be treated equally, and he believed that a busy and strict regimen was the best way to run a prison. Prison life included lockstep marching and very rigid discipline. It is at this time that the classic white-and-black striped uniforms appeared. All inmates were expected to work, read the Bible, and pray each day. The idea was that through hard work, religious instruction, penitence, and obedience, the inmate would change from criminal behavior to law-abiding behavior (Carlson & Garrett, 2008).

The Auburn system of prison operation initially had economic success due to several factors. First, the proceeds generated from inmate labor aided in offsetting the costs of housing the inmates. Second, the use of the congregate system allowed more productive work to take place—work that often required group effort. Third, other innovations of the Auburn system ensured its profitability. One of these was the use of inmate labor for

Contract labor system:
Utilized inmate labor through state-negotiated contracts with private manufacturers.

profit through a **contract labor system**, which eventually became a mainstay feature of the Auburn system. The contract labor system utilized inmate labor through state-negotiated contracts with private manufacturers who provided the prison with raw materials so that prison labor could refine those materials (Roth, 2011). Items such as footwear, carpets, furniture, and clothing were produced through this system.

Two American Prototypes in Conflict

Both the Pennsylvania system and the Auburn system of prison construction and management had achieved attention in Europe by the late 1830s and were seen as unique models of prison management that were distinctly American in thought and innovation (Carlson et al., 2008). It was not long, however, until questions regarding the superiority of one system over the other began to emerge. Both the Pennsylvania system and the Auburn system had potential benefits and drawbacks.

Ultimately, the Auburn system was the model that states adopted due to the economic advantages that were quickly realized. In addition, the political climate of the time favored an emphasis on separation, obedience, labor, and silence since sentiments toward crime and criminals were less forgiving during this era. Maintaining a daily routine of hard work was seen as the key to reform. Idleness, according to many advocates of this more stern system, provided convicts with time to teach one another how to commit future crimes. Thus, it was important to keep convicts busy so that they did not have the time or energy to dwell on the commission of criminal activity.

The Southern System of Penology: Before and After the Civil War

The climate and philosophy of southern penology has been captured on the silver screen in several classic prison movies, such as *Cool Hand Luke* and *Brubaker*. Indeed, more modern films, such as *O Brother, Where Art Thou?*, portray southern penology in a manner that is similar to its predecessors. When examining southern penology, it is important to understand the different cultural and economic characteristics of the region, particularly when comparing this type of prison system with the Pennsylvania and New York systems. From a historical, social, and cultural standpoint, students should keep in mind that the slave era took place during the early to mid-1800s (up until 1864 or so), and this impacted the manner in which corrections was handled in the South.

Prior to the Civil War, separate laws were required for slaves and free men who turned criminal. These laws were referred to as **Black Codes**, and they included harsher punishments for crimes than were given to white offenders (Browne, 2010). What is notable is that black slaves were not usually given prison sentences because this interfered with the ability of plantation owners to get labor out of the slave, a commodity desperately needed in the plantation system (Browne, 2010; Roth, 2011). Thus, during the pre–Civil War era, prisons typically had populations that included mostly white inmates with only a few free blacks (Browne, 2010).

After the Civil War, the economy was in ruin, and the social climate was chaotic throughout the southern United States. In a time when things were very uncertain, there were few resources of any sort, and ideas as to how the inmate population should be dealt with were scarce. Because there were not sufficient prison resources, the lease system continued to be implemented and expanded. It is interesting to point out that after the Civil War, over 90% of all leased inmates were in the South (McShane, 1996a, 1996b; Roth, 2011). This was largely due to the political and economic characteristics of the region as well as the termination of slavery that occurred with the South's defeat.

Eventually, southern states abolished the leasing system and created large prison farms that were reminiscent of the old plantations of the South (Roth, 2006). These farms operated to maximize profits and reduce the costs associated with incarceration of the inmate population. During this time, some major southern penal farms, such as Angola in Louisiana and Cummins in Arkansas, developed a sense of notoriety (Roth, 2006).

Since the majority of the law-abiding citizenry had no concern for the welfare of convicts, both of these systems proved to be lucrative and workable arrangements for businesses and state systems. With this in mind, it is perhaps accurate to say that southern penology took a step backward in correctional advancement and did so in a manner that maximized profit at the expense of long-term reform and crime reduction. Because these systems were profitable, there was no incentive to eliminate abuses.

Black Codes: Separate laws were required for slaves and free men who turned criminal.

PHOTO 1.6 Louisiana State Penitentiary Angola is a sprawling, farm-like state prison that was built on the grounds of a plantation in the South. This prison is now modern and sophisticated in the programming that is offered.

Lomax, A., photographer. (1934) Prison compound no. 1, Angola, Louisiana. Leadbelly Huddie Ledbetter in the foreground. Angola Louisiana United States. 1934. July. [Photograph] Retrieved from the Library of Congress. https://www.loc.gov/item/2007660073/.

PHOTO 1.7 Yuma Prison, pictured here, is reflective of the southwestern style of penology.

Larry Mayer/Stockbyte/Getty Images

The Chain Gang and the South

Chain gangs were a common feature within the southern penal system. This type of labor arrangement was primarily used by counties and states to build railroads and levees and to maintain county roads and state highways (Carroll, 1996). Most jurisdictions viewed this type of labor as a way to make money and also reduce overhead in housing inmates. The shackles were never removed from inmates on many chain gangs, and the men would usually sleep chained together in cages (Carroll, 1996).

In addition, the overseers of this system were poorly paid and often illiterate. This meant that, in a manner of speaking, the guard staff became dependent upon this system in which they settled for the substandard wage given as they furthered the cause of a system that exploited even them, though to a lesser extent when compared with the convict (Carroll, 1996). Given these circumstances and the limited skills of the guard staff, the use of brute force and clumsy tactics of inmate control prevailed.

The Western System of Penology

As crime rose in the Wild West, settlers responded by building crude jails in the towns that lay scattered across the desert terrain. These jails were not very secure and typically did resemble how they are often portrayed on American television (Carlson & Garrett, 2008). For the most part, they were used as holding cells, and long-term housing simply did not exist. During these years, most western states were territories that had not achieved statehood, and inmates were usually held in territorial facilities or in federal military facilities (Johnson et al., 2008).

As the need for space became greater, most western states found it more economical and easy to simply contract with other states and with the federal government to take custody of their inmates (Carlson & Garrett, 2008). The western states paid a set cost each year and simply shipped their offenders elsewhere; given the social landscape at the time, this was perhaps the most viable of options that these states could choose. According to Carlson and Garrett (2008), western states paid for other states to maintain custody of their offenders. This allowed western states to avoid the costs of building and maintaining large prisons and/or plantations. As time went on, state governments in the West developed, and the region became more settled. Once this occurred, western states began to build their own prisons. These prisons were designed along the lines of the Auburn system with an emphasis on labor.

THE AGE OF THE REFORMATORY IN AMERICA

In 1870, prison reformers met in Cincinnati and ultimately established the National Prison Association (NPA). This organization was responsible for many changes in prison operations during the late 1800s, which were listed in its Declaration of Principles (Wooldredge, 1996). This declaration advocated for a philosophy of reformation rather than the mere use of punishment, progressive classification of inmates, the use of indeterminate sentences, and the cultivation of the inmate's sense of self-respect—perhaps synonymous with self-efficacy in today's manner of speaking. These innovations eventually became themes in the evolution of American corrections. This meeting and the recommendations that emanated from it were actually quite remarkable for the time period in which this occurred. It was only a handful of years after the Civil War, and the cattle drives and Old West tales had not yet become legend.

Elmira Reformatory:
The first reformatory prison.

The first reformatory, **Elmira Reformatory**, was opened in July 1876 when the facility's first inmates arrived from Auburn Prison. Ironically, the site of the Elmira Reformatory had at one time been a prisoner-of-war camp for captured Confederate soldiers during the Civil War (Brockway, 1912; Wooldredge, 1996). The camp had a vile history, and thousands of southern soldiers died in the squalid, harsh, and brutal environment. However, the use of Elmira in 1876 was one of reform (thus the word *reformatory*), and this ushered in a new era in the field of penology.

The warden of Elmira Reformatory was a man by the name of **Zebulon Brockway**, who started his career in corrections as a prison guard in a state prison in Connecticut (Brockway, 1912). Brockway contended that imprisonment was designed to reform inmates, and he advocated for individualized plans of reform. During his term as warden, Brockway embarked on perhaps the most ambitious attempts to have the Declaration of Principles implemented within a correctional facility (Wooldredge, 1996). Judges, working within the framework of these principles and adopting an indeterminate sentencing approach, would sentence first-time offenders with modified indeterminate sentences. When serving these sentences, the reform of the offender was monitored, and, if successfully reformed, the offender was released prior to the expiration of the sentence. If the offender did not demonstrate sufficient proof of reform, he simply served the maximum term.

The Elmira Reformatory used a system of classification that had been produced due to Brockway's admiration of the work of Alexander Maconochie, a captain in the British Royal Navy who in 1837 was placed in command over the English penal colony at Norfolk Island. While serving in this command, Maconochie proposed a system where the duration of the sentence was determined by the inmate's work habits and righteous conduct. Called a mark system because "marks" were provided to the convict for each day of successful toil, this system was quite well organized and thought out (Brockway, 1912).

Under this plan, convicts were given marks and were moved through phases of supervision until they finally earned full release. Because of this, Maconochie's system is considered indeterminate in nature, with convicts progressing through five specific phases of classification. **Indeterminate sentences** include a range of years that will be potentially served by the offender. The offender is released during some point in the range of years that are assigned by the sentencing judge. Both the minimum and maximum times can be modified by a number of factors, such as offender behavior and offender work ethic. The indeterminate sentence stands in contrast to the use of **determinate sentences**, which consist of fixed periods of incarceration imposed on the offender with no later flexibility in the term that is served. Brockway was a strong advocate of the indeterminate concept and believed that it was critical to turning punishment into a corrective and reformative tool. Ultimately, it was found that these institutions were actually no more successful at molding inmates into law-abiding and productive citizens than were prisons, and by 1910, the reformatory movement began to decline in use.

Zebulon Brockway: The warden of Elmira Reformatory.

Indeterminate sentences: Sentences that include a range of years that will be potentially served by the offender.

Determinate sentences: Consist of fixed periods of incarceration with no later flexibility in the term that is served.

PRISONS IN AMERICA: 1900s TO THE END OF WORLD WAR II

Prison Farming Systems

The prison farm concept was one that began in Mississippi and then extended throughout a number of southern states. The use of this type of prison operation lasted until well after World War II. As was noted earlier, prison farms were profit driven and based on agricultural production. Even though their particular market was agricultural, much of their operation was similar in approach to industrial prisons; the key difference was simply in the product that was manufactured. Two systems in particular capture the essence of southern prison farming: Arkansas and Texas.

The Arkansas System: Worst of the Worst

The conditions within the Arkansas prison system are thought to be the worst of all those among the southern prison farm era. The Arkansas system actually only consisted of two prison plantations, the Cummins Farm, which covered approximately 16,000 acres of territory, and the Tucker Farm, which spanned about 4,500 acres of territory. Each of these facilities produced rice, cotton, vegetables, and livestock. What made this prison system so particularly terrible was the corruption, brutality, and completely inhumane means of operation that existed.

The Arkansas prison system, similar to the Mississippi prison system, placed inmates in charge of other inmates. In Arkansas, these inmates were referred to as trusties and were at the top of the inmate hierarchy. Civilian employees in the prisons in Arkansas were scarce, meaning that trusties were responsible for most of the day-to-day order on the farm. The trusties served as guards over the other inmates and carried weapons. They also controlled and operated critical services, such as food and medical services. Trusties had their own dormitory to themselves, more freedom than other inmates, and the best food, and they were free to extort other inmates for money, goods, or services. As one might expect, such extortion happened quite frequently.

The overall supervisor of this system was the superintendent, whose primary role was to ensure that the prison farm operated at a profit. This meant that the superintendent tended to provide all authority to the trusties, so long as they made the prison a profit. The control of desperate, underfed, exhausted, and often ill inmates was maintained through a process of constant punishment. Some of these punishments were nothing less than the use of torture. Punishments included whipping; the inmate's fingers, nose, ears, or genitals being pinched with pliers; and even inserting needles under the inmate's fingernails. One of the most infamous forms of torture used was the "Tucker Telephone." The Tucker Telephone consisted of an old-fashioned crank telephone wired in sequence with two batteries. Electrodes coming from it were attached to a prisoner's big toe and genitals. The electrical components of the phone were modified so that cranking the telephone sent an electric shock through the prisoner's body.

The Progressive Era

Progressive Era: A period of extraordinary urban and industrial growth and unprecedented social problems.

From 1900 to 1920, numerous reforms took place across the United States, and this led to some dubbing this period the Age of Reform. For prison operations, the Age of Reform reflected an era of change and attention to humane treatment of inmates. During the **Progressive Era**, a particularly influential group, known as the Progressives, cast attention on social problems throughout the nation and sought to improve the welfare of the underprivileged. The members of this group remained steadfast in the belief that understanding deviant behavior lay with social and psychological causes, and they also contended that social and psychological treatment programs were the key to offender reform. Due to this line of thought and the influence of the Progressives, the field of penology eventually included psychologists, social workers, and psychiatrists in addition to lawyers and security staff.

The Era of the "Big House"

Big House prisons: Typically large stone structures with brick walls, guard towers, and checkpoints throughout the facility.

The Big House era lasted from the early 1900s to just before the emergence of the civil rights movement.

Big House prisons were typically large stone structures with brick walls, guard towers, and checkpoints throughout the facility. The key architectural feature to Big House prisons was the use of concrete and steel. The cell blocks sometimes had up to six levels, making the entire structure large and foreboding. The interior of each cell block often was extremely hot and humid during the summer months and cold during the winter months. In addition, these structures magnified noise levels, creating echoes throughout as steel doors and keys clanged open and shut, announcements were made, and machinery operated within the facility.

The Medical Model

During the 1930s, another perspective emerged regarding inmate treatment and the likelihood for reform. The medical model developed in tandem with the rise of the behavioral sciences in the field of corrections (Carlson et al., 2008). The **medical model** can be described as correctional treatment that utilizes a type of mental health approach incorporating fields such as psychology and biology; criminality is viewed as the result of internal deficiencies that can be treated. The key to the medical model is understanding that it is rehabilitative in nature.

The medical model was officially implemented in 1929 when the U.S. Congress authorized the Federal Bureau of Prisons to open correctional institutions that would use standardized processes of classification and treatment regimens within their programming. One early proponent of the medical model and its clinical approach to rehabilitation was Sanford Bates, who was the first director of the Bureau of Prisons and had also served as a past president of the American Correctional Association (students will recall that this was originally named the National Prison Association in 1870).

At the heart of the medical model was the classification process; everything in the medical model that followed hinged on the accuracy and effectiveness of this process. The developers of the process believed that such a systematic approach would improve treatment outcomes and overall recidivism among offenders. However, as Carlson et al. (2008) note, "Although classification was one of the greatest concepts invented during this period, it became at best a management process rather than a reliable tool to aid in rehabilitation" (p. 13). This, unfortunately, emerged as the truth across the nation, and classification ultimately became a systematic process for housing and to aid institutional and community-based professionals in managing the inmate population rather than for changing the inmates' behavior.

Medical model:
An approach to correctional treatment that utilizes a type of mental health approach incorporating fields such as psychology and biology.

The Reintegration Model

The **reintegration model** evolved during the last few years that the medical model was still in vogue. The term *reintegration* was used to identify programs that looked to the external environment for causes of crime and the means by which criminality could be reduced. This model was commonly used during the 1960s and 1970s as an alternative to punitive approaches that were gaining momentum. However, as crime continued to rise, strong skepticism of both the medical model and the reintegration model became commonplace. One of the sharpest and most distinctive blows to both of these models "was a rather infamous negative report produced in the early 1970s by a researcher studying rehabilitation programs across the country" (Carlson et al., 2008, p. 16). This report was the work of Robert Martinson, who had conducted a thorough analysis of research programs on behalf of the New York State Governor's Special Committee on Criminal Offenders.

Martinson (1974) examined a number of various programs that included educational and vocational assistance, mental health treatment, medical treatment, and early release. In his report, often referred to as the **Martinson Report**, he noted that "with few and isolated exceptions, the rehabilitative efforts that have been reported so far have had no appreciable effect on recidivism" (Martinson, 1974, p. 22). Martinson's work was widely disseminated and used as ammunition for persons opposed to treatment, whether individual- or community-based. Thus, skepticism of rehabilitation and/or reintegration rose to its pinnacle as practitioners cited (often in an inaccurate manner) the work of Robert Martinson.

Reintegration model:
Used to identify programs that looked to the external environment for causes of crime and the means to reduce criminality.

Martinson Report:
An examination of a number of various prison treatment programs.

The Crime Control Model

The **crime control model** emerged during a "get tough" era on crime. The use of longer sentences, more frequent use of the death penalty, and an increased use of intensive supervised probation all were indicative of this era's approach to crime. The use of determinate sentencing laws took the discretion from many judges so that, like it or not, sentences were awarded at a set level regardless of the circumstances associated with the charge. Increasingly, states and the federal government are realizing that the approach of the crime control era may have been a bit too ambitious, particularly since states cannot afford, in the current state of the economy, to pay the bills for the long-term incarceration that has been invoked under this approach.

Crime control model:
An approach to crime that increased the use of longer sentences, the death penalty, and intensive supervision probation.

MODERN-DAY SYSTEMS: FEDERAL AND STATE INMATE CHARACTERISTICS

The Federal Bureau of Prisons (BOP) was initially established by Congress in 1930 and has since that time become a highly centralized organization with over 33,000 employees who supervise more than 209,000 inmates. The federal system has over 100 facilities that include maximum-security prisons, supermax facilities, detention centers, prison camps, and even halfway houses. The variety of correctional services provided by this system is much greater than what most state systems provide (Federal Bureau of Prisons, 2010b).

Since the War on Drugs that occurred during the 1980s, the proportion of drug offenders has remained high, constituting more than half of the BOP population (Carson, 2014). However, unlike state prisoners, most federal offenders are not violent, and their drug crimes are also not usually associated with violence. Also interesting is that roughly 12% of all federal inmates are citizens of other countries (Carson, 2014). As an indicator of the types of crimes and the types of criminals that tend to be included in the federal system, consider that 54% of federal inmates are classified as being either a low- or minimum-security risk, with the average time served for BOP inmates being around 6.5 years in length (Federal Bureau of Prisons, 2010b).

Within state correctional systems, there is quite a bit of variety, in terms of both their operation and the inmates that they house. The size of prisons within one state can have a great degree of variability. The Louisiana State Penitentiary (Angola) houses over 5,000 inmates (more than the total inmate count for the entire state of North Dakota), while other prisons in other states may house fewer than 1,000 inmates. A wide variety of types of facilities may be included in a state system, just as with the federal system described previously. Additionally, working for one state prison system can be quite different from working for another in terms of salary, training, opportunities, and so forth.

In late 2016, national statistics indicated that more than half (54%) of all state prison inmates were violent offenders, while nearly half (47%) of federal inmates were drug offenders (Carson, 2018). To make matters worse for state systems that house these difficult populations, consider that state budgets tend to not be as large as the federal budget, so funding is often an issue that keeps state systems from operating as effectively as the BOP. This also means that working conditions, salaries, and training among state prison staff tend to vary, though the American Correctional Association has been very influential in professionalizing the field of corrections throughout numerous states. All in all, state corrections tends to be the most common form of corrections, but, despite advances, these systems do not fare as well as the BOP.

It is also important to note that the majority of inmates are housed in state prison systems. Among these, most are in custody in one of the seven largest prison systems. The largest three systems each have populations near to or over a 100,000 inmate head count and include Texas (with 163,703 inmates), California (with 130,390 inmates), and Florida (with 99,974 inmates). All three of these state prison populations are significantly larger than the other 47 state systems to which they can be compared (Carson, 2018). According to Carson (2018), the remaining four of the largest seven systems each house between 49,000 and 54,000 inmates and include the states of Georgia (53,267 inmates), Ohio (52,175 inmates), New York (50,716 inmates), and Pennsylvania (49,244 inmates). Collectively, these four prison systems house just under 205,000 inmates (Carson, 2018). All of the other states were reported to house fewer than 49,000 inmates, with most housing substantially less than this number.

The Emergence of the Top Three in Corrections

This term the *Top Three in corrections* is an apt description of the three largest state correctional systems in the United States. Texas is the largest system, California is the next largest, and Florida is third (Carson, 2018). These systems are referred to as the Top Three due to the fact that they are the largest three systems according to inmate count. We will not discuss each state individually. Rather, students should understand that the Top Three in corrections are important for a number of reasons that go beyond their mere head count.

First, these three states have large overall free-world populations as well as prison populations. This means that each of these states has a large population that is likely to be more representative of the overall U.S. population than would be the case for numerous other states. When taken in total, these three states should be considered somewhat representative of the overall U.S. population. Because they are representative, this means that research conducted from samples taken from these three states will, collectively, be likely to yield results that generalize to the rest of the United States.

Second, each of these states has had to grapple with immigration issues and the constant ingress and egress of legal and illegal persons within its borders. This is a unique characteristic that is not shared by a majority of the states. While other states may also struggle with this issue, the Top Three do so on a large-scale basis. This makes a difference because of the type of crime problems that are encountered (i.e., more drug trafficking, smuggling issues, and organized crime activity) as well as the factors that are associated with those problems (more drug use, cultural clashes, and more complicated crime problems). Third, these states all possess a truly diverse array of racial and cultural groups. The history of each of the Top Three reflects exchanges between various cultures. In all three states, the Latino population is well represented, as are the African American and Asian American populations. Other racial and cultural groups are likewise represented in each of these three states, partially due to routine immigration and also due to the unique histories of the states.

Fourth and lastly, each of these three states tends to have a fairly robust economy. The market conditions in all are active and vibrant due to their locations (all have extensive coastlines) and due to a sufficient number of urban areas within their borders. The fact that these three states tend to have more stable economies (at least throughout most of their history) impacts how well they are able to fund their correctional programs. This can make a considerable difference in the overall approach to a correctional agency's response in processing the offender population.

CONCLUSION

Corrections is a term that has origins in the need and/or desire to punish those who commit an aberrant behavior that is proscribed by society. Indeed, the terms *punishment* and *corrections* have shared common meanings throughout history. This text presents the term *corrections* as a process whereby practitioners from a variety of agencies and programs use tools, techniques, and facilities to engage in organized security and treatment functions intended to correct criminal tendencies among the offender population.

In ancient times, the ability of an aggrieved party to gain retribution for a crime required some form of retaliation. In most cases, individuals or groups only achieved retribution if they were able to personally extract it from the offender. Later, over time, rulers of various groups organized processes of achieving retribution, thereby reducing the likelihood that conflicts between individuals and groups would escalate. Regardless of the type of customs that existed in various areas of Europe, the use of physically humiliating punishments and crippling punishments was still widespread. When examining the history of punishment and corrections, it is clear that early forms of punishment were quite barbaric when compared with those today.

The rise of the Enlightenment and the writings of a variety of scholars and philosophers helped shape the use of simple punishments from barbaric cruelty to corrective mechanisms intended to reduce problematic behaviors. Further, a distinct sense of rationality was used in administering punishments, and new concepts were introduced. One of the premiere figures who advocated the use of reason was Cesare Beccaria. It was Beccaria who advocated for proportionality between the crime committed and the punishment received. Beccaria also contended that it was the certainty of punishment, not the severity, that would be more likely to deter crime. These novel concepts, as well as the contention that offenders should be treated humanely, marked the Enlightenment and the emergence of prison reform in Europe and the United States.

As prison development in America began, two competing mindsets emerged: the Pennsylvania and Auburn systems of prison operation. Numerous dichotomies and disagreements in philosophy as to the rightful goal of prisons emerged as the Pennsylvania system and the Auburn system competed. The intent of the Pennsylvania system was strictly to reform offenders. On the other hand, the primary motive behind the Auburn system had a business-model perspective—prisons should be self-sufficient or as close to self-sufficient as possible. Ultimately, the money-making option was more compatible with the capitalist notions of the United States, and the Auburn system gave way to the penal farm, particularly in the southern United States.

The profit motive ultimately drove southern states to implement the farming prison, while the northeastern areas of the nation adopted the use of prison industries. In both cases, inmates were leased out to private businesses that could make a profit off of inmate labor. This again highlighted the impact of the Auburn system. In the South, the use of prison farms became reminiscent of the old plantation era prior to the Civil War, and, in fact, some prisons were built right on the grounds of prior plantations. The traditions in the South, along with racial discrimination and disparity and poor economic circumstances, served to replicate many of the injustices that occurred in the prior slave era, just under a different guise.

The Big House era emerged from the prison industry model, but, unlike the prison industry or the prison farming approach, inmates in the Big House were not put through grueling labor, and they were not subjected to the same level of rule setting as inmates in the past. Eventually, the Big House era, the prison industry model, and the prison farm model gave way to the state and federal systems that we now have in place. In 1930, the Federal Bureau of Prisons was established and has emerged as a premiere correctional agency. Among state prisons systems, three states (Texas, California, and Florida) are by far the largest of the state systems, with each at or exceeding 100,000 inmates. These three states collectively include nearly one third of the entire state inmate population throughout the nation. Because of this, any research or other generalization made about corrections in the United States should, at least for the most part, include each of these states as an object of interest. Going further, these systems, along with four others that combined include another approximately 210,000 inmates, house most of the violent offenders throughout the United States despite the tight correctional budgets with which they must operate.

Want a Better Grade?

Get the tools you need to sharpen your study skills. Access practice quizzes, eFlashcards, video, and multimedia at **edge.sagepub.com/hanserbrief**

Interactive eBook

Visit the interactive eBook to watch SAGE premium videos. Learn more at **edge.sagepub.com/hanserbrief/access**.

 Career Video 1.1: Jail Administrator

 Criminal Justice in Practice Video 1.1: Overview of the Criminal Justice System

 Prison Tour Video 1.1: Punishment Reform and Living and Working Conditions

DISCUSSION QUESTIONS

Test your understanding of chapter content. Take the practice quiz at edge.sagepub.com/hanserbrief.

1. Identify *punishment* and identify *corrections*. How does each differ from the other, and why are they often confused with one another?

2. How has punishment progressed from ancient and medieval times to current practices? Are there still similarities in thought, and, if so, what are they?

3. Identify key thinkers and persons of influence who have impacted the field of corrections. For each, be sure to highlight their particular contribution(s) to the field.

4. What is the significance of Old Newgate Prison? What distinguishes this structure from the penitentiary wing added to the Walnut Street Jail? Why is Old Newgate Prison important to correctional history in the United States?

5. Explain how the classical school of criminology, behavioral psychology, and the field of corrections can be interrelated in reforming offender behavior.

6. What are some key differences between the Pennsylvania and Auburn prison systems?

7. How did different regions vary in their approaches to prison operations? Compare at least two regions.

8. What is meant by the Top Three in American corrections, and why is this important?

KEY TERMS

Review key terms with eFlashcards at edge.sagepub.com/hanserbrief.

Auburn system, 16

Banishment, 8

Big House prisons, 20

Black Codes, 17

Branding, 7

Brutalization hypothesis, 4

Classical criminology, 11

Code of Hammurabi, 4

Contract labor system, 16

Corrections, 2

Crime control model, 21

Determinate sentences, 19

Eastern State Penitentiary, 14

Elmira Reformatory, 18

Great Law, 10

Hedonistic calculus, 12

Indeterminate
  sentences, 19

Lex talionis, 4

Martinson Report, 21

Medical model, 21

Old Newgate Prison, 12

Private wrongs, 6

Progressive Era, 20

Public wrongs, 6

Reintegration model, 21

Sanctuary, 5

Trial by ordeal, 5

Walnut Street Jail, 13

Western State
  Penitentiary, 14

Zebulon Brockway, 19

APPLIED EXERCISE 1.1

Student Debate

Many people in society believe that incarcerated offenders should be made to work as a means of paying for their crime and supporting their stay while in prison. This, in and of itself, is not a problematic notion. However, inmates must be given humane working conditions, and, as a result, there are limits to the type of work they can do and the circumstances under which it is done.

The ancient Romans essentially considered the inmate to be civilly dead and to also be a slave of the state. Though modern-day thinking by prison management does not advocate for inmates to hold such an arbitrary classification, some might say that such a classification is appropriate for offenders.

For this exercise, have half the room or forum argue for classifying offenders as slaves of the state and the other half argue against categorizing inmates in such a way.

Students should keep in mind some of the counterintuitive findings when punishment is too severe, and they should also consider the thoughts of Cesare Beccaria and other philosophers on corrections. Both teams of students should come up with at least three substantial points to argue for the side they have been assigned.

> **Group 1:** Half of the students in the classroom (or the online forum) provide tangible and logical reasons for why inmates should be treated as slaves of the state.

> **Group 2:** Have the other half of the room or forum argue against categorizing inmates as slaves of the state.

The instructor should regulate the debate and encourage students to find specific examples from the text and/or their own independent research.

WHAT WOULD YOU DO?

You are a judge in Old England, the year is 1798, and the Crown has given you some very explicit instructions for this week. It appears that there is no room onboard the hulks that float in the River Thames, and, due to the traitorous rebellion of the American colonies in the New World, there is nowhere to transport criminals for banishment. With this in mind, the Crown is desperate to reduce criminal acts and has recently decided that the best means to do this is by setting some very strong and severe examples to the public. Thus, you have been told that you must use one of two sentences today: provide the death penalty for anyone found guilty of any crime that is eligible for it or find those persons innocent of their charges and thereby make invalid the need for any punishment whatsoever. In other words, you must either rule innocence or give all the offenders in your court the death penalty.

This is during a time when England's criminal code has been given the nickname of the "Bloody Code" among the commoners of the British Empire. You are well aware that there is serious discontent among the peasantry in your area with this code and that the Crown has previously approached the extensive use of the gallows with trepidation; such circumstances can breed riots and, in very extreme times, rebellion. On the other hand, you know that many of the wealthy in the area are typically supportive of harsh penalties against the working poor (such penalties discourage theft of their own property). You sit at your bench, waiting to make your determination regarding three offenders who are accused of different crimes. All of the crimes for which they are accused would entail the use of the death penalty. These three offenders and their circumstances are noted as follows:

> Offender 1: Mr. Drake Dravies, a brigand and a buccaneer who deflowered a 10-year-old girl against her will and attempted to kill her but was caught before doing so. *You know for a fact that this man committed this crime.*

> Offender 2: Ms. Eliza Goodberry, a single spinster maid who worked in the fish market. She was found guilty of being a witch and consorting with demons. It is rumored that she gave secret birth to a demon child. *You know for a fact that this woman did not commit this crime, and you know that she is not a witch.*

> Offender 3: Mr. John McGraw, a general laborer who stole food in the open market (and almost got away with it) to feed his family. Labor shortages and tough economic times have left him with few other options. *You know for a fact that this man committed this crime, and you also know that it is true that he committed this crime simply to feed his family.*

You must make a decision: Either you must declare all three innocent of the crimes as charged, or you must give all three the death penalty by hanging at the gallows. There is no option to try these persons for these crimes at a later date.

What do you do?

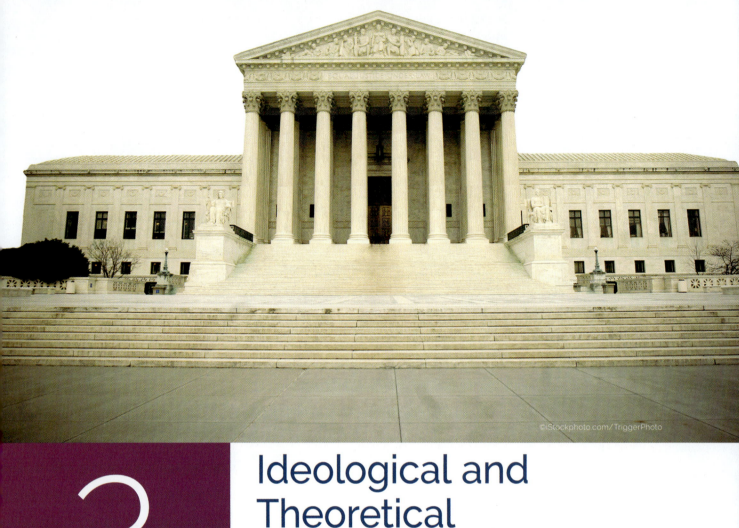

2

Ideological and Theoretical Underpinnings to Sentencing and Correctional Policy

Learning Objectives

1. Identify and discuss the philosophical underpinnings associated with correctional processes.
2. Identify and discuss different types of sanctions used in correctional operations.
3. Evaluate the outcomes of different sentencing schemes.
4. Apply criminological theories to different correctional processes.
5. Integrate philosophical underpinnings, types of sanctions, sentencing schemes, and criminological theories to develop a multifaceted understanding of corrections.

The Original Gangsta

It was about 3:00 a.m., and Desmond's cellie, Tederick, was up on the top bunk, keeping watch. Desmond took out the shank hidden behind the toilet and worked it back and forth against his metal bunk. The noise from the continuous friction was loud enough to be heard in the cell but not so loud as to resonate throughout the entire cellblock.

Tederick continued to look out the cell onto the run below to see if the guard or anyone else was listening or aware of what was happening. All was quiet, including the cells next to theirs. The other inmates knew what was going on and minded their own business if they were awake; others slept through the noise.

Tederick asked Desmond, "So, you gonna get him in the rec yard or in the dayroom?"

Desmond replied, "I'm going to hit him in the rec yard."

Desmond thought about the situation and how he was going to get the blade to the rec yard. The inmate that he was going to "hit," Cedric Jackson, was a member of an opposing gang who had been talking smack. Both Desmond and Cedric were members of small, local gangs in New Orleans; neither was affiliated with large gangs like the Crips or the Bloods.

Desmond had lived a life of poverty in New Orleans, and his father had died in prison. His mother did her best, working odd jobs and raising three kids as a single mom. Desmond's cousin, Nate, always had a strong influence on Desmond. Nate had a car with really fly rims, he had women, he had dope, and he had respect on the streets. But now Nate was at Angola, doing real time, and he was writing letters to Desmond to take care of some "business" for him.

Tederick looked down at Desmond and said, "I thought that you wanted to go to school and hook up with that girl?"

"Yeah, that's what I wanna do," Desmond responded.

"Then if you make this hit on Cedric to get even for Nate, you gonna be in here for a long time; maybe you should forget about ole girl and just go to school a few years from now."

Desmond thought about this. He considered how Nate had always had "stuff" when on the streets but was now stuck in Angola for at least another 15 years.

Desmond also thought about his "ole girl," Angela, and all the letters she had sent him. His mom thought well of Angela, and they both had seen to it that when he got out he would be able to get settled and get a job with a nearby warehouse (his mom worked there in the administrative office). Angela would even help him get started in school.

Tederick spoke again. "You know that they are really looking at giving more good time for the drug treatment programs, don't you? The feds and the state are reducing sentences, giving more good time, and closing down prisons. . . . This is a good time to be doing time 'cause you can get out early, right?"

Desmond frowned. "Yeah, I guess so. What are you trying to say?"

Tederick shook his head. "Man, I am saying to hell with Nate and to hell with doing time for Nate. . . . You gotta do you, man, and get on with your life. Let Nate take care of his own business. Let Cedric run his mouth. He is getting shipped soon, anyway, and you can be out in 6 months if you play your cards right." Locking eyes with Desmond, Tederick urged, "Look man, you got a girl on the outside who cares about you, a mom who can get you a job, and you might be in school within a year, or you can sit and rot some more in here. That might be what you want, but it ain't for me, no sir!" His voice rose and he slapped his hand down on his mattress. "I am not lettin' THE MAN take my life from me, and I ain't letting all the wrong learning I got from the streets decide my life for me! You shouldn't either, homie!"

Desmond snorted. "Fine, so I let him make it; I give him a break.... What then; what about the gang?"

"You give me the steel and I can get rid of it. I got 3 more years, but you can be out in 6 months. Do the drug program that just accepted you and do it *for real*; then get a real life, not this hell hole. Forget the gang; right now you do not owe them.... It is all square business at this point. Go further with it, and you won't ever get out of it."

Desmond turned the shank over in his hands as he considered Tederick's words. He stood, looked at his cellie, and said, "You know, I ain't never had many real friends." After only a moment's hesitation, he extended the homemade blade, handle first, and dropped it onto Tederick's open palm.

Tederick smiled. "My brother, you are doing the right thing; trust me."

INTRODUCTION

This chapter focuses on the reasons for providing correctional services in today's society. In considering these reasons, it is important to understand two key aspects related to corrections. First, it is helpful to be familiar with the historical developments related to punishment and corrections. Chapter 1 provided information about how our current views on correctional practices have evolved. Understanding the history of corrections helps us to make sense of today's correctional system. This is true for legal precedent that shapes correctional policies as well as philosophical and/or political motives behind our use of correctional resources. Thus, it is the rich history of corrections that has shaped it into what we know today.

The second aspect is the need for a clear definition of the term *corrections*. As demonstrated in Chapter 1, this term can have many different meanings to many different practitioners, scholars, and researchers around the world. Nevertheless, it is important to be able to define the term in a clear and succinct manner so that one can correctly connect it with the means by which correctional practices are implemented and the reasons for implementing them. This is essential since this is what will provide clarity in purpose, which, in turn, should lead to clarity in action.

PHILOSOPHICAL UNDERPINNINGS

Within the field of corrections itself, four goals or philosophical orientations of punishment are generally recognized. These are retribution, deterrence, incapacitation, and treatment (rehabilitation). Two of these orientations focus on the offender (treatment and specific deterrence), while the others (general deterrence, retribution, and incapacitation) are thought to focus more on the crime that was committed. The intent of this section of the chapter is to present philosophical bases related to the correctional process. In doing this, it is useful to first provide a quick and general overview of the four primary philosophical bases of punishment (see Table 2.1). These bases were touched upon in Chapter 1 but are now provided in more detail and with the purpose of elucidating the true purposes and rationales behind the correctional process.

Retribution: Offenders committing a crime should be punished in a way that is equal to the severity of the crime they committed.

Retribution

Retribution is often referred to as the "eye for an eye" mentality, and it simply implies that offenders committing a crime should be punished in a like fashion or in a manner

TABLE 2.1

Philosophical Underpinnings in Corrections

PHILOSOPHICAL UNDERPINNING	PREMISE
Retribution	Implies that offenders committing a crime should be punished in a like fashion or in a manner that is commensurate with the severity of the crime.
Incapacitation	Deprives offenders of their liberty and removes them from society with the intent of ensuring that society cannot be further victimized.
Deterrence (general and specific)	General deterrence occurs when observers see that offenders are punished for a given crime and are themselves discouraged from committing crime. Specific deterrence is punishment upon a specific offender in the hope that the offender will be discouraged from committing future crimes.
Rehabilitation	Offenders will be deterred from reoffending due to their having worthwhile stakes in legitimate society.
Restorative justice	Interventions that focus on restoring the health of the community, repairing the harm done, meeting victims' needs, and emphasizing that the offender can and must contribute to those repairs.
Reintegration	Focused on the reentry of the offender into society.

that is commensurate with the severity of the crime that they have committed. As discussed in Chapter 1, retribution is the justification for punishment by the concept of *lex talionis*. It is a "just deserts" model that demands that punishments match the degree of harm that criminals have inflicted on their victims (Stohr, Walsh, & Hemmens, 2013). Thus, those who commit minor crimes deserve minor sentences, and those who commit serious crimes deserve more severe punishments (Stohr et al., 2013). This model of punishment is grounded in the idea that, regardless of any secondary purpose that punishment might be intended to serve, it is right to punish offenders because justice demands it. In

PHOTO 2.1 One example of retribution would be having someone who vandalized property work to repair the damage he or she caused.

Joe Sohm Visions of America/Newscom

essence, society has an ethical duty and obligation to enforce the prescribed punishment; otherwise the sentencing process is based on lies and exceptions.

It is important that students not equate retribution with the mere practice of primitive revenge; retribution has many distinctions that set it apart from such a simplistic understanding. Retribution is constrained revenge that is tempered with proportionality and enacted by a neutral party. This neutral party is required to stay within the bounds of laws that afford offenders certain rights despite the fact that they are to be punished. As we have seen in Chapter 1, the use of this formalized method of punishment emerged out of the chaotic times where blood feuds and retaliation for private wrongs abounded. Retribution was grounded in the notion that the offender (or the offender's family) must pay for the crime committed. The need to keep feuds from escalating between aggrieved families was important among the ruling class. Thus, retribution was designed to adhere to a rational process of progressive sanctions, separating it from mere retaliation.

In addition, when we hold offenders accountable for their actions, we make the statement that we (as a society) believe that offenders are free moral agents who have self-will.

PHOTO 2.2 It is hoped that placing offenders in such noxious circumstances will deter them from further criminal behavior. There is no actual empirical proof that this is the case.

©iStockphoto.com/ISignature Collection

Incapacitation: Deprives offenders of their liberty and removes them from society, ensuring that they cannot further victimize society for a time.

Selective incapacitation: Identifying inmates who are of particular concern to public safety and providing them with much longer sentences.

General deterrence: Punishing an offender in public so other observers will refrain from criminal behavior.

Specific deterrence: The infliction of a punishment upon a specific offender in the hope that he or she will be discouraged from committing future crimes.

It is the responsibility of the offender, not society, to pay for the crime that has been committed. Once this payment (whatever the sanction might be) has been made, there is no further need for punishment. While this type of approach works well in justifying punishment of offenders who are culpable and cognizant of their crime, it is not appropriate for offenders who have mental deficiencies and/or mitigating circumstances that remove fault from them. It is in these cases where retribution loses its logical application within the punishment or correctional process.

Incapacitation

Incapacitation simply deprives offenders of their liberty and removes them from society with the intent of ensuring that society cannot be further victimized by them during their term of incarceration. The widespread use of incapacitation techniques during the 1990s is purported by some experts to be the cause for the drop in crime that was witnessed after the year 2000. Though this has not been proven, the argument does seem to possess some potential validity. Regardless, it has become increasingly clear that the use of mass incarceration efforts simply cannot be afforded by most state budgets. This has led to more increased use of community corrections techniques and techniques of selective incapacitation.

Selective incapacitation is implemented by identifying inmates who are of particular concern to public safety and by providing those specific offenders with much longer sentences than would be given to other inmates. The idea is to improve the use of incapacitation through more accurate identification of those offenders who present the greatest risk to society. This then maximizes the use of prison space and likely creates the most cost-effective reduction in crime since monies are not spent housing less dangerous inmates.

Deterrence

Deterrence is the prevention of crime by the threat of punishment (Stohr et al., 2013). Deterrence can be general or specific. **General deterrence** is intended to cause vicarious learning whereby observers see that offenders are punished for a given crime and so they are discouraged from committing a similar crime due to fear of punishment. **Specific deterrence** is simply the infliction of a punishment upon a specific offender in the hope that that particular offender will be discouraged from committing future crimes. For specific deterrence to be effective, it is necessary that a punished offender make a conscious connection between an intended criminal act and the punishment suffered as a result of similar acts committed in the past.

Stohr and Walsh (2011) note that the effect of punishment on future behavior also must account for the contrast effect, a notion that distinguishes between the circumstances of the possible punishment and the life experience of the person who is likely to get punished. As they explain it,

> For people with little or nothing to lose, arrest and punishment may be perceived as merely an inconvenient occupational hazard, an opportunity for a little rest and recreation, and a chance to renew old friendships. But for those who enjoy a loving family and the security of a valued career, the prospect of incarceration is a nightmarish contrast. Like so many other things in life, deterrence works least for those who need it the most. (p. 10)

Thus, it appears that deterrence has as much to do with *who* is being deterred as it does with *how* deterrence is being implemented. However, research on the effectiveness of deterrence has generally been mixed, even during the mid-1990s to about 2006 (Kohen & Jolly, 2006), when the use of increased incarceration was touted to be the primary cause for lowered crime rates. It is still seemingly impossible to determine whether a deterrent effect, or simply an incapacitation effect, was being observed.

Rehabilitation

Rehabilitation implies that an offender should be provided the means to achieve a constructive level of functioning in society, with an implicit expectation that such offenders will be deterred from reoffending due to their having worthwhile stakes in legitimate society—stakes that they will not wish to lose as a consequence of criminal offending. Vocational training, educational attainment, and/or therapeutic interventions are used to improve the offender's stakes in prosocial behavior. The primary purpose of rehabilitation is solely the recovery of the offender, regardless of the crime that was committed. In other words, if it is deemed that offenders are treatable, and they are successfully treated to refrain from future criminal behavior, rehabilitation is considered a success, and concern over the severity of the past crime is not considered important. With this approach, it is feasible that offenders with lesser crimes may end up serving more time behind bars than a person with a more serious crime if it is determined that they are not amenable to rehabilitative efforts.

The rehabilitative approach is based on the notion that offenders are provided treatment rather than punishment. Punitive techniques are completely alien to the rehabilitative model; the goal is to cure the offenders of their criminal behavior, much as would be done with a medical or mental health issue. As a result, sentencing schemes under a rehabilitation orientation would be *indeterminate*, a term that will be discussed in more detail later in this chapter. Indeterminate sentences have no specific amount of time provided upon which offenders are released from custody. Rather, a minimum and maximum amount of time is awarded, and, based on offenders' treatment progress, they are released prior to the maximum duration of their sentence once rehabilitative efforts have been determined a success.

Rehabilitation: Offenders will be deterred from reoffending due to their having worthwhile stakes in legitimate society.

Restorative Justice

Restorative justice is a term for interventions that focus on restoring the health of the community, repairing the harm done, meeting victims' needs, and emphasizing that the offender can and must contribute to those repairs. This definition was adapted from restorative justice advocate Thomas Quinn during his interview with the National Institute of Justice in 1998. More specifically, restorative justice considers the victims, communities, and offenders (in that order) as participants in the justice process. These participants are placed in active roles to work together to do the following: (1) empower victims in their search for closure, (2) impress upon offenders the real human impact of their behavior, and (3) promote restitution to victims and communities.

Dialogue and negotiation are central to restorative justice, and problem solving for the future is seen as more important than simply establishing blame for past behavior. Another key factor to this type of correctional processing is that the victim is included in the process. Indeed, the victim is given priority consideration, yet, at the same time, the process is correctional in nature, as offenders must face the person whom they victimized and the offenders must be accountable for the crimes that they committed against the victim.

Restorative justice: Interventions that focus on restoring the community and the victim with involvement from the offender.

Reintegration

Reintegration is focused on the reentry of the offender into society. The ultimate goal of reintegration programs is to connect offenders to legitimate areas of society in a manner that is gainful and productive. When used inside correctional institutions, this approach emphasizes continued contact between offenders and their families, their friends, and even the community. This approach is set against the backdrop realization that the overwhelming majority of offenders will ultimately return to society. While reintegration efforts do emphasize offender accountability, the use of reintegration processes is focused on ensuring that the offender has a maximal set of circumstances that, at least initially, diminish the need or desire to engage in crime by cultivating the connections

Reintegration: Focused on the reentry of the offender into society by connecting offenders to legitimate areas of society that are gainful and productive.

PHOTO 2.3 No-contact visitation often consists of a glass partition between both parties. Each person uses the phone receiver as a means of communicating with one another.

Thinkstock Images/Stockbyte/Thinkstock

United States v. Booker:
Determined judges no longer had to follow the sentencing guidelines that had been in place since 1987.

that the offender has to legitimate society. Reintegration efforts are intended to reduce recidivism among offenders. During the past few years, there has been an upsurge in national interest in offender reentry programs, which, inherently, are all reintegrative in nature.

TYPES OF SANCTIONS

It is through the use of intermediate (graduated) sanctions, various types of probation, incarceration, and the death penalty that various types of punishments (also known as sanctions) are meted out. While the public perhaps identifies prison as the final outcome for criminal offenders, the reality is that few offenders go to prison. Rather, the overwhelming majority are placed on probation or on some type of community supervision. Indeed, prisons and jails tend to hold only one fifth to one fourth of the entire offender population. However, chronic offenders and those who commit serious crimes tend to be given some period of incarceration.

Problems in determining the appropriate sentence for offenders are noted in the literature and have been the focus of at least one influential Supreme Court ruling. In 2005, the Court held in **United States v. Booker** that federal judges no longer were required to follow the sentencing guidelines that had been in effect since 1987. The Court held that federal judges now must only consider these guidelines with certain other sentencing criteria when deciding a defendant's punishment. Because of this ruling, and because of the trend toward alternative sanctions, there has been an observed trend toward more use of indeterminate sentencing (Debro, 2008). This also is consistent with much of the push for reintegrative efforts that has been observed throughout the nation.

The Continuum of Sanctions

The continuum of sanctions refers to a broad array of sentencing and punishment options that range from simple fines to incarceration and ultimately end with the death penalty. Between each of these visible points in the sentencing/sanctioning process (fines, incarceration, and the death penalty) is a variety of options that are used throughout the United States. The reasons for this variety of sanctions are manifold. Perhaps chief among them is the desire to calibrate the sanction in a manner that is commensurate with the type of criminal behavior.

When using the term *calibrate*, it is meant that sanctions can be selected in such a manner that allows us to, through an additive process, weigh the seriousness and number of the sanction(s) that are given so that the punishment effect is as proportional to the crime as can be arranged. The desire to establish proportionality harkens back to the thinking of classical criminologists, and this should not be surprising. Classical criminologists appealed to the use of reason in applying punishments, and that is precisely what a continuum seeks to achieve as well: a reasonable, commensurate, and gradual progression of sanctions that can be consistently additive in nature so as to be logically proportional to the frequency and seriousness of the criminal behavior in question (Lilly, Cullen, & Ball, 2014).

In addition to the desire for proportionality, there is another reason for the use of varied sanctions, particularly intermediate sanctions: the desire to save beds in prisons. As noted earlier in this chapter, there is a push for reintegration efforts in the federal government and in many states throughout the nation. The reason for this has to do with both a shift in ideologies and, more specifically, the rising costs of imprisonment. The national and international economic crisis that began in 2008 negatively impacted numerous state budgets throughout the United States. A slow recovery is underway, but many states are still more cash-strapped than usual, making the use of alternatives in sentencing all the more appealing.

Another rationale for this continuum is associated with treatment purposes. While we have noted that rehabilitation efforts are typically not contingent on the sentence that is imposed, the fact that indeterminate sentences tend to be used with a rehabilitation orientation demonstrates the need for incentives to exist so that offenders will change their behavior. Without an indeterminate sentence, offenders might not find their efforts toward reform to have any substantive reward; thus, early release provides a strong incentive that encourages offenders to actively work toward behavior change. The use of alternative sanctions follows this same logic, where lesser sanctions can be given to those offenders who show progress in treatment, and more serious sanctions can be administered to offenders who prove to be dangerous or a nuisance to a given facility.

From this point, we move to a description of some of the more common versions of sanctions. In providing these descriptions, we will progress from the least severe to the most severe types of sanctions that are usually encountered. The list of sanctions that follows is not all-encompassing but simply is intended to provide the student with an understanding of the types used and the means by which they are categorized. We begin with sanctions that involve fines or monetary penalties and progress to the ultimate form of punishment: the death penalty.

PRACTITIONER'S PERSPECTIVE

"Diversity is huge. It's all around us, every day."

Visit the IEB to watch Maxine Cortes's video on her career as a court administrator.

Maxine Cortes
Court Administrator

Monetary

Most monetary sanctions come in the form of fines. Most offenders convicted of a criminal offense are assessed a fine as a punishment for committing the offense. A **fine** can be defined as a monetary penalty imposed by a judge or magistrate as a punishment for being convicted of an offense. In most cases, the fine is a certain dollar amount established either by the judge or according to a set schedule dependent upon the offense committed. The logic behind the fine is that it will deter the offender from committing another offense in the future for fear of being fined again. In most jurisdictions, the fines are assessed and paid in monthly payments to the receiving agency.

Fine: A monetary penalty imposed as a punishment for having committed an offense.

Probation and Intermediate Sanctions

The use of probation and other community-based sanctions accounts for all the varied types of sentencing punishments available short of a jail or prison sentence. When on probation, offenders will report to a probation officer (in most cases) on a scheduled routine that varies with the seriousness of their crime and their expected risk of recidivism. Additional community-based sanctions, tacked on to a probation sentence, further allow for the calibration of the sentence with respect to the crime that was committed and the offender who is on supervision. Intermediate sanctions are a range of sentencing options that fall between incarceration and probation and are designed to allow for the crafting of sentences that respond to the offender or the offense, with the intended outcome of the case being a primary consideration. The purpose of intermediate sanctions is to make available a continuum of sanctions scaled around one or more sanctioning goals. Such a continuum permits the court or corrections authority to tailor sanctions that are meaningful with respect both to their purposes and to the kinds of offenders that come before them.

Incarceration

Though imprisonment is the most visible penalty to the public eye, its ability to deter crime is questionable at best (National Institute of Justice [NIJ], 2016). Though the majority of offenders under supervision are on community supervision rather than in prisons or jails, the incarcerative type of sentence still draws public interest due to its ominous nature. This punishment remains the most commonly used for serious offenders. This remains true despite research that has found that in many respects the likelihood of recidivism increases once an offender is incarcerated (NIJ, 2016). Thus, the effectiveness of incarceration to change potential criminal behavior is questionable. Because of this, it is recommended that incarceration be viewed as best suited for meeting the goals of incapacitation (and perhaps retribution) rather than rehabilitation, deterrence, or crime reduction.

Incarceration Options

Among incarceration options, the jail facility is considered the first stage of incarceration for the offender. Jail facilities come in a variety of sizes and designs, but all are generally intended to hold offenders for sentences that are short. Aside from those persons who are held for only brief periods (such as immediately after an arrest), jails tend to hold offenders who are sentenced to a year or less of incarceration. In most cases, jail facilities are the first point at which an offender is officially classified as being in the correctional component of the criminal justice system. In simple terms, a jail is a confinement facility, usually operated and controlled by county-level law enforcement, that is designed to hold persons charged with a crime who are either awaiting adjudication or serving a short sentence of 1 year or less after the point of adjudication. Similarly, the Bureau of Justice Statistics (2019) defines jails as "locally-operated correctional facilities that confine persons before or after adjudication. Inmates sentenced to jails usually have a sentence of a year or less, but jails also incarcerate persons in a wide variety of other categories." Thus, there is some degree of variance in the means by which jails are utilized, but they tend to be short-term facilities in most cases.

On the other extreme, consider the use of the supermax prison. The supermax prison is perhaps the epitome of incarceration-based sentences. There are some prison administrators who contend that supermax facilities have a general deterrent effect. However, this is unlikely because inmates in supermax facilities do not form bonds with persons in the prison or outside of the prison. Further, the disruptive inmates who will be kept in supermax facilities are least likely to care about the consequences of their actions and/or their ability to bond with other people. Deterrence as a philosophical orientation targets those inmates who would engage in antisocial behavior if not for the deterring mechanism. However, the inmates typically channeled into a supermax facility are those who have not been deterred when incarcerated in less secure environments, such as minimum-, medium-, and maximum-security facilities. Thus, these inmates are unlikely to be among those who would commit crimes were it not for the penalty of incarceration; they are impervious to the threat of incarceration and the deprivations that this sanction entails. Thus, supermax facilities act as simple holding spaces for the most incorrigible of inmates and are devoid of any deterrent and/or therapeutic value.

Because much of this text later involves coverage of the prison environment, further discussion related to specific aspects of incarceration schemes will not be provided at this time. It is sufficient to say that incarceration, while accounting for no more than 30% of the entire correctional population, tends to draw substantial public and media attention. Further, the offenders who are kept incarcerated are among those who are either repetitive or violent, or both. Therefore, the correctional process within institutions is one that deals with harder-core offenders than might be encountered among community supervision personnel.

The Death Penalty

The most extreme outcome when offenders are at the end of the correctional process: the death penalty, which is also referred to as **capital punishment**. Obviously, the death penalty results in the end of the offender's journey through the correctional process and entails no true rehabilitative efforts on the part of the correctional facility.

Capital punishment:
Putting the offender to death.

ARGUMENTS FOR AND AGAINST THE DEATH PENALTY

The debate over the death penalty has been active in the United States for generations. Generally, it is not difficult to find people who have strong views regarding this sanction, both pro and con. These arguments generally focus on one of three common philosophical perspectives on punishment: deterrence, retribution, and arbitrariness. Each of these three have been discussed in this chapter but are independently discussed in the following sections in reference to the death penalty itself.

Deterrence

Opponents of the death penalty who argue against deterrence as a rationale for using the death penalty note that while numerous statistical studies have been conducted, there is no conclusive evidence that the death penalty lowers crime. Table 2.2 provides an overview of findings regarding the deterrent effect of the death penalty. The outcomes, as one will see, are mixed, at best, and sometimes indicate that the death penalty *increases* the likelihood of future acts of homicide. Support for this can be seen when comparing states that do not employ the death penalty with those that do; generally crime rates and murder rates are lower in states that do not have the death penalty. Interestingly, the United States, an ardent proponent of the death penalty, has a higher murder rate than do countries in Europe and Canada, which do not have the death penalty.

Further, most people who commit murders do not usually plan on being caught, and most commit their crimes due to fits of anger when in impaired states, such as when they are drunk or high on drugs. These types of circumstances do not allow for an offender to contemplate the outcome of his or her actions, and, depending on the offender's emotional framework at the time of the crime commission, it may be doubtful that the knowledge of this sanction would be a deterrent. Since these factors—the unpremeditated nature of the crime and the offender's altered state of mind due to the substance abuse—are often cited as reasons to mitigate the punishment an offender may receive, it is clear that such circumstances may not truly justify the death penalty, at least not on a logical basis.

Lastly, life sentences without the possibility of parole are just as effective as death sentences. Both can arguably deter crime in a general and specific manner. However, the life sentence allows for remediation in cases where it may later be found that an offender was, in fact, innocent. In addition, life sentences tend to be less expensive for prison administrators (and taxpayers) than death sentences. It is also worth noting that most murderers on death row are very well mannered and do not represent an institutional hazard.

Retribution

While retribution has been couched as a logical approach, there is an emotional component that is also addressed: Families of the victim can see that, if nothing else, there is some connection between the action and the consequences received. In addition, the death penalty can help to facilitate the grieving process as some families may desire reciprocation commensurate to the loss that they have incurred. To ask for a payment of

TABLE 2.2

Selected Studies of the Deterrent Effect of the Death Penalty

STUDY	UNIT OF ANALYSIS	PERIOD	RESULT
Sellin, 1959	Matched state comparison	1920–1962	No deterrent
Ehrlich, 1975	U.S. (aggregate)	1933–1969	7–8 fewer murders per execution (C.I. 0–24)
Bowers and Pierce, 1980	New York state	1907–1963	2 more homicides per month after an execution
Mocan and Gittings, 2003	State-level	1977–1997	5 fewer homicides per execution
Katz, Levitt, and Shustorovich, 2003	State-level	1950–1990	No systematic evidence of a deferent (+3.1 to -5.6)
Dezhbakhsh, Rubin, and Shepherd, 2003	Country-level	1977–1996	16 fewer homicides per execution
Shepherd, 2004	State-level	1977–1999	3 fewer murders per execution
Zimmerman, 2004	State-level	1978–1997	14 fewer murders per execution
Shepherd, 2005	Country-level	1977–1996	21 states have brutalization effect 6 states have deterrent effect 23 states have no effect Overall, 4.5 fewer murders per execution
Donohue and Wolfers, 2005	Canada vs. U.S.	1950–2003	No deterrent
Martin, 2016	Country-level	2016	No deterrent

Source: Kohen, A., & Jolly, S. J. (2006). *Deterrence reconsidered: A theoretical and empirical case against the death penalty.* Paper presented at the annual meeting of the Midwest Political Science Association, Chicago, IL.

anything less than the offender's life would seem to indicate that the life of the victim was somehow less valuable. Thus, there may indeed be an emotional sense of justice that is fulfilled. While many critiques of retribution might not find this adequate as a rationale, supporters may contend that families of murder victims have a right to feel as they do and, correspondingly, are entitled to seek relief as they are allowed within the law.

Opponents of retribution may hold that at its base, retribution is simply a form of revenge. The contention is that retribution simply provides a reason and rationale behind the pursuit of unbridled revenge. In fact, opponents of retribution tend to believe that the use of the death penalty itself contradicts the "evolving standards of decency" espoused by the Supreme Court. The mark of a civilization is how it aids those who are troubled, which many believe should not be through the eradication of their existence. This is even truer when the offender has a diminished capacity and/or acted in an altered state of mind. Critics often claim the use of the death penalty is simply barbaric and that justifications based on retribution do not make this penalty more civilized.

Arbitrariness

Many opponents of the death penalty point to the arbitrary nature of the application of the death penalty. Supporters of the death penalty argue that it is not arbitrarily applied and even note that more Caucasian offenders are executed than minority offenders. On the other hand, the number of African Americans on death row tends to be disproportionately high when compared to their overall population numbers. Thus, this racial disparity can seem to point to some degree of arbitrariness in the use of this sanction.

Regardless of whether racial disparities do indicate arbitrariness, there is general consensus that no matter how much we try, we can never perfectly calibrate a sanction to be exactly commensurate with the crime that is committed. The inability to ensure proportionality, therefore, undermines the argument for retribution and instead further illustrates how we, as a society, are at risk of enabling arbitrary practices.

BRUTALIZATION HYPOTHESIS

There is some evidence that the death penalty may not only be a failure at deterring crime but may actually increase homicide levels in areas where executions occur. This observation is often based on the notion that violence begets violence and is referred to as the brutalization hypothesis. The brutalization hypothesis, first introduced in Chapter 1, contends that the death penalty may actually cause an increase in murders because it reinforces the use of violence. Because of this, researchers such as Bowers and Pierce (1980) contend that "the lesson of the execution then, may be to devalue life by the example of human sacrifice" (p. 457).

Cochran, Chamlin, and Seth (1994) examined the reinstatement of the death penalty in Oklahoma. They found "no evidence that Oklahoma's reintroduction of the execution produced a significant decrease in the level of criminal homicides during the period under investigation" (p. 129). Even further, they noted that the death penalty seemed to produce a brutalizing effect that further encouraged offenders to commit murders if they had feelings that their own life or circumstances were fundamentally unfair.

Further still, the recent work of Mann (2017) shows that not only is it questionable as to whether the death penalty deters crime, but there may be some evidence that it actually aggravates the likelihood of future crime. Using a 5-year longitudinal design, Mann compared death penalty states that actively utilized the death penalty with other states that did not utilize the death penalty. He found that those states using the death penalty "on average over the five-year period demonstrated a statistically significantly higher violent crime rate, per capita" than states not using the death penalty (p. 48). These findings led Mann to provide the following additional comments:

> Law makers should consider alternate theories focused on social issues, economics, opportunity and other control theories in crime control policy. Based on this study, it is possible that the death penalty has an opposite effect to deterrence. (p. 49)

What this means is that the legal system and our society should discard our deterrence rationale for the death penalty because deterrence theory has questionable validity. Rather, we should instead simply acknowledge that we keep the death penalty simply because we like it. Simply put, we keep the death penalty for emotional reasons, not logical ones.

Regardless of the argument that one believes, no one can seem to prove whether "it works" at doing anything in the manner that is intended, and both advocates and opponents are able to generate evidence for their views. This makes the entire issue difficult to resolve. Work by Kohen and Jolly (2006) demonstrates the mixed state of affairs in researching the efficacy of the death penalty. When looking at Table 2.2, it can be seen that there are numerous studies that have had different results. Some studies find support for deterrence, and some do not; others occasionally find support for the brutalization effect. While Kohen and Jolly were making a case against the death penalty in their review of the research, the key point for this chapter is to simply understand that it seems that there is no airtight case for *or* against the death penalty despite years of debate.

SENTENCING MODELS

Sentencing involves a two-stage decision-making process. After the offender is convicted of a crime, the initial decision is made as to whether probation should or should not be granted. The chief probation officer or his or her designee will typically make this decision based on the presentence investigation (PSI). The presentence investigation report is a thorough file that includes a wide range of background information on the offender.

PHOTO 2.4 For judges, deciding the length of a sentence requires difficult calculations that take into account the possibility for rehabilitation, the need to protect society, the need to fulfill the demand of retribution, and the implementation of deterrence strategies.

©iStockphoto.com/ISignature Collection

Mitigating factors:
Circumstances that make a crime more understandable and help to reduce the level of culpability that an offender might have.

Aggravating circumstances: Magnify the offensive nature of a crime and tend to result in longer sentences.

This file will typically include demographic, vocational, educational, and personal information on the offender as well as records on his or her prior offending patterns and the probation department's recommendation as to the appropriate type of sentencing for the offender.

If incarceration is chosen, the second decision involves determining the length of the sentence. For many judges, deciding the length of the sentence (when they are required to do so) is not an easy task. They must consider several factors, such as the possibility for rehabilitation, the need to protect society, the need to fulfill the demand of retribution, and the implementation of deterrence strategies. The most important factor in deciding on a sanction is the seriousness of the crime. Sentencing on the basis of seriousness is one key way that courts attempt to arrive at consistent sentences. Once the seriousness of the crime has been determined, the next factor to consider is the prior record of the offender. The worse the prior record, the more likely the offender will receive a lengthy sentence. The last few issues considered in the sentencing process are mitigating and aggravating factors. **Mitigating factors** do not exonerate an offender but do make the commission of the crime more understandable and also help to reduce the level of culpability that the offender might have had. **Aggravating circumstances**, on the other hand, magnify the offensive nature of the crime and tend to result in longer sentences. Each of these factors can impact the outcome of the sentence. It is with this in mind that we turn our attention to the two types of sentencing: indeterminate and determinate sentencing.

Indeterminate Sentences

Indeterminate sentencing is sentencing that includes a range of years that will be potentially served by the offender. The offender is released during some point in the range of years that are assigned by the sentencing judge. Both the minimum and maximum times can be modified by a number of factors, such as offender behavior and offender work ethic. Under the most liberal of approaches using indeterminate sentences, judges will assign custody of the offender to the department of corrections, and the release of the offender is completely dependent on the agency's determination if he or she is ready to function appropriately in society. This type of sentence is typically associated with treatment-based programming and community supervision objectives. In such cases, indeterminate sentencing provides correctional officials a good deal of control over the amount of time that an offender will serve.

Penal codes with indeterminate sentencing stipulate minimum and maximum sentences that must be served in prison (2 to 9 years, 3 to 5 years, and so forth). At the time of sentencing, the judge will explain to the offender the time frame that the offender may potentially be in prison. The offender is also informed of any potential eligibility for parole once the minimum amount of time has been served. However, the actual release date is determined by the parole board, not the judge. Note that this particular sentence is different from the determinate *discretionary* sentence that will be described in the following subsection. The difference is that while the determinate discretionary sentence has a range of time to be served, the specific sentence to be served within that range is decided by the judge at the point of initial sentencing. Once this specific amount of time has been decided, there is no further modification to the sentence, regardless of the offender's progress within the institution.

Determinate Sentences

Determinate sentencing consists of fixed periods of incarceration with no later flexibility in the term that is served. This type of sentencing is grounded in notions of retribution, just deserts, and incapacitation. These types of sentences came into vogue due to disappointments with the use of rehabilitation and due to increased support for retribution. When offenders are given a determinate sentence, they are imprisoned for a specific period of time. Once that time has expired, the inmate is released from prison.

It should be pointed out that in many states inmates may be given "good time" if they maintain good behavior while in the correctional facility (Schriro, 2009). Generally, this entails a willingness to work in the prison, engage in educational and therapeutic programs, and participate in other prosocial activities. Good time earned is taken off the total sentence that inmates must serve, thereby allowing them to be released early from prison. While this does add some degree of variability to the total time that offenders serve in the institution, the actual sentence given to the inmates is not connected to their level of participation in treatment or to the likelihood of parole or early release (Schriro, 2009).

One variant of the determinate sentence is the determinate presumptive sentence. The **determinate presumptive sentence** specifies the exact length of the sentence to be served by the inmate. Judges are required to impose these sentences unless there are aggravating or mitigating circumstances, in which case they may lengthen or shorten the sentences within narrow boundaries and with written justification. This type of sentence is perhaps more realistic than a pure determinate sentencing model because it accounts for the variety of circumstances that are different from one case to another. In fact, very few criminal cases are exactly alike even when the charge is the same. The circumstances associated with each type of criminal case (e.g., theft) tend to vary, with different motivations, different outcomes, and different issues, and this may make the crime seem more or less severe in nature, especially on a human level (Carter, 1996).

To further demonstrate the potential complexity of sentencing schemes, consider also the determinate discretionary sentence. The **determinate discretionary sentence** (discussed briefly in the prior subsection) sets a stated range of time that must be served. This range of time (e.g., 3 to 5 years) is not subject to modification by judges who impose a sentence under this model (Carter, 1996). However, the judge is able to use his or her own discretion in determining the exact sentence so long as it falls within the range that has been predetermined by legislative bodies. Thus, the sentence is determinate in nature with parameters being set (in our example, a minimum of 3 years and a maximum of 5 years), but it is also discretionary since it allows the judge to select the exact time that will be served. Note that this sentence is different from the indeterminate sentence that was presented in the prior subsection. The difference is that while the indeterminate sentence often has a range of time to be served, the eventual date of release for the offender is decided by correctional officials who work with the offender and determine his or her progress toward and suitability for reintegration into society. Thus, the exact amount of time served depends on an offender's progress within the correctional treatment regimen.

Mandatory Minimum Sentences

Mandatory minimum sentences require that some minimum length of incarceration be served by offenders who commit certain specified crimes, such as drug-related crimes. In these cases, judges are extremely limited in their consideration of the offender's background or circumstances, and the use of community-based sanctions is out of the question. One type of mandatory minimum sentence is the "three strikes and you're out" law. This law requires that judges award a long-term prison sentence (in some cases life in prison) to offenders who have three felony convictions. This has resulted in the growth of prison populations around the nation and has also resulted in a graying of the prison population in the United States. As more and more inmates serve lengthy mandatory minimum sentences, the proportion of inmates who are elderly continues to climb. Since elderly inmates are more costly to house than younger inmates (due to medical care and other related costs), this has proven to be a serious drain on many state-level prison systems.

Indeed, this issue has been given considerable attention in recent years, with Texas, California, Florida, New York, and Louisiana all experiencing a rise in per capita elderly inmates that are incarcerated. Each of the states just mentioned has either one of the largest prison populations or one of the highest rates of incarceration in the United States. In all cases, the costs that are associated with the elderly inmate are exponentially

Determinate presumptive sentence:
This type of sentence specifies the exact length of the sentence to be served by the inmate.

Determinate discretionary sentence:
Type of sentence with a range of time to be served; the specific sentence to be served within that range is decided by the judge.

Mandatory minimum:
A minimum amount of time or a minimum percentage of a sentence must be served with no good time or early release modifications.

PHOTO 2.5 When sentencing occurs, the result can be disastrous for the defendant, who now must cope with the reality of the punishment that will be meted out.

©iStockphoto.com/RichLegg

higher than those associated with the average inmate. This has led to other issues for administrators to consider, such as the possibility of early release of inmates who are expected to die, the implementation of human caregiver programs such as hospice, and accountability to the public. It is this accountability that places prison administrators in a dilemma since public safety is the primary concern for all custodial programs. Thus, in one generation, mandatory minimum, three strikes, habitual offender, and other enhanced sentences have created a new crisis that looms on the correctional horizon of the United States. This outcome is likely to affect sentencing patterns in the future, which, in turn, will impact the state of corrections.

Sentencing Has Become More Indeterminate in Nature

At the time that the first edition of this text was written, evolution in sentencing practices pointed to the possibility of a more indeterminate nature. While it was not clear then, and is perhaps no clearer now, whether this was the best approach from a public safety perspective, it was obvious that the 1990s had been reflective of a crime control model of criminal justice with an emphasis on mandatory minimums for sentencing and purely determinate sentencing schemes. This led to a swelling of the offender population behind bars. Many of these inmates were drug offenders rather than violent offenders, calling into question for many whether this type of mass incarceration was truly warranted.

Crime has not increased during the past few years and, in fact, has gone down. This further begs the question, is this level of mass imprisonment really necessary? It would appear that the federal government has started to ask this question of itself as well, as can be seen with recent recommendations from the U.S. Sentencing Commission. The U.S. Sentencing Commission is an independent agency in the judicial branch of government intended to establish sentencing policies and practices for the federal courts, including guidelines to be consulted regarding the appropriate form and severity of punishment for offenders convicted of federal crimes. It also advises Congress and the executive branch on effective crime policy and sentencing issues.

In 2014, the commission unanimously voted to reduce sentencing terms for drug traffickers who are already in prison. This meant that approximately 46,000 drug offenders would be eligible for early release (Greenblatt, 2014). Before any other discussion is provided, it should be pointed out that this recommendation included *drug traffickers*, not just drug users. Typically, drug traffickers are given stiffer sentences because they often are viewed as part of the cause of drug use. Given that this is a more serious charge than possession or consumption of drugs, this makes the recommendation by the commission even more noteworthy.

The commission intends for these recommendations to be indeterminate in nature, depending on the facts and circumstances of each case. Indeed, Greenblatt (2014) notes that not all offenders will be released. Rather, petitions for each will be considered on an individual basis by federal judges, reflecting the indeterminate feature to these recommendations. In addition, this process has not moved quickly; the reduction was instituted in 2014, but none of the affected offenders were released until November 2015, which allowed an 18-month process for judges to review offender petitions before any were released.

Interestingly, the commission cited "fundamental fairness" as the primary motivation behind these sentencing changes (Greenblatt, 2014). Indeed, it has become a goal of the Department of Justice to seek leniency with nonviolent drug offenders as a means of reducing the sentencing disparities that date back to the mass incarceration of crack cocaine users in the 1980s and 1990s (Greenblatt, 2014). The Sentencing Commission, on the other hand, has gone forward with a more aggressive plan for sentence reduction that is fully retroactive, going beyond the initial efforts of the Department of Justice in reducing drug-related sentences (U.S. Sentencing Commission, 2014a).

It is important for students to understand that aside from theoretical and philosophical reasons for reducing sentencing disparities, there is a more pragmatic and mercenary reason for the commission to make such aggressive recommendations. As cited in the commission's official news release, one key priority is to reduce the inmate population in the nation's federal prisons (U.S. Sentencing Commission, 2014a). Indeed, Judge Patti Saris, the chair of the commission, noted that "this modest reduction in drug penalties is an important step toward reducing the problem of prison overcrowding at the federal level in a proportionate and fair manner" (U.S. Sentencing Commission, 2014a, p. 1). She went on to add that "reducing the federal prison population has become urgent, with that population almost three times where it was in 1991" (p. 1).

What is important to understand is that regardless of the moral, philosophical, or theoretical reasons given for many criminal justice policies, the reality is that economics always plays a strong role in how the system can and does operate. In fact, the economic circumstances of the times may not only shape sentencing processes and correctional system operations, but also are important in determining what law enforcement agencies can and will enforce. Students should understand that despite the philosophical and theoretical perspectives on punishment that may come under consideration, as discussed in Chapter 1, economics is often the "trump card" variable in determining correctional policy. This has historically been true in American corrections, as we saw when the Pennsylvania system (grounded in theories of reformation) competed with the Auburn system (grounded in terms of profit and loss). The author wants to make clear to students that in the world of the practitioner, theoretical perspectives and philosophy often take a back seat to economic pressures.

Sentencing Disparities

In the previous subsection, we referred to the term *disparities* and noted that the U.S. Department of Justice has made efforts to reduce these incongruous elements within sentencing policies and among the incarcerated population. The term *disparity* should be held distinct from the better-known term *discrimination*. **Disparity** refers to inconsistencies in sentencing and/or sanctions that result from the decision-making process. This typically results when the criminal justice system provides an unequal response toward one group as compared with the response given to other groups. Distinct from this is **discrimination**, which focuses on attributes of offenders when providing a given sentence. This usually results in a differential response toward a group without providing any legally legitimate reference to the reasons for that differential response. According to Neubauer (2019), the most commonly cited forms of disparity in sentencing involve geography and judicial attitudes. We now proceed with a discussion of these two types of disparity and will further explore how disparities impact corrections throughout the United States.

Geographical disparity in sentencing patterns has been tied to various areas of the United States and reflects the cultural and historical development of correctional thought in those regions. Neubauer (2019) notes that geographical differences in justice are the product of a variety of factors, such as the amount of crime, the types of crime affecting a given area, the effectiveness of police enforcement, and media attention given to criminal activity in the region. Overall, it is clear that the South imposes more harsh sentences than other areas of the nation, and the western part of the United States seems to follow suit. Interestingly, executions are concentrated in these regions as well. When one considers our discussions in Chapter 1 regarding southern penology and the development of corrections in the West, this observation may not be too surprising.

Lastly, a discussion regarding disparity in sentencing would not be complete without at least some reference to observed disparity in death penalty sentences. The use of this sentence and problems regarding racial disparity in its application help to illustrate why disparity in sentencing is an important issue to the field of corrections. This also helps to illustrate a philosophical or ideological influence on the sentencing process that is at least perceived to be true by many in the public arena. This then may undermine punishment schemes that are intended to rehabilitate and/or deter offenders. Rather,

Disparity: Inconsistencies in sentencing and/or sanctions that result from the decision-making process.

Discrimination: A differential response toward a group without providing any legally legitimate reasons for that response.

the cultural impact and/or influence from individuals of influence in the justice system may obscure and impair the intended outcome of various sentencing and punishment schemes. In regard to rehabilitation, this can create additional distrust of helping professionals from a given group that has been marginalized. From a deterrence viewpoint, these factors may only deter one group while giving the impression that criminal activity will be tolerated among other groups. It is clear that these impressions undermine the correctional process and, as a result, further complicate the process as a whole.

Smarter Sentencing Act:
Sentence Leniency to Relieve Disparities

Smarter Sentencing Act of 2014: A bill that adjusts federal mandatory sentencing guidelines in an effort to reduce the size of the U.S. prison population.

Increased political support in recent years to reduce many of the mandatory minimums that were enacted during the 1990s led to the **Smarter Sentencing Act of 2014**. According to GovTrack.us (2015, p. 2), the act adjusts federal mandatory sentencing guidelines for a variety of crimes in an effort to reduce the size of the current U.S. prison population and costs associated with it. This bill has spawned substantial discussion. While the reasons for this are many, for academic purposes, the author will point out once again that this legislation is simply a reflection of the "pendulum" of justice whereby criminal justice policy goes back and forth between harder and softer approaches to crime. As we saw in Chapter 1, different eras in modern corrections reflect an ebb and flow between more stern approaches to criminal offending followed after a time by more humane approaches.

One of the main advocates of this legislation is Senator Ted Cruz of Texas, who in 2015 gave a speech that captures the essence of what is current and common sentiment in regard to sentencing reform in general and the Smarter Sentencing Act of 2014 specifically. According to Cruz (2005),

> The issue that brings us together today is fairness. What brings us together is justice. What brings us together is common sense. This is as diverse and bipartisan array of members of Congress as you will see on any topic and yet we are all unified in saying commonsense reforms need to be enacted to our criminal justice system. Right now today far too many young men, in particular African American young men, find their lives drawn in with the criminal justice system, find themselves subject to sentences of many decades for relatively minor non-violent drug infractions. (p. 1)

Again, what is important for students to understand is that despite comments such as Cruz's plea for justice, fairness, and common sense, there is nevertheless an ulterior purpose grounded in economics behind much of this proposed sentencing reform. For example, consider that this legislation affirms that the proposed changes are consistent with the U.S. Sentencing Commission mandate to *minimize the likelihood that the federal prison population will exceed the capacity of the federal prisons*. This legislation also directed the Justice Department to issue a report outlining the reduced expenditures and cost savings as a result of the Smarter Sentencing Act within a 6-month period of its enactment. It should be clear from these facts that the undergirding concern for the federal government is money. Indeed, it would appear that the push for smarter sentencing comes at a time when the economy has been ailing, and public leaders do not consider it a smart practice to spend money on prisons—money that society does not have. Whether a criminal behavior is more or less wrong today than it was in prior years and/or whether racial disparities exist within our sentencing processes and prison systems is of secondary concern. The primary concern revolves around money and the economic conditions of the time.

CRIMINOLOGICAL THEORIES AND CORRECTIONS

If a correctional program is to be effective it must have a clear theoretical and philosophical grounding. Numerous criminological theories exist that are used to explain why crime occurs and how one might predict crime. The ability to predict crime allows us to find those factors that lead to crime and therefore give us guidance on what should be done to prevent crime. Further, if we are able to explain why crime occurs, we can better determine those factors that must be addressed to correct aberrant tendencies toward criminal behavior. Thus, appropriate grounding in theoretical underpinnings to criminal behavior can improve any correctional effort. However, theoretical applications may not always be quite so clear in the day-to-day practice of corrections.

The specific theoretical applications to institutional corrections may not be clear to many students. Therefore, a discussion on some of the more common theories of criminal behavior is presented so that students can connect philosophies on correctional intervention, the use of different sentencing characteristics, and theories on criminal behavior as a means of understanding the many different bases to the field. Further, the philosophical underpinnings behind punishment are important to understand since this will often shape official reactions to criminal offending. Both sociological and psychological theories are important for this process and are thus included in the pages that follow.

Individual Traits

According to some theorists, criminal behavior can be directly connected to specific personality characteristics that offenders tend to possess. **Individual personality traits** that are associated with criminal behavior include defiance, self-assertiveness, extroversion, impulsivity, narcissism, mental instability, a tendency toward hostility, a lack of concern for others, resentment, and a distrust of authority. These traits are, quite naturally, psychological in nature, and, presuming that these characteristics are the root causal factor behind an offender's criminal behavior, correctional interventions along the line of the medical model would be most appropriate. In such circumstances, criminal behavior would be treated as a form of pathology and would be treated with a correctional scheme that integrates mental health interventions (Lilly et al., 2014).

Individual personality traits: Traits associated with criminal behavior.

Classical Theory and Behavioral Psychology

Students will recall from Chapter 1 that classical criminologists contend that punishment must be proportional, purposeful, and reasonable. Beccaria, in advocating this shift in offender processing, contended humans were hedonistic—seeking pleasure while wishing to avoid pain—and that this required an appropriate amount of punishment to counterbalance the rewards derived from criminal behavior. It will become clear in subsequent pages that this emphasis on proportional rewards and punishments dovetails well with behavioral psychology's views on the use of reinforcements (rewards) and punishments.

Though our correctional system today is much more complicated than in times past, classical criminology serves as the basic underlying theoretical foundation of our criminal justice system in the United States, including the correctional components. It is indeed presumed that offenders can (and do) learn from their transgressions through a variety of reinforcement and punishment schedules that corrections may provide.

Operant Conditioning

One primary theoretical orientation used in nearly all programs associated with correctional treatment is operant conditioning. This form of behavioral modification is based on the notion that certain environmental consequences occur that strengthen the likelihood of a given behavior, and other consequences tend to lessen the likelihood that a given behavior is repeated. We now turn our attention to those consequences that impact human behavior, for better or worse.

Reinforcers and Punishments

Those consequences that strengthen a given behavior are called reinforcers. Reinforcers can be both positive and negative, with **positive reinforcers** being a reward for a desired behavior (Davis, Palladino, & Christopherson, 2012). An example might be if we provided a certificate of achievement for offenders who completed a life skills program. **Negative reinforcers**

Positive reinforcers: Rewards for a desired behavior.

Negative reinforcers: Unpleasant stimuli that are removed when a desired behavior occurs.

PHOTO 2.6 Carlos Valdez, district attorney in Nueces County, Texas, holds one of the signs that a judge ordered posted outside the homes of registered sex offenders.

AP Photo/Corpus Christi Caller-Times, Paul Iverson

Positive punishment:
Punishment where a stimulus is applied to the offender when the offender commits an undesired behavior.

Negative punishment:
The removal of a valued stimulus when the offender commits an undesired behavior.

Social learning theory:
Contends that offenders learn to engage in crime through exposure to and adoption of definitions that are favorable to the commission of crime.

Strain theory/ institutional anomie:
Denotes that when individuals cannot obtain success goals, they will tend to experience a sense of pressure often called *strain*.

Labeling theory:
Contends that individuals become stabilized in criminal roles when they are labeled as criminals.

are unpleasant stimuli that are removed when a desired behavior occurs (Davis et al., 2012). An example might be if we agreed to remove the requirement of wearing electronic monitoring devices when offenders successfully maintained their scheduled meetings and appointments for a full year without any lapse in attendance.

Consequences that weaken a given behavior are known as punishments. Punishments, as odd as this may sound, can also be either positive or negative. A **positive punishment** is one where a stimulus is applied to the offender when the offender commits an undesired behavior (Davis et al., 2012). For instance, we might require offenders to pay an additional fee if they are late in paying restitution to the victim of their crime. A **negative punishment** is the removal of a valued stimulus when the offender commits an undesired behavior (Davis et al., 2012). An example might be when we remove offenders' ability to leave their domicile for recreational or personal purposes (i.e., place them on house arrest) if they miss any of their scheduled appointments or meetings.

The key in distinguishing between reinforcers and punishments to keep in mind is that reinforcers are intended to *increase* the likelihood of a *desired* behavior whereas punishments are intended to *decrease* the likelihood of an *undesired* behavior. In operant conditioning, the term *positive* refers to the addition of a stimulus rather than the notion that something is good or beneficial. Likewise, the term *negative* refers to the removal of a stimulus rather than being used to denote something that is bad or harmful.

Social Learning

Social learning theory and differential association theory are presented together because they have a common history and because many of their basic precepts are similar (Ronald Akers's social learning theory was spawned from Edwin Sutherland's differential association theory). As with differential association theory, **social learning theory** contends that offenders learn to engage in crime through exposure to and adoption of definitions that are favorable to the commission of crime (Lilly et al., 2014). While both theories contend that exposure to normative definitions that are favorable to crime commission can influence others to commit crime (through vicarious learning and/or reinforcement for repeating similar acts), social learning explicitly articulates the manner by which such definitions are learned by criminals. Differential association, on the other hand, does not clarify this point, and this is the primary distinction between the two theories.

Anomie/Strain

The next theory to be examined is **strain theory/institutional anomie**. This theory denotes that when individuals cannot obtain success goals (money, status, and so forth), they will tend to experience a sense of pressure often called *strain*. Under certain conditions, they are likely to respond to this strain by engaging in criminal behavior. Merton (1938) and Messner and Rosenfeld (2001) note that this is often aggravated in American society by the continued emphasis on material (monetary) success and the corresponding lack of emphasis on the means by which such material accumulation is obtained. In other words, these authors contend that society in the United States emphasizes winning the game (of life) much more than how the game (of life) is played.

Labeling and Social Reaction

Another theoretical application that is relevant to the correctional process is **labeling theory**. This theory contends that individuals become stabilized in criminal roles when they are labeled as criminals, are stigmatized, develop criminal identities, are sent to prison, and are excluded from conventional roles (Cullen & Agnew, 2006). In essence, the label of "criminal offender" or "convict" stands in the way of the offender reintegrating back into society. Such labels impair the offender's ability to obtain employment, housing, and/or other goods or services necessary to achieve success. Tracking and labeling often result from the need to ensure public safety (as with pedophiles) and thus are simply a necessary aspect of the punishment, incapacitation, and public safety objectives of many community corrections programs. However, it may be that these functions can be achieved in a manner that aids public safety but does not prevent the offender from achieving reintegration.

The desire to allow for an offender's past errors to be public information (due to a need to achieve public safety) without undo blockage of the offender's ability to reintegrate has been directly addressed by labeling theory scholars. One particular labeling theorist, John Braithwaite, provided a particularly insightful addition to the labeling theory literature that is specifically suited for the field of community supervision. In his work *Crime, Shame, and Reintegration*, Braithwaite (1989) holds that crime is higher when shaming is stigmatizing and criminal activity is lower when shaming effects serve a reintegrative purpose.

According to Braithwaite (1989), the negative effects of stigmatization are most pronounced among offenders who have few prosocial bonds to conventional society (such as family, religious institutions, and civic activities). This would place young males who are unmarried and unemployed at the greatest risk of being thrust further into criminality due to shaming effects. Due to their lack of resources, connections, and general social capital, these offenders find themselves further removed from effective participation in legitimate society. Over time, these offenders will find that it is much easier to join criminal subcultures where tangible reinforcements for their activities can be found. Thus, a cycle is created where a given segment of the offender population is further encouraged to repeat criminal activity simply due to the fact that other options have essentially been knifed away from them.

Conflict Criminology

According to **conflict theory**, the concepts of inequality and power are the central issues underlying crime and its control. This theory is derived from the work of Karl Marx (Lilly et al., 2014). Conflict criminologists note that capitalism perpetuates a system that benefits the rich. In the process, the poor are denied access to economic opportunities and are therefore prevented from improving their social standing. Thus, the wide economic gap between the social classes is increased and perpetuated with each successive generation. In a similar vein, the state—which includes the criminal law and the criminal justice system—operates to protect social arrangements that benefit those profiting from capitalism (Lilly et al., 2014). In general, the injurious acts committed by the poor and powerless are defined as crime, but the injurious acts committed by the rich and powerful are not brought within the reach of the criminal law. One can see this in sentencing practices that tend to mete out harsher terms to those groups who lack wealth and the ability to hire expensive defense attorneys but assign comparatively light sentences to those who are wealthy. Thus, critical criminologists point at the social system itself as the chief cause of America's growing prison population.

Conflict theory: Maintains that concepts of inequality and power are the central issues underlying crime and its control.

CONCLUSION

This chapter began with a review of the purpose of corrections as a process whereby practitioners from a variety of agencies and programs use tools, techniques, and facilities to engage in organized security and treatment functions intended to correct criminal tendencies among the offender population. It is with this purpose in mind that a variety of philosophical underpinnings were presented, including retribution, incapacitation, deterrence, rehabilitation, restorative justice, and reintegration. As can be seen, each of these philosophical approaches has held sway throughout the history of corrections at one time or another. However, it is clear from the definition of corrections, as provided in this text, that the ultimate and modern philosophy of corrections is one that likely includes elements of rehabilitation, restorative justice, and reintegration more than it does retribution, incapacitation, and deterrence.

Though modern corrections is considered more reintegrative in nature, the reality cannot be ignored: Prisons are not effective instruments of rehabilitation or reintegration. Research demonstrates that incarceration is not likely to lower recidivism and, in some cases, may actually increase it. Thus, other philosophical uses of prisons, such as incapacitation and deterrence, continue to proliferate among correctional agencies.

Similarly, the most serious sentence that can be meted out, the death penalty, also seems to not be a very effective form of punishment. Just as with prisons, there is serious doubt as to the deterrent value of the death penalty. Also, just as with prisons, there is some research that shows that the death penalty may actually increase the commission of future crime, not lower it. As with prisons, the philosophical views on the use of the death penalty were discussed, providing pros and cons associated with this sanction.

Because one criminal offense is not always equal to another, there has emerged the need for a continuum of sanctions. This continuum provides for a number of punishments (sanctions) that have varying levels of severity. Monetary fines are perhaps the least serious of sanctions, followed by a very wide range of intermediate community-based sanctions. Community-based sanctions are given extensive coverage due to their variability in administration and their effectiveness in calibrating the punishment to the criminal offense and the criminal offender. Discussion regarding the use of incarceration as a primary tool of punishment was provided, as was an explanation of the different types of custody arrangements when using incarceration with serious offenders. Next, three types of sentencing models were presented: indeterminate, determinate, and mandatory minimum sentences. The reasons for using these types of sentences were provided, as were the pitfalls to each one. While intentions may be good, the outcomes of each of these types of sentencing schemes have not necessarily been effective in achieving the desired goal of their application. Further still, despite the use of complicated sentencing approaches and philosophical approaches to administering punishments, it is clear that sentencing disparities exist throughout America. Disparities were noted to be especially problematic in the southern and western parts of the United States, and it has been found that disparities with punishments exist with both prison sentences and the application of the death penalty. In discussion of the issue of disparity, the distinction between disparity and discrimination was made clear.

It would seem that a degree of fervor has developed regarding the sentencing schemes used in recent decades. In particular, the federal system is overcrowded with drug offenders, which has prompted the development of policies to release these offenders early from prison. To a lesser extent, this has been true in some state systems as well, where effects of the War on Drugs that have resulted in widespread racial disparities in the sentencing of African American men is promoted as the rationale for the reduction of time served. While this may seem like an altruistic concern, the reality is that these prison systems are financially broke and the use of sentence reductions can help alleviate overcrowding.

Lastly, a number of criminological theories were presented with an emphasis on their application to the field of corrections. An understanding of the theoretical bases of the criminal justice discipline in general and the correctional system in particular will aid in the correctional process. Indeed, if we are able to explain why different types of crime occur, we can then determine those factors that should be addressed to eliminate the likelihood of criminal behavior. This means that an understanding of the theoretical underpinnings to criminal behavior can improve any correctional effort. Though a diverse number of theories were presented, each provides its own vantage on how and why crime exists, and each provides a framework from which correctional agents can approach the task of providing organized security and treatment functions intended to correct criminal tendencies among the offender population and, in the process, enhance public safety.

Want a Better Grade?

Get the tools you need to sharpen your study skills. Access practice quizzes, eFlashcards, video, and multimedia at **edge.sagepub.com/hanserbrief**

Interactive eBook

Visit the interactive eBook to watch SAGE premium videos. Learn more at **edge.sagepub.com/hanserbrief/ access**.

 Career Video 2.1: Court Administrator

 Criminal Justice in Practice Video 2.1: Judge Sentencing Guidelines

 Prison Tour Video 2.1: Death Row Housing

 Prison Tour Video 2.2: Preparing for Execution

DISCUSSION QUESTIONS

Test your understanding of chapter content. Take the practice quiz at edge.sagepub.com/hanserbrief.

1. Compare and contrast the two philosophical orientations of *incapacitation* and *deterrence*. In your opinion, which one is the better approach to correctional practice?

2. Compare and contrast the two philosophical orientations of *rehabilitation* and *retribution*. In your opinion, which one is the better approach to correctional practice?

3. Provide a topical overview of the different types of informal sanctions discussed in this chapter. Also, explain what is meant by the "continuum of sanctions" when talking about informal sanctions.

4. Compare and contrast the terms *disparity* and *discrimination*. Which one do you think is most appropriate to explaining the overwhelming proportion of minority inmates behind bars?

5. Compare and contrast indeterminate and determinate sentencing. What are the pros and cons of each?

6. What is restorative justice and how is it unique from many other perspectives on the resolution of crime?

7. According to the chapter, in what ways are classical criminology and operant conditioning similar to one another in their orientations on shaping human behavior?

KEY TERMS

Review key terms with eFlashcards at edge.sagepub.com/hanserbrief.

Aggravating circumstances, 40

Capital punishment, 37

Conflict theory, 47

Determinate discretionary sentence, 41

Determinate presumptive sentence, 41

Discrimination, 43

Disparity, 43

Fine, 35

General deterrence, 32

Incapacitation, 32

Individual personality traits, 45

Labeling theory, 46

Mandatory minimum, 41

Mitigating factors, 40

Negative punishment, 46

Negative reinforcers, 45

Positive punishment, 46

Positive reinforcers, 45

Rehabilitation, 33

Reintegration, 33

Restorative justice, 33

Retribution, 30

Selective incapacitation, 32

Smarter Sentencing Act of 2014, 44

Social learning theory, 46

Specific deterrence, 32

Strain theory/institutional anomie, 46

KEY CASE

United States v. Booker (2005), 34

APPLIED EXERCISE 2.1

Match each of the following modern-day programs with its appropriate philosophical underpinning, sentencing scheme, or theoretical orientation.

PROGRAM	IDEOLOGY OR PHILOSOPHY OF ORIGIN
1. You are sentenced to 10 years in prison and must serve no less than 80% of that time (8 years) without the benefit of early release or parole	A. Restorative justice
2. Laws that select specific types of offenders and provide enhanced penalties to ensure that they are effectively removed from society (habitual offender laws, three-strikes laws, etc.)	B. General deterrence
3. Punishing an offender in public so other observers will refrain from criminal behavior	C. Negative punishment
4. Providing the offender an opportunity to restore damages done to the victim and minimizing stigma/shame for the offender	D. Treatment
5. Removal of visitation privileges because an offender commits the undesired criminal behavior of child abuse	E. Indeterminate sentencing
6. Exacting a fine for undesired behavior	F. Mandatory minimum
7. Treats crime similar to a mental health issue or along the medical model perspective	G. Determinate sentencing
8. Providing substance abusers with certificates of graduation when completing an addiction treatment program	H. Positive reinforcement
9. Sentencing has no flexibility in terms	I. Positive punishment
10. Sentencing with variable terms, affected by the context of the crime and later behavior of offenders while serving their sentence	J. Incapacitation

WHAT WOULD YOU DO?

You are the judge in a small-town court. In this town, everybody knows each other, and things are usually fairly informal. Your position and title carry a great deal of respect throughout the town, and because of this you take your role in the community very seriously. You have recently had a case appear on your docket that is very troubling. A mentally challenged man, 19 years old, is in the county jail; he bludgeoned another man to death with a ball-peen hammer, hitting the victim repeatedly across the head to the point that the deceased victim was barely recognizable. The defendant, Lenny Gratzowskowitz, was in a fit of fury during the crime and continued pounding the skull of the deceased well beyond the point of death.

The police who arrested Lenny were careful to ensure their behavior was well within ethical boundaries, and the agency ensured that legal representation was present before any questions were asked of Lenny. In fact, many of the police officers (including the chief of police) know Lenny on a semipersonal basis because of the tight-knit nature of the town. Generally, Lenny is not problematic, and he has never been known to be violent. However, throughout his history, from childhood on up, he has been subjected to ridicule and embarrassment by a handful of town residents who are of a fairly unsavory disposition.

In fact, the victim, Butch Wurstenberger, had been a childhood bully in grade school, and he had terrorized Lenny on numerous occasions. Now, as an adult, Butch was known to be an abrupt and sarcastic man, but not violent. Both Butch and Lenny had obtained jobs with a general contractor to complete construction of a Walmart supercenter that was slated for a grand opening during the upcoming year. Each had worked in the construction field: Butch had become known for his skill with foundation work and drywall setting and his experience with industrial air-conditioning and refrigeration systems; Lenny had been hired due to his routine dependability on other job sites and his willingness to work, regardless of the circumstances.

Once both arrived on the job site, Butch heckled Lenny on a few but sparing occasions. Most other members of the work crew were from out of town and were not aware of the history between Butch and Lenny. None of them had noticed any serious problems between the two men—that is, not until they reported to work one morning to find that both Butch and Lenny had arrived at the work scene early, and one of them (Butch) was dead while the other (Lenny) was bloody from the act of violence that he had committed.

Consider this situation and determine which philosophical orientation you would use when sentencing Lenny. Select only *one* of the following philosophical orientations: retribution, deterrence, incapacitation, treatment, restorative justice, or reintegration. Consider why you selected that orientation and why each of the other orientations might not be as appropriate as the one that you chose. Write this down as an essay that answers the following question:

What would you do?

3

Correctional Law and Legal Liabilities

Learning Objectives

1. Describe the hands-off doctrine and its relevance to corrections.

2. Identify key rights inmates possess.

3. Evaluate the application of the First, Fourth, Eighth, and Fourteenth Amendments of the Constitution to corrections.

4. Discuss the shift to a more restrained, hands-on approach and the impact of the Prison Litigation Reform Act of 1995.

5. Identify and discuss legal liabilities associated with correctional staff.

6. Explain how prisons have had to change to comply with judicial orders.

7. Apply legal principles to challenges in the field of corrections.

The Writ Writer

"The 1960s were a time when 'rights' was a nebulous concept with many different interpretations in America, depending on who you were," Professor Schwin noted. "Back then, rights for African Americans and Caucasians were distinctly different, a woman's right to decide the fate of her own body was hotly debated, and Latinos were simply looked at as a third world source of cheap labor." The professor swept his gaze across the classroom. "And though it was not a popular topic, the rights of those who were incarcerated became an area of intense legal scrutiny."

Professor Schwin went on to note that the odd thing about prison law is that it centers on parameters of treatment that affect, paradoxically, those who have themselves broken the law. He also noted that the heroes of prison litigation are not always those who are free from blemish. Often they are criminals, which is precisely why few people sympathize with the plight of plaintiffs in prison litigation cases.

In 1960, a young Latino American man named Fred Cruz was arrested in Texas for robbery. Despite contesting his guilt, Cruz could not afford an attorney for his appeals while in the Texas Department of Corrections (TDC). Although he only had an eighth-grade education, he refined his reading skills while serving time and read what law books were available to him. Over time, he became the quintessential jailhouse lawyer, and he fought and eventually won a legal battle with Texas to secure several constitutional rights for inmates not only in Texas but also throughout the United States. He became, as they call it inside the institution, a **writ writer**.

Writ writer: An inmate who becomes skilled at generating legal complaints and grievances within the prison system.

At the time of his arrest, Cruz was, by all appearances, your garden variety criminal and was sentenced to 50 years for robbery. His sentence required that he be committed to hard labor, so he would pick cotton during the hot spring and summer months in the never-ending fields of eastern Texas. But at night and when the planting and harvesting were slow, he would study law books, learning the processes to file lawsuits. He initiated numerous suits that challenged the back-breaking working conditions of TDC field labor, the brutal beatings and physical torture that occurred on prison farms, and the capricious nature of the "kangaroo courts" inside prisons that addressed disciplinary charges and processed inmate complaints. He also challenged the Texas building tender system, which allowed other inmates to act as guards and get away with abusing and exploiting other, less fortunate inmates. As punishment for his suits, Cruz was eventually sent to the Ellis Unit, which was at that time known as the Alcatraz of Texas. The Ellis Unit was headed by Warden C. L. McAdams, one of the roughest and most calloused wardens in the state.

While at the Ellis Unit, Cruz was placed in solitary confinement for long periods of time, made to live on bread and brackish water, and was sometimes beaten. His legal work was confiscated, he was restricted from writing or filing paperwork, and he was not allowed to see an attorney on numerous occasions. Even with these terrible obstacles, Cruz managed to provide assistance to other inmates within the system through means that violated institutional rules and procedures. One of these instances was discovered when an inmate was caught with legal work drafted with Cruz's guidance. The work was related to the violation of rights for Islamic inmates. The prison's reaction led to an inmate uprising that included both Islamic and non-Islamic offenders and was so difficult to contain that outside involvement became necessary. This outside involvement eventually included attorneys who took interest in the issues presented and assisted Cruz in reaching the U.S. Supreme Court with the case.

In *Cruz v. Beto*, the Court held that inmates must be given reasonable opportunities to exercise their religious beliefs.

A student raised his hand, and Professor Schwin nodded at him to speak.

"Was Cruz Muslim, or was he just helping them out?"

"No, Cruz was not Muslim," the professor responded. "But often, when fighting for matters of justice, we must be willing to fight for the rights of others as well as our own, even if we do not necessarily see eye to eye with their beliefs or lifestyle. It's a matter of principle. And it never hurts to make friends with the enemy of your enemy."

Professor Schwin then concluded by saying, "Cruz filed litigation related to a number of issues and at the behest of numerous individuals. Sometimes he was successful; other times he was not. But the key point is to understand that regardless of the times, legal principles and rights inherently secured by the Constitution are intended to extend to all of us, both large and small in social standing. It is intended that we all be equal in the eyes of the law."

INTRODUCTION

From our previous two chapters, it is clear that the field of corrections has gone through many transformations throughout the ages. But as Chapter 2 demonstrates, some of these changes took place due to courthouse intervention into prison operations. The period when the hands-off doctrine ended marked the beginning of case law that has permanently impacted correctional operations in the United States. An understanding of this case law is critical to both the practitioner and the student. It is with this in mind that this chapter is presented, which will allow students to develop an understanding of the legal issues associated with correctional processes and practices.

During the early history of corrections in the United States, the courts were not typically involved in prison operations. This is generally because, as noted previously, inmates were seen as slaves of the state. With such a philosophy, concern for prisoners' rights seemed alien to most people at the time. As a result of the culture and the period, there was little reason for the Supreme Court to be involved with inmate issues, and, at that time, inmates would not have considered the possibility of suing the Court; such an option did not exist (Branham & Hamden, 2009).

In addition, there was a general belief that the public and even the courts should not be concerned about the goings-on inside the prisons. Since inmates were considered slaves of the state and since prisons were not intended to be pleasant places of existence, it was thought that prisons were best left in the hands of those who operated them (Branham & Hamden, 2009). There was no need to meddle, and, essentially, the "no news is good news" mentality prevailed with the public and the Court in regard to prison operations.

Hands-off doctrine:
The policy of the courts of avoiding intervention in prison operations.

PRISON TOUR VIDEO:
Many legal rulings related to prisons are in regard to the conditions of a facility. Go to the IEB to watch two wardens describe laws and their impacts on prison.

THE HANDS-OFF DOCTRINE

The policy of the Supreme Court and the lower courts of avoiding intervention in prison operations is generally known as the **hands-off doctrine** and was based on two primary premises. First was the premise that under the separation of powers inherent in the U.S. Constitution, the judicial branch was not justified to interfere with prisons, which were operated by the executive branch. The second premise was, simply put, that judges should leave prison administration to the prison experts. With these two premises in place, states tended to operate their prisons with impunity; they were free to do as they pleased with no fear of outside scrutiny (Branham & Hamden, 2009).

As we have seen in the prior chapters, courts did eventually begin to intervene in prison operations during the early to mid-1900s. This was, in part, due to reform-minded persons who were active in educating the public about inmate issues. As public sentiment moved toward programs that were more rehabilitative in nature, prisons began to catch the eye of the courts more routinely. Further, the civil rights movement of the 1960s highlighted the abuses that occurred in prisons as well as the complete lack of legal protections for inmates. As civil rights issues in society took center stage, civil rights issues in prisons also drew more attention from both the public and the courts.

PHOTO 3.1 Front row, left to right: Associate Justice Stephen G. Breyer, Associate Justice Clarence Thomas, Chief Justice John G. Roberts, Jr., Associate Justice Ruth Bader Ginsburg, Associate Justice Samuel A. Alito. Back row: Associate Justice Neil M. Gorsuch, Associate Justice Sonia Sotomayor, Associate Justice Elena Kagan, Associate Justice Brett M. Kavanaugh.

Fred Schilling, Collection of the Supreme Court of the United States

The Beginning of Judicial Involvement

Perhaps the clear beginning of the end for the hands-off doctrine occurred in the 1941 Supreme Court case *Ex parte Hull*. Prior to this case, it was common for prison officials to screen inmate mail, including legal mail. As a result, prison staff were known to misplace petitions and even deny the inmate the opportunity to mail them. In *Hull*, the Supreme Court held that no state or its officers could legally interfere with a prisoner's right to apply to a federal court for writs of habeas corpus. The term *habeas corpus* refers to a challenge of the legality of confinement and is a Latin term that means "you have the body." A writ of habeas corpus is a court order requiring that an arrested person be brought forward to determine the legality of his or her arrest. In *Hull*, the Supreme Court ruled that inmates had the right to unrestricted access to federal courts to challenge the legality of their confinement (Branham & Hamden, 2009; Stohr, Walsh, & Hemmens, 2009). This was important because writs of habeas corpus were the primary mechanism that inmates had to challenge unlawful incarceration. Without the ability to use this avenue of redress, inmates were powerless to make legal challenges to their confinement.

One case truly opened the door for inmate civil litigation. *Cooper v. Pate* (1964) validated and made clear the right of inmates to sue prison systems and prison staff. In this case, the Court ruled that state prison inmates could sue state officials in federal courts under the Civil Rights Act of 1871. This act was initially enacted to protect southern African Americans from state officials (such as judges) who were often members of the subversive group the Ku Klux Klan. This act is now codified and known as 42 U.S.C. Section 1983, or simply Section 1983 in day-to-day conversation among legal experts and/or practitioners. This act holds that

> every person who under color of law of any statute, ordinance, regulation, custom, or usage of any state or territory, subjects or causes to be subject, any citizen of the United States or other person within the jurisdiction thereof to the deprivation of any rights, privileges, or immunities secured by the Constitution and laws, shall be liable to the party injured in an action at law. (Branham & Hamden, 2009, p. 553)

Because it had been determined that the Court was responsible for protecting against misuses of power possessed by those vested with state authority, it became clear that the Court was within its purview to rule regarding any number of issues related to civil rights of the inmate. Further, it was determined that officials "clothed with the

Ex parte Hull (1941): Ruling that marked the beginning of the end for the hands-off doctrine.

Cooper v. Pate (1964): Ruling that state prison inmates could sue state officials in federal courts.

authority of state law" did include prison staff and security personnel. Therefore, state prison personnel who violated an inmate's constitutional rights while performing their duties under state law could be held liable for their actions in federal court. The clarity provided in *Cooper v. Pate* and the route of litigation opened for inmates essentially created what some have called the hands-on doctrine of correctional case law, which we now consider in the following subsection.

THE EMERGENCE OF INMATE RIGHTS

As noted earlier, much of the public concern with the rights of inmates dovetailed with the civil rights movement during the 1960s. During this time, the National Association for the Advancement of Colored People (NAACP) Legal Defense and Educational Fund and the National Prison Project of the American Civil Liberties Union (ACLU) began to advocate for the rights of inmates. At the same time, legal protections for inmates were given substantial public and political attention. In determining inmate rights, the balance between humane treatment and the need to maintain public safety remained an issue among the courts. Many courts used contradictory rulings to navigate the myriad complications that emerged from balancing these two overarching concerns.

Turner v. Safley (1987): A prison regulation that impinges on inmates' constitutional rights is valid if it is reasonably related to legitimate penological interests.

Rational basis test: Sets guidelines for the rights of inmates that still allow correctional agencies to maintain security.

The difficulty in setting this balance led to further involvement by the Supreme Court in the landmark case of *Turner v. Safley* (1987). In this case, the Court ruled on a Missouri ban against correspondence sent among inmates in different institutions within the state's jurisdiction (del Carmen, Ritter, & Witt, 2005). The Court upheld the ban, noting that such forms of regulations are valid if they are "reasonably related to legitimate penological interests." Further, the Court enunciated four key elements of what is known as the **rational basis test**. These four elements are as follows:

1. There must be a rational and clear connection between the regulation and the reason that is given for that regulation's existence.

2. Inmates must be given alternative means to practice a given right that has been restricted, when feasible.

3. The means by which prison staff and inmates are affected must be kept as minimal as realistically possible.

4. When less restrictive alternative means of impeding upon an inmate's rights are available, prison personnel must utilize those alternative means.

This rational basis test has provided a good degree of clarity in resolving the conflict between inmate rights and the need for institutional and public safety. It is within these guidelines that the rights of inmates have emerged but, at the same time, been held in check sufficiently to allow correctional agencies to maintain security. Thus, inmate rights became a permanent fixture in the U.S. correctional system, but these rights were tempered with some degree of feasible pragmatism.

Access to Courts and Attorneys

Johnson v. Avery (1969): Held that prison authorities cannot prohibit inmates from aiding other inmates in preparing legal documents.

In tandem with the newfound rights of inmates was a corresponding recognition that such rights were useless unless inmates were afforded access to the courts. The primary case that deals with inmate access to the courts is *Johnson v. Avery* (1969). In this case, the Supreme Court held that prison authorities cannot prohibit inmates from aiding other inmates in preparing legal documents unless they also provide alternatives by which inmates may access the courts. More importantly, the Court further made it clear that, as a result, prison systems have an obligation to provide some form of documented access to the courts, to attorneys, or to some sort of legitimate legal aid (Anderson, Mangels, & Dyson, 2010).

Access to Law Libraries

Bounds v. Smith (1977): Determined that prison systems must provide inmates with law libraries or professional legal assistance.

The primary case that addresses the issue of law libraries is *Bounds v. Smith* (1977). In this case, the Court held that even when prison policies allow jailhouse lawyers to provide assistance to inmates, prison systems must still provide inmates with either adequate law libraries or adequate legal assistance from persons trained in the law (Anderson et al., 2010). As with *Johnson v. Avery*, the opinion of the Court in *Bounds v. Smith* was far from specific in explaining what would constitute legal services and materials. However, most lower courts, in interpreting the ruling by the Supreme Court, have concluded that the requirements in *Bounds* are satisfied when states provide inmates with adequate law libraries and access to materials with some quasi-professional help (del Carmen et al., 2005).

While the specific requirements under *Bounds* are not necessarily clear, it would appear that, at a minimum, most courts currently require either an adequate law library or assistance from persons who have some sort of verifiable legal training. When determining whether the inmates of a prison have the appropriate assistance for filing court documents, courts generally consider the following four factors:

PHOTO 3.2 Many times when inmates sue state agencies, they may use resources such as the law library.

AP Photo/*Daytona Beach News-Journal*/Nigel Cook

1. The number of inmates entitled to legal assistance

2. The types of claims these inmates are entitled to bring

3. The number of persons rendering assistance

4. The training and credentials of those who provide legal assistance

Although *Bounds* still remains the primary case regarding inmates' right to access to courts, another case, *Lewis v. Casey* (1996), has since narrowed the scope of this right in a manner that provides prison officials with much more leeway when establishing legal assistance that is likely to be constitutionally permissible. In *Lewis*, the Court held that any inmate who alleges a violation of *Bounds* must show that shortcomings in the prison's library or services for assistance caused an actual harm or injury that was directly attributable to the inmate's inability to pursue legitimate legal claims.

CORRECTIONS AND THE CONSTITUTION

First Amendment Cases in Corrections

The First Amendment states that

> Congress shall make no law respecting an establishment of religion, or prohibiting the free exercise of religion thereof; or abridging the freedom of speech, or of the press; or the right of the people peaceably to assemble, and to petition the government for a redress of grievances.

This amendment essentially safeguards the very right to speak, communicate, and act freely. Given the strict confines of the prison setting, it is no surprise that this amendment has been the legal basis behind numerous forms of inmate redress. Generally, case law has revolved around inmate access to publications, mail censoring processes, correspondence with family, and religious practices within the institution (see Table 3.1).

Turner v. Safley, which set the rational basis test, involved prison staff's basis for censoring different types of mail among inmates. Prison staff may censure mail and other forms of written communications and engage in oversight so long as there is a legitimate penological interest at stake. Some of the reasons that might be given for this include the safety and security of the institution, rehabilitative concerns, and prison

TABLE 3.1

First Amendment Prison Law Cases From the Supreme Court

First Amendment issues include inmate rights to freedom of speech and freedom of religion. Cases below are organized in chronological order.	
COURT CASE	**RULING**
***Fulwood v. Clemmer* (1962)**	The Muslim faith is a valid religion that requires prison officials to allow Muslim inmates to engage in their respective religious activities.
Gittlemacker v. Prasse (1970)	Inmates must be allowed to practice their religion, but states are not required to provide a clergy member in such cases.
***Cruz v. Beto* (1972)**	Inmates must be given reasonable opportunities to exercise their religious beliefs.
***Procunier v. Martinez* (1974)**	Prison officials may censor inmate mail only to the extent necessary to ensure security of the institution.
Kahane v. Carlson (1975)	Jewish inmates have a right to their kosher diet as a bona fide part of their religion. If this cannot be provided, prison officials must show why such a diet cannot be provided.
Theriault v. Carlson (1977)	Bogus or fake religious sects that are not genuinely practiced with sincerity are not protected under the First Amendment.
Turner v. Safley (1987)	A prison regulation that impinges on inmates' constitutional rights is valid if it is reasonably related to legitimate penological interests.
O'Lone v. Estate of Shabazz (1987)	Prison policies that in effect prevent inmates from exercising freedom of religion are constitutional because they are reasonably related to legitimate penological interests.
Beard v. Banks (2006)	Prisons may implement policies that restrict access to magazines, publications, and photographs for inmates who are classified as higher risk to the safety and security of the institution.
Holt v. Hobbs (2015)	In cases of legitimate religious actions, the government must show that substantially burdening the religious exercise of an individual is "the least restrictive means of furthering that compelling governmental interest."

Source: del Carmen, R. V., Ritter, S. E., & Witt, B. A. (2005). *Briefs of leading cases in corrections* (4th ed.). Lexington, KY: Anderson..

Fulwood v. Clemmer (1962): Ruled that correctional officials must recognize the Muslim faith as a legitimate religion and allow inmates to hold services.

Procunier v. Martinez (1974): Prison officials may censor inmate mail only to the extent necessary to ensure security of the institution.

Cruz v. Beto (1972): Ruling that inmates must be given reasonable opportunities to exercise their religious beliefs.

O'Lone v. Estate of Shabazz (1987): Held that depriving an inmate of attending a religious service for "legitimate penological interests" was not a violation of the inmate's First Amendment rights.

order. Courts tend to defer to the judgment of prison administrators when considering First Amendment issues, as it is considered that they are the best suited to determine security and safety needs.

In the realm of religion in prison, the Court has made at least two significant determinations. In **Cruz v. Beto (1972)**, the Court ruled that inmates must be given reasonable opportunities to exercise their religious beliefs (see Table 3.1). In this case, Cruz was a member of the Buddhist faith and an inmate in the Texas Department of Corrections. He had been placed in solitary confinement and, during that time, was not allowed access to religious services for his faith although other inmates of other religions (e.g., Catholic, Jewish, and Protestant faiths) who were in solitary confinement were given such benefits. The Court held that inmates with unconventional beliefs must be afforded reasonable opportunities to exercise their religious beliefs, especially if this is afforded to inmates of other faiths (Anderson et al., 2010; Branham & Hamden, 2009).

In another case, **O'Lone v. Estate of Shabazz (1987)**, the Court held that the free exercise rights of a Muslim inmate were not violated when prison officials would not adjust the time of his work schedule (see Table 3.1). The reason for this ruling was that the prison system's justification for avoiding alterations in the work schedule was grounded in security and logistic concerns, which, according to the Court, were sufficient and legitimate penological interests. Since this ruling, various religious minorities among the inmate population have gained ground in securing the means for practicing their religions. As an example, lower court decisions have supported specialized dietary requirements for different religious faiths and the right to assemble for services, to contact religious leaders of the inmate's respective faith, and even to wear a beard if one's religion legitimately calls for it. Indeed, this was upheld in *Holt v. Hobbs* (2015), when the Supreme Court ruled that Muslim inmates may wear beards within institutions. A failure to allow this on the part of prison administrators is a violation of the Religious Land Use and Institutionalized Persons Act of 2000, as explained in the paragraph that follows.

The right to practice religion in prison was strengthened through the **Religious Land Use and Institutionalized Persons Act of 2000**. This act prohibits a state or local government from taking any action that substantially burdens the religious exercise of an institutionalized person unless the government demonstrates that the action constitutes the least restrictive means of furthering a compelling governmental interest. When a compelling governmental interest does exist, the government must use the least restrictive alternative to further that interest. Specifically, Section 3 contends that governments cannot impose a "substantial burden on a person residing in or confined to an institution," which of course includes inmates. The issue of religion has been under consideration for over 30 years, and it seems that just as the American cultural and religious landscape has become more diverse in society, so too has the world of corrections and correctional religious services.

Fourth Amendment Cases in Corrections

According to the Fourth Amendment,

> the right of the people to be secure in their persons, houses, papers, and effects, against unreasonable searches and seizures, shall not be violated, and no Warrants shall issue, but upon probable cause, supported by Oath or affirmation, and particularly describing the place to be searched, and the persons or things to be seized.

In prison settings, the Fourth Amendment has limited applicability because, for the most part, inmates do not have a legitimate expectation of privacy while serving their sentence. Indeed, the fact that they are incarcerated limits such an expectation given that prison officials must assume responsibility for their safety and security while in custody. Students should understand that inmates have no expectation of privacy when in the institution (whether in a cell or in a dorm-like setting), nor do they have such an expectation upon their person. Whenever prison staff determine appropriate, inmates are subject to search without any justification being necessary beyond a security rationale (see Table 3.2).

One area that has generated substantial litigation is that of cross-gendered searches of the person. However, when correctional staff of the opposite gender conduct a search, there is no constitutional violation. So long as the search is conducted according to policy, in a professional manner, and without sexual connotations, the inmate will have no standing for a grievance. Though this is the case, it is generally considered prudent for prison administrators to consider an inmate's privacy as much as is reasonable, especially in areas such as showers and dressing rooms. Though personal searches are not considered problematic, the use of opposite-gendered strip searches is usually prohibited by most courts. Thus, prison and jail systems with inmates of both genders will find it wise to have staff of both genders readily available as well.

Searches of the person are conducted at varying levels of intrusiveness. Some consist of the simple pat search of the outer clothing, others require strip searching of the inmate, and the most invasive form of search is the body cavity search. Policies should clearly demonstrate the need for more invasive searches and connect them to a legitimate

PRISON TOUR VIDEO: Accommodating an inmate's religion in prison presents a unique set of challenges for prison officials. Go to the IEB to watch a warden discuss these challenges about religion in prison.

Holt v. Hobbs (2015): In cases of legitimate religious actions, the government must show that substantially burdening the religious exercise of an individual is "the least restrictive means of furthering that compelling governmental interest."

Religious Land Use and Institutionalized Persons Act of 2000: Prohibits the government from substantially burdening an inmate's religious exercise.

TABLE 3.2

Fourth Amendment Prison Law Cases From the Supreme Court

Fourth Amendment issues address unreasonable searches and seizures of persons and their things. In prisons, this tends to involve invasive searches of inmates and/or their property that inmates believe should be protected.	
COURT CASE	**RULING**
Lanza v. New York (1962)	Verbal and written conversations in jail (and prison) visitation rooms do not enjoy any Fourth Amendment privacy safeguards.
United States v. Hitchcock (1972)	Searches of inmate cells and dormitories are not subject to Fourth Amendment considerations. Prison staff may conduct searches of inmate living quarters as the need arises.
Bell v. Wolfish (1979)	The use of body cavity searches of inmates after contact visits is permissible. Prison staff may search inmates' quarters in their absence. Double bunking does not deprive inmates of their liberty without due process of law.
Hudson v. Palmer (1984)	A prison cell may be searched without a warrant and without probable cause. Prison cells are not protected by the Fourth Amendment.
Albert W. Florence v. Board of Chosen Freeholders of the County of Burlington, et al. (2012)	Prison staff may routinely strip search minor offenders and detainees (e.g., traffic violation offenders) when they are arrested and detained within a jail or detention facility.

Source: del Carmen, R. V., Ritter, S. E., & Witt. B. A. (2005). *Briefs of leading cases in corrections* (4th ed.). Lexington, KY: Anderson.

PHOTO 3.3 Muslims attend Juma, Friday's group prayer, in a gathering room reserved for prisoner activities at a state prison in Virginia.

Andrew Lichtenstein/Getty Images

***Hudson v. Palmer* (1984):** Held that prison cells may be searched without the need of a warrant and without probable cause.

***Albert W. Florence v. Board of Chosen Freeholders of the County of Burlington, et al.* (2012):** Prison staff may strip search minor offenders and detainees within a jail or detention facility.

institutional need. Thus, the more invasive the procedure, the more clear the evidence must be that a legitimate institutional interest was at stake, and the more important it is that the search be documented and justified (Branham & Hamden, 2009).

For offenders who are on probation and/or parole, the issue of searches can be a bit more complicated. In such circumstances, police officers and other external enforcement officials need to adhere to Fourth Amendment considerations when dealing with these offenders. However, community supervision staff (e.g., probation and parole officers) are not under those same restrictions, particularly if the inmate is on their own caseload. In such circumstances, police have been known to work in partnership with community supervision personnel to avoid Fourth Amendment restrictions; such partnerships exist throughout the nation and have been found to be constitutionally permissible.

Eighth Amendment Cases in Corrections

The Eighth Amendment states that "excessive bail shall not be required, nor excessive fines imposed, nor cruel and unusual punishments inflicted." This amendment is cited in most cases involving excessive use of force or other physical injuries where inmates file suit (see Table 3.3). But in addition to such overt acts where civil rights are in question, inmates also use this amendment as the basis for a number of other issues, such as the conditions of confinement, medical care, and other minimal standards of living.

TABLE 3.3

Eighth Amendment Prison Law Cases From the Supreme Court

Eighth Amendment issues address whether treatment of inmates entails cruel or unusual punishment.	
COURT CASE	**RULING**
Estelle v. Gamble (1976)	Deliberate indifference to inmate medical needs constitutes cruel and unusual punishment and is therefore unconstitutional.
Gregg v. Georgia (1976)	Death penalty statutes that contain sufficient safeguards against arbitrary and capricious imposition are constitutional.
***Ruiz v. Estelle* (1980)**	The conditions in the prison system of Texas were found to be unconstitutional.
Rhodes v. Chapman (1981)	Double celling of inmates does not, unto itself, constitute cruel and unusual punishment.
Whitley v. Albers (1986)	The shooting of an inmate without prior verbal warning in order to suppress a prison riot does not violate that inmate's right against cruel or unusual punishment.
***Wilson v. Seiter* (1991)**	Deliberate indifference is required for liability to be attached to condition of confinement cases. This means that a culpable state of mind on the part of prison administrators must be demonstrated.
Overton v. Bazzetta (2003)	Prison staff may restrict prison visitations so long as their actions are related to legitimate penological interests.
Hill v. McDonough (2006)	Methods of implementing the death penalty are subject to suit under Section 1983 litigation.
Brown v. Plata (2011)	A court-mandated population limit is sometimes necessary to remedy violations of prisoners' Eighth Amendment constitutional rights.

Source: del Carmen, R. V., Ritter, S. E., & Witt, B. A. (2005). *Briefs of leading cases in corrections* (4th ed.). Lexington, KY: Anderson.

PRACTITIONER'S PERSPECTIVE

"Studies have shown there is a link between education and violence; the more education is present, the less violence there is."

Go to the IEB to watch Ray Bynam's video on his career as a corrections sergeant.

Ray Bynam
Retired Corrections Sargeant

In general, when determining if acts or conditions are violations under the Eighth Amendment, the courts have utilized three basic considerations:

1. Whether the treatment shocks the general conscience of a civilized society

2. Whether the treatment is cruel beyond necessity

3. Whether the treatment is within the scope of legitimate penological interests

Ruiz v. Estelle (1980):
Ruled that the Texas prison system was in violation of the prohibition against cruel and unusual punishments.

Wilson v. Seiter (1991):
Deliberate indifference is required for liability to be attached for condition of confinement cases.

Totality of the conditions: A standard used to determine if conditions in an institution are in violation of the Eighth Amendment.

Hutto v. Finney (1978): Held that courts can set time limits on prison use of solitary confinement.

To further aid in determining claims against prison conditions under the Eighth Amendment, federal courts have implemented a standard that takes into account various factors. This standard, which was born out of Supreme Court case law, is known as the totality of circumstances in general precedent by the Court. The totality of circumstances has been modified in prison law to consist of the **totality of the conditions**, to determine if conditions in an institution are in violation of the Eighth Amendment. In numerous cases, the Court has made a point to effect positive change in prison operations throughout the nation and, in doing so, has laid out three principles that should be implemented when mandating change. The three principles, as articulated in ***Hutto v. Finney* (1978)**, consist of the following:

1. Courts should consider the totality of the conditions of confinement.

2. Courts should make a point to specify each individual factor that contributed to this totality of conditions that were found to be unconstitutional, with clear orders for remediation and changes.

3. When and where possible, courts should articulate the minimal standards necessary for an institution to remedy the constitutional violation.

As noted just previously, inmates also use the Eighth Amendment as the legal basis to challenge force used against them. For instance, in *Hudson v. McMillian* (1992), the Supreme Court ruled that inmates do not need to suffer a severe physical injury in order to file an Eighth Amendment claim as long as the use of force or abusive treatment involved the "wanton and unnecessary infliction of pain," or if the force was used "maliciously or sadistically for the very purpose of causing pain." Naturally, such forms of treatment serve no legitimate penological interest and simply aggravate criminogenic mindsets among the offender population. More recently, the case of *Brown v. Plata* (2011) serves as modern-day confirmation of the criteria set in *Hutto*. In *Brown*, it was determined that overcrowding in the California Department of Corrections and Rehabilitation (CDCR) was unconstitutional. Prior to this Supreme Court ruling, the CDCR, as a system, was designed to hold 80,000 inmates. A three-judge district court ruling required the CDCR to submit a plan to reduce the 150,000 inmates within that system by 40,000 to bring it to 110,000, which would have still been a population that was 37% above capacity for that system. The district judges noted that California had failed to obey previous orders to improve prison crowding conditions, making the reduction necessary to deal with overcrowding and very poor health care, the result of which had been at least one preventable inmate death each week, on average (Liptak, 2011). This case spurred substantial debate among the justices, particularly over concern about public safety when releasing so many inmates in such a short period of time. Nevertheless, the Court upheld the ruling of the district judges, affirming that when conditions are so bad as to be unconstitutional, states must take corrective action.

There is one other legal area that falls under Eighth Amendment consideration: the death penalty. While the death penalty has been held as constitutional since the ruling of ***Gregg v. Georgia* (1976)**, the specific means by which that penalty is administered are still subject to potential legal challenge (see Table 3.3). There is an expectation that the application of this penalty will not be arbitrary or capricious in nature. Further, as with *Hudson v. McMillian*, the imposition of the death penalty must not consist of the unnecessary infliction of pain or be malicious or sadistic in nature.

Gregg v. Georgia (1976): Held that death penalty statutes that contain sufficient safeguards against arbitrary and capricious imposition are constitutional.

CONSTITUTIONALITY OF THE DEATH PENALTY

The death penalty has always been the source of substantive debate in the United States, but it was most hotly deliberated within the courts during the civil rights movement of the 1960s. Prior to this time, most proponents of the death penalty interpreted it as conforming to the Fifth, Eighth, and Fourteenth Amendments of the Constitution. However, the notion that this sanction was perhaps a violation of the "cruel and unusual" punishment clause of the Eighth Amendment led some of these proponents to reconsider their stance, and numerous cases began to appear that challenged the legality of the death penalty.

In 1958, the Supreme Court made a ruling in the case *Trop v. Dulles* that was seemingly unrelated to concerns with the death penalty. In *Trop*, the Court developed a phrase that would be cited in future cases because it fit so well with many compelling arguments in favor of reform within the correctional system. According to the Court, there existed "evolving standards of decency that marked the progress of a maturing society" (*Trop v. Dulles*, 1958, p. 86). This statement has been used by opponents of the death penalty to justify why the death penalty should be eliminated from the range of sentencing options.

In 1968, the Supreme Court made an important decision regarding the jury selection process in cases where the death penalty might be given. The Court held that it was not sufficient to strike a potential juror from serving if he or she had doubts or reservations about the use of the death penalty (*Witherspoon v. Illinois,* 1968). Rather, it was determined that jurors could only be disqualified if it could be shown that they were incapable of being impartial in their decision making while on the jury.

Perhaps one of the most important cases, at least from a historical standpoint, was *Furman v. Georgia* (1972). In *Furman*, it was determined that the death penalty had been administered in an arbitrary and capricious manner. The Court, in ruling on Georgia's death penalty statute, noted the wide discretion given to juries, who were largely unguided in the application of death penalty decisions. This was considered arbitrary in nature and a violation of the Eighth Amendment prohibition against cruel and unusual punishments. This decision resulted in the commutation of hundreds of death row sentences to life imprisonment.

In discussing the *Furman* decision, it is important to emphasize that the Supreme Court did not rule that the use of the death penalty, in and of itself, was unconstitutional. Rather, the Court simply ruled that the manner in which the death penalty was applied was unconstitutional.

Once states were clear on the point made in *Furman*, many quickly acted to revise their laws so that their own use of the death penalty would not be stricken down. According to the Death Penalty Information Center (2004), Florida was the first state to rewrite its death penalty statute, just 5 months after the *Furman* ruling. Shortly after this, another 34 states proceeded to enact new death penalty statutes. Nevertheless, as we will see in the next subsection, the ability to apply the death penalty has been eroding over time due to increasingly challenging restrictions on its use. Later in *Glossip v. Gross* (2015), the Court added that when considering the method of execution further noted that the chosen method of execution is be considered constitutional unless the convicted can show that a substantial risk of harm was increased through a given method when compared with another known and available alternative method. In other words, implementation of the death penalty does not include, as a goal, the infliction of unnecessary pain. We now turn to several U.S. Supreme Court decisions that have shaped and tempered the use of the death penalty since the ruling in *Gregg v. Georgia* (1976; see Chapter 3).

Constitutional Limits on the Death Penalty

Numerous Supreme Court decisions followed the ruling in *Gregg v. Georgia* that have greatly limited the use of the death penalty. For instance, in 1977, just 1 year after the *Gregg* decision, the Supreme Court ruled in *Coker v. Georgia* that the death penalty was unconstitutional when used for the rape of an adult woman if she had not been killed during the offense. In other words, the commission of a rape alone was found to not be sufficient grounds for the death penalty. Another case, *Kennedy v. Louisiana*, affirmed this years later in 2008, wherein the Court affirmed this ruling even if the rape involves a child. Later, in 1986, the Court further restricted the use of the death penalty for those who were insane in *Ford v. Wainwright* (1986). Thus, defendants who could successfully use the insanity defense could avoid the death penalty. Much later still, in *Atkins v. Virginia* (2002), the Court ruled that the execution of the mentally retarded was also unconstitutional. This issue was a controversial one, with earlier Court rulings allowing mental retardation to be used as a mitigating factor to reduce sentence severity of offenders eligible for the death penalty. With the rulings in *Ford* and *Atkins*, it was clear that the death penalty was not likely to be considered legal if the offender suffered from some serious mental defect that

Trop v. Dulles (1958): Developed a phrase that would be cited in future cases because it fit with many compelling arguments in favor of correctional reform.

Witherspoon v. Illinois (1968): Held that it was not constitutional to strike a potential juror from serving if the juror had doubts or reservations about the use of the death penalty.

Furman v. Georgia (1972): Ruling that the death penalty was arbitrary and capricious and violated the prohibition against cruel and unusual punishment.

Glossip v. Gross (2015): Supreme Court case that determined that the Eighth Amendment does not require that a constitutional method of execution be free of any risk of pain.

Coker v. Georgia (1977): Ruling that the death penalty was unconstitutional for the rape of an adult woman if she had not been killed during the offense.

Kennedy v. Louisiana (2008): A Supreme Court case where it was determined that it is a violation of the Eighth Amendment to impose the death penalty for the rape of a child when the crime did not result, and was not intended to result, in the death of the child.

Ford v. Wainwright (1986): Ruling that defendants who could successfully invoke the insanity defense could avoid the death penalty.

Atkins v. Virginia (2002): Held that the execution of the mentally retarded is unconstitutional.

PHOTO 3.4 This death row inmate is being escorted by two officers down a prison hallway.

AP Photo/*Rapid City Journal,* Steve McEnroe

Hall v. Florida (2014): Supreme Court case that held that executing an intellectually disabled person violates the Eighth Amendment's protection against cruel and unusual punishment.

Roper v. Simmons (2005): Ruled that the death penalty was unconstitutional when used with persons who were under 18 years of age at the time of their offense.

Thompson v. Oklahoma (1988): Held that the Eighth and Fourteenth Amendments prohibited the execution of a person who is under 16 years of age at the time of his or her offense.

impacted his or her knowing intent to commit the crime. Nevertheless, because some states sought to circumvent these rulings so as to optimize their ability to implement the death penalty, the Court further ruled in **Hall v. Florida (2014)**, that cases where it is difficult to distinguish mental disability cannot rely solely on intelligence test scores, alone, to determine whether a death row inmate is to be put to death. Because these tests have margins of error, those with scores that are near to normal or average must be allowed to provide other evidence of their mental disability. This ruling helped to solidify the intent to the U.S. Supreme Court rulings in the prior *Ford* and *Atkins* cases.

Later, the issue of juveniles and the death penalty was finally resolved after years of continued Supreme Court rulings that had slowly minimized the application of the death penalty to this offender population. In the 2005 case of **Roper v. Simmons**, the Supreme Court ruled that the death penalty was unconstitutional when used with persons who were under 18 years of age at the time of their offense. The reason for this was simply that juveniles are not considered culpable in the same manner that adults are and therefore cannot have the intent necessary to qualify for a death penalty charge.

Prior to this, in **Thompson v. Oklahoma (1988)**, the Supreme Court found that the Eighth and Fourteenth Amendments prohibited the execution of a person who was under 16 years of age at the time of his or her offense, though only four of the justices fully concurred with this ruling. In *Stanford v. Kentucky* (1989) and *Wilkins v. Missouri* (1989), the Supreme Court sanctioned the imposition of the death penalty on offenders who were at least 16 years of age at the time of the crime. The decision in *Roper* overturned these prior judgments.

Further, the Court has consistently kept restricting the means by which death penalty cases are tried. Again in Florida, another case in 2014, *Hurst v. Florida*, upheld the idea that death penalty cases should only be tried by juries, not judges. In this case, the Court struck down a Florida law that allowed judges to decide on facts during sentencing as a violation of the Sixth Amendment right to a trial by jury.

Key U.S. Supreme Court Decisions

One area of concern with the death penalty has been the potential for racial disparity, both in how the case is treated and in the outcome of the case. Two key cases come to mind when considering racial factors and the death penalty. The first case is *Batson v. Kentucky* (1986), in which the Court addressed the manner by which juries are formed in death penalty cases. In *Batson*, the Court ruled that prosecutors had to provide nonracial reasons for eliminating potential jurors from serving on the jury. Essentially, race is not allowed to be a factor in the jury selection process, and, when prosecutors remove a disproportionate number of citizens with the same racial identity of the defendant (the person facing the death penalty), they may be required to explain their actions.

Another important case regarding racial issues and the death penalty was *McClesky v. Kemp* (1987). In *McClesky*, the Court held that statistical data used to demonstrate disparities in the use of the death penalty were not sufficient evidence to invalidate its use. In other words, if it is shown that a higher-than-expected number of minority offenders are put to death when compared with Caucasians, this is not grounds against the use of the death penalty. In this case, the Court contended that Georgia's application of the death penalty was discriminatory, and statistical analyses were used to show a pattern of racial disparity determined by the race of the victim. In particular, when the victim was Caucasian and the defendant was African American, the death penalty was much more likely to be given. However, the Court held that racial disparities would not be recognized as a constitutional violation of "equal protection of the law" unless deliberate racial discrimination against the defendant could be shown.

Aside from racial disparities, other questions and concerns have arisen with regard to the death penalty. One of these is whether judges should be allowed to determine if the death penalty will be given with no input from a jury. In *Ring v. Arizona* (2002), the Court held that juries, rather than judges, are to be the body that determines if the death penalty will be given to a convicted murderer. This ruling overturned the death penalty processes of several states that had previously allowed judges to make this determination on their own. Another stipulation of the *Ring* decision is that any and all aggravating factors that would increase the likelihood of a defendant incurring the death penalty must be included in the initial indictment. Due to this requirement, the federal government itself had to revise some of its death penalty laws.

Two cases that address legal representation and/or procedural issues during death penalty proceedings are also worthy of mention. In *Strickland v. Washington* (1984), the Supreme Court ruled that defendants in capital cases have a right to representation that is objectively reasonable. To demonstrate that their representation was not sufficient, defendants must show that the ruling against them would be appreciably different if they had not used the specific legal counsel that had been afforded them; this is a very difficult standard to meet. In a very recent case, the Court in **McCoy v. Louisiana (2018)** noted that the Sixth Amendment guarantee of the assistance of counsel does not mean that the accused concedes control of his or her case to assigned counsel. Decisions over whether to plead guilty, waive his or her right to trial by jury, pass on an appeal, or testify in person are all rights that are reserved by the accused individual. In this case, Robert McCoy had repeatedly claimed his innocence, but his appointed counsel nevertheless entered guilty pleas, despite McCoy's vocal objections (see Focus Topic 3.1).

In *Uttecht v. Brown* (2007), the Court made a ruling that seemed to undermine its earlier ruling in *Witherspoon v. Illinois* (1968). In the *Uttecht* ruling, the Court held that it was, in fact, acceptable to remove potential jurors from service if they expressed mere doubts about the use of the death penalty. In this case, the trial court had removed a potential juror who had expressed doubts about, but not absolute opposition to, the death penalty.

Lastly, there is one case that is most unusual because it cited trends and legal rulings in other countries and also addressed the consequences when a defendant who is a citizen of another country is put to death in the United States. In *Medellin v. Texas* (2008), a Mexican citizen who raped two young girls in Texas had been given a death sentence. The nation of Mexico sued the United States because the Mexican national consulate had not been notified about Medellin's case, a requirement of the Vienna Convention on Consular Relations. The state of Texas rejected this notion and continued to pursue the death penalty for Medellin. What is most interesting about this case is that the Bush administration sought to influence the Supreme Court, entering the case on behalf of Medellin in an attempt to get the Supreme Court to overturn the Texas court's ruling.

McCoy v. Louisiana **(2018):** Supreme Court case that ruled that it is a violation of an accused's Sixth Amendment rights if his or her defense counsel pleads guilty when the defendant objects and does not wish to concede guilt.

FOCUS TOPIC 3.1

The Sixth Amendment Right to Counsel Does Not Transcend the Will of the Client

In a 6–3 ruling, the U.S. Supreme Court sided with Louisiana death row inmate Robert McCoy, and granted him a new trial when they found his attorney's entry of a guilty plea violated McCoy's Sixth Amendment rights. In August 2011, McCoy was convicted of three counts of first-degree murder and sentenced to death in January 2012. McCoy adamantly maintained his innocence throughout his trial proceedings and ultimately sought removal of his appointed public defenders because, according to McCoy, they were not attempting to prove his innocence but were instead more focused on obtaining a plea bargain. McCoy's family borrowed money to gain an attorney to represent McCoy. Later, this attorney informed McCoy of his intent to concede McCoy's guilt to avoid the likelihood of a death sentence. McCoy vehemently disagreed with the idea and asserted his innocence. McCoy also tried to remove his attorney from the case so that he could defend himself, but the court would not

(Continued)

(Continued)

grant his requests. During trial, McCoy's counsel repeatedly stated that McCoy killed the victims, even though McCoy continued to assert his innocence and explained that a drug-trafficking ring was responsible for the murders. McCoy raised the constitutionality of his lawyer's actions on appeal to the Louisiana Supreme Court. That court held that defense counsel had authority to determine whether to concede guilt, even over McCoy's objections.

The question raised to the U.S. Supreme Court was whether it was unconstitutional for defense counsel to concede their client's guilt over the client's express objection.

The Court held that it is, indeed, unconstitutional, and explained that the Sixth Amendment guarantees an accused the assistance of counsel for his or her defense, which includes directing the objective of the case. Thus, when an individual's liberty, or life as was

the case, is at stake, it is the defendant's prerogative, not counsel's, to decide whether to admit guilt in the hope of gaining mercy at the sentencing stage, or to maintain innocence, leaving it to the jury to decide his or her guilt or innocence. The assistance of counsel that is guaranteed by the Sixth Amendment does not require a defendant to give up his or her own decision-making power to counsel. Rather, many decisions must be reserved for the client, such as whether to plead guilty, waive the right to a jury trial, testify in one's own behalf, and forgo an appeal. In this case, the Court found that McCoy's objective—to maintain that he was innocent of murdering his family—was irreconcilable with trial counsel's objective—to avoid a death sentence. Therefore, "when a client expressly asserts that the objective of his defense is to maintain innocence of the charged criminal acts, his attorney is required to abide by that objective and may not override it by conceding guilt." ●

Fourteenth Amendment Cases in Corrections

According to the Fourteenth Amendment,

> All persons born or naturalized in the United States, and subject to the jurisdiction thereof, are citizens of the United States and of the state wherein they reside. No state shall make or enforce any law which shall abridge the privileges or immunities of citizens without due process of law, nor deny to any person within its jurisdiction the equal protection of the laws.

The primary application of the Fourteenth Amendment to prison law issues has to do with procedural due process issues and issues related to equal protection (see Table 3.4). The Fourteenth Amendment often comes into play when inmates file suit regarding parole release, intraprison transfers, and disciplinary hearings. The other aspect of the Fourteenth Amendment that is relevant to prison law is equal protection. Inmates using this aspect of the amendment may file suit regarding equal protection pertaining to issues based on race, gender, or religious discrimination (Anderson et al., 2010).

Wolff v. McDonnell (1974): Inmates are entitled to due process in prison disciplinary proceedings that can result in the loss of good time credits or in punitive segregation.

The case of ***Wolff v. McDonnell* (1974)** provided guidelines for minimal due process rights that should be afforded inmates facing disciplinary proceedings (see Table 3.4). The Court held that when a prisoner faces serious disciplinary action resulting in the loss of good time or in some form of housing custody change, the procedures used should provide the following:

1. The inmate must be given 24-hour written notice of the charges.

2. There must be a written statement by the fact finders as to the evidence relied on and reasons for the disciplinary action.

3. The inmate should be allowed to call witnesses and present documentary evidence in his or her defense so long as this does not jeopardize institutional safety and security.

4. Counsel substitute must be permitted when the inmate is illiterate or when the complexity of the issues makes it unlikely that the inmate will be able to collect and present the evidence for an adequate comprehension of the case.

5. The prison disciplinary board must be impartial.

TABLE 3.4

Fourteenth Amendment Prison Law Cases From the Supreme Court

The Fourteenth Amendment addresses due process of law and equal protection under the law.	
COURT CASE	**RULING**
Wolff v. McDonnell (1974)	Inmates are entitled to due process in prison disciplinary proceedings that can result in the loss of good time credits or in punitive segregation.
Baxter v. Palmigiano (1976)	Inmates are not entitled to counsel or cross-examination in prison disciplinary hearings. In addition, silence by inmates in a disciplinary proceeding may be used as adverse evidence against them.
Vitek v. Jones (1980)	Inmates are entitled to due process in involuntary transfers from prison to a mental hospital.
Superintendent, Walpole v. Hill (1985)	Disciplinary board findings that result in loss of good time credits must be supported by a "modicum" of evidence to satisfy due process requirements.
Kingsley v. Hendrickson (2015)	For claims of excessive force brought by pretrial detainees, it is only necessary to show that the force used was objectively unreasonable, not that the officer subjectively intended to injure the inmate. These protections for detainees are thus similar to those for persons who are officially convicted and incarcerated.

Source: del Carmen, R. V., Ritter, S. E., & Witt, B. A. (2005). *Briefs of leading cases in corrections* (4th ed.). Lexington, KY: Anderson.

While the above conditions must be provided, the Court has made it clear that disciplinary hearings do not rise to the standard of a true courthouse proceeding but are designed to provide some measure of protection against arbitrariness.

In ***Baxter v. Palmigiano (1976)***, the distinction between prison disciplinary proceedings and genuine courtroom affairs was directly addressed (Anderson et al., 2010). The Court noted that inmates do not have the right to either retained or appointed counsel for disciplinary hearings that are not part of an actual criminal prosecution (see Table 3.4). Further, inmates are not entitled to confront and cross-examine witnesses at all times. In addition, and unlike courtroom proceedings, an inmate's decision to remain silent when faced with disciplinary proceedings can be used as adverse evidence of the inmate's guilt by disciplinary decision makers. Thus, disciplinary hearings simply require fundamental fairness and nondiscriminatory application; they are not to be confused with regular court proceedings.

A RESTRAINED HANDS-ON PERSPECTIVE AND COURT DEFERENCE TO PRISONS

In changing to a more restrictive interpretation of inmate rights, two cases, ***Bell v. Wolfish (1979)*** and *Rhodes v. Chapman* (1981), served as major turning points. These cases reflect an attempt of the Court to balance institutional operations with the need to protect inmates' constitutional rights from oppressive government actions. In *Bell v. Wolfish*, the Court noted that

> maintaining institutional security and preserving internal order and discipline are essential guides that may require limitation or retractions of the retained constitutional rights of both convicted prisoners and pretrial detainees.... Prison officials must be free to take appropriate action to ensure safety of inmates and correctional personnel and to prevent escape or unauthorized entry. (p. 1878)

Thus, it is clear that the Court has been receptive to the challenges that face prison staff and administrators. However, in the case of *Rhodes*, the Court, in addressing the totality of conditions regarding prison facilities, noted very clearly that "the constitution

Vitek v. Jones **(1980):** Inmates are entitled to due process in involuntary transfers from prison to a mental hospital.

***Kingsley v. Hendrickson* (2015):** For claims of excessive force brought by pretrial detainees, it is only necessary to show that the force used was objectively unreasonable.

***Baxter v. Palmigiano* (1976):** Determined that inmates do not have the right to counsel for disciplinary hearings that are not part of a criminal prosecution.

***Bell v. Wolfish* (1979):** Determined that body cavity searches of inmates after contact visits is permissible, as are searches of inmates' quarters in their absence. Double bunking does not deprive inmates of their liberty without due process of law.

does not mandate comfortable prisons," nor should prisons aspire to reach such an ideal standard of living.

From 1980 onward, the Court has taken a more balanced approach to inmate litigation. This balanced approach is known as the **one hand on, one hand off doctrine**. This doctrine contends that while incarcerated, (1) inmates do not forfeit their constitutional rights, (2) inmate rights are not as broad and encompassing as are free persons', and (3) prison officials need to maintain order, security, and discipline in their facilities.

Over time, the Supreme Court has become even more conservative regarding inmates' rights and meddling in the affairs of prison operations. While the passage of the Prison Litigation Reform Act is seen as indication that the Court now gives deference to prison officials, rulings such as that found in *Brown v. Plata* (2011) demonstrate that the Court will become involved and rule against prison administrators when prison conditions seem to be egregious.

The Prison Litigation Reform Act of 1995

The **Prison Litigation Reform Act (PLRA)** was initially passed through Congress with the intent of preventing the wave of frivolous lawsuits that had been filed by inmates in federal court. Essentially, the PLRA was designed to be a screening mechanism to eliminate those cases that were found to be malicious in nature, failed to state a bona fide claim that qualified for legal relief, or sought damages from agency personnel who had immunity from suit. More specifically, Carlson and Garrett (2008) note that the PLRA contains these provisions:

1. Limits inmates' ability to file lawsuits

2. Requires inmates to exhaust all available administrative remedies prior to filing suits

3. Requires the payment of full filing fees in some instances

4. Imposes harsh sanctions, including the loss of good time credit, for filing frivolous or malicious lawsuits

5. Requires that any damages awarded to inmates be used to satisfy pending restitution orders

Further, the PLRA sets a limit on the relief that an inmate may secure from the courts, the restriction being that courts may only require of prison systems a minimum of modification so as to most expediently and in a Spartan manner address a noted deficiency. Even with this, courts must keep their requirements narrow in focus, and they must consider the potential impact that their ruling may have on overall institutional and public safety. Further, oversight from federal judges regarding prison operations was given an expiration of 2 years unless another successful hearing proves that constitutional violations continue to persist within that prison facility or system. Thus, the PLRA has provided state governments with a mechanism by which they can limit the effects of suits regarding prison conditions. This is good news for state prison systems that were, in their earlier history, under the constant scrutiny of federal courts. Figure 3.1 demonstrates the impact that the PLRA has had on the overall number of state inmates who have filed petitions in U.S. district courts.

It is noteworthy that the PLRA has been challenged in court but that the Supreme Court has continued to support and uphold the provisions of this act. For example, in *Booth v. Churner* (2001), the Court ruled that inmates must exhaust all of their available administrative processes for a grievance before filing suit, even if those provisions do not allow for monetary damages. This is an important point because up until this time, inmates would sometimes utilize legal remedies to simply gain monetary compensation rather than using available tools within the prison to rectify wrongs.

The PLRA and its general stance on inmate litigation are reflective of a more conservative approach with prison rulings. In fact, it appears that the Court is leaning more toward a "one hand on, one hand off" doctrine regarding intervention in prison operations. More conservative rulings are being handed down from the Court, and this reflects an eclipse of the hands-off doctrine in American corrections.

One hand on, one hand off doctrine: More conservative rulings are being handed down from the Court, reflecting an eclipse of the hands-off doctrine.

Prison Litigation Reform Act (PLRA): Limits an inmate's ability to file lawsuits and the compensation that he or she can receive.

FIGURE 3.1

The Effect of the PLRA on Petitions Filed by State Inmates From 1990 to 2006*

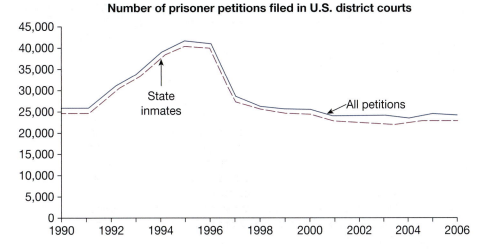

Number of prisoner petitions filed in U.S. district courts

Source: U.S. Department of Justice. (2009). *Civil rights complaints in U.S. district courts, 1990–2006.* Washington, DC: Author.

Note: Most recent data available.

STATE AND FEDERAL LEGAL LIABILITIES

Liability is a term used to note that when a person commits a wrongful action or fails to act when they had a duty to do so, they can be held legally accountable. There are many means by which liability may ensue. First, liability can attach at both the state and the federal levels of government. Second, liability, though most often civil, can be criminal as well. In most cases, the sources of liability are typically not restricted to community supervision officers but often serve as the bases of liability for any officer of the state. Nevertheless, just as with other practitioners who act under "color of law" (a term we will expound upon later in this chapter), community supervision officers usually incur potential liability due to their role as agents of the state (*state* being used in a general sense to cover local, state, and federal government). In addition, the types of liability may actually apply in a variety of forms and from multiple sources. For instance, civil and even criminal liability can emerge, and a correctional officer can be subject to both state and federal levels of civil liability, depending upon the circumstances. For purposes of this chapter, we will begin with state-level forms of liability and then progress toward federal levels of liability (students should also see Table 3.5 for additional details).

State Levels of Liability

Civil liability under state law often is referred to as tort law. A **tort** is a legal injury in which the action of one person causes injury to the person or property of another as the result of a violation of one's duty that has been established by law (del Carmen, Barnhill, Bonham, Hignite, & Jermstad, 2001). Torts can be either deliberate or accidental in nature. A deliberate tort is considered intentional and refers to acts that are intended to have a certain outcome or to cause some form of harm to the aggrieved. An accidental tort would generally be one that is committed out of negligence with the intent of harm being nonexistent.

Tort: A legal injury in which a person causes injury as the result of a violation of one's duty as established by law.

Torts

According to *Black's Law Dictionary*, an **intentional tort** is one in which the actor was judged to have possessed intent or purpose to injury, whether expressed or implied. This means that an intentional tort has numerous components, all of which must be proven

Intentional tort: The actor, whether expressed or implied, was judged to have possessed intent or purpose to cause an injury.

TABLE 3.5

Comparing State and Federal Lawsuits Against Community Supervision Officers

STATE TORT CASES	FEDERAL SECTION 1983 CASES
Based on state law	Based on federal law
Plaintiff seeks money for damages	Plaintiff seeks money for damages and/or policy change
Usually based on decided cases	Law was passed in 1871
Usually tried in state court	Usually tried in federal court
Public officials and private persons can be sued	Only public officials can be sued
Basis for liability is injury to person or property of another in violation of a duty imposed by state law	Basis for liability is violation of a constitutional right or of a right secured by federal law
Good faith defense usually means the officer acted in the honest belief that the action taken was appropriate under the circumstances	Good faith defense means the officer did not violate a clearly established constitutional or federal right of which a reasonable person should have been aware

Source: Adapted from del Carmen, R. V., Barnhill, M. B., Bonham, G., Hignite, L., & Jermstad, T. (2001). *Civil liabilities and other legal issues for probation/parole officers and supervisors.* Washington, DC: National Institute of Corrections.

by a plaintiff if the suit is to prevail. The components of an intentional tort that must be proven are as follows:

1. An act was committed by the defendant.

2. The act was deliberate and can be shown to be such due to the fact that the defendant had to have known of the potential consequences of the act.

3. The resulting harm was actually caused by the act.

4. Clear damages can be shown to have resulted from the act.

A hypothetical example might be a scenario in which Correctional Officer X conducts a search of an inmate's cell. While in the cell, the correctional officer assaults the inmate and says, "I can reach out and touch you anytime, and if you say anything about it, I will see to it that you get written up and lose your good time." The inmate is injured, has to go to the infirmary for internal injuries, and decides to disclose the source of his injuries. In this case, the correctional officer has obviously committed an intentional tort. The assault was committed by Correctional Officer X (the defendant in this case), the assault was obviously on purpose and Correctional Officer X was clearly cognizant of his actions, and clear damages in the way of physical injuries occurred as a direct result of the assault.

With regard to nonphysical torts, several specific types of offenses or damages might occur. These include defamation, acts that cause harm to a person's emotional well-being, and malicious prosecution, among others. Because this area of liability is so broad, it is necessary to address many of these concepts specifically to ensure that the student has a clear understanding of each potential source of liability. Though the following discussion will not be exhaustive, it will highlight the more common areas of liability that correctional officers may face.

First, **defamation** is an invasion of a person's interest through his or her reputation. In order for this to occur, some form of slander or libel must have occurred against the aggrieved individual. Many people use the term *slander* loosely but often are unclear about its meaning, though they may not realize this. To be clear, **slander** is any oral communication provided to another party (aside from the aggrieved person) that lowers the reputation of the person discussed and where people in the community would find such facts to actually be damaging to that person's reputation. **Libel**, on the other hand, is the written version of slander.

Defamation: Some form of slander or libel that damages a person's reputation.

Slander: Verbal communication intended to lower the reputation of a person where such facts would actually be damaging to a reputation.

Libel: Written communication intended to lower the reputation of a person where such facts would actually be damaging to a reputation.

The next category, infliction of **emotional distress**, refers to acts (either intentional or negligent) that lead to emotional distress of the client. Emotional distress can occur due to words or gestures, and of course the conduct of correctional staff. For instance, tactics used to bully or abuse an inmate would fall within this category. However, one simple incident, though not likely to be in adherence to agency policy, typically does not incur liability unless the situation can be shown to be "extreme" and "outrageous" (del Carmen et al., 2001).

The last area of nonphysical tort that will be discussed is malicious prosecution. **Malicious prosecution** occurs when a criminal accusation is made by someone who has no probable cause and who generates such actions for improper reasons. In such cases, the accused must be, as a matter of material fact, innocent of the charges that were made.

PHOTO 3.5 It is from the local or federal courthouse that legal suits are filed. The courthouse has the role of determining inmate rights as well as the rights of offenders on probation or parole.
©iStockphoto.com/OlegAlbinsky

The next primary category of state tort includes acts of negligence. In the vast majority of cases, negligence suits are not successful, particularly if the officer had adhered to agency policy. Most agencies do have sufficient policy safeguards to ensure that an adequate good faith attempt at public safety is made, and, presuming that the officer follows policy, liability is not likely to be incurred. For purposes of this text, **negligence** is defined as doing what a reasonably prudent person would not have done in similar circumstances or failing to do what a reasonably prudent person would have done in similar circumstances. The following minimal conditions are typically required to establish a case of negligence:

1. A legal duty is owed to the aggrieved person (in prisons, such a duty exists between custodial staff and inmates).

2. A breach of that duty must have occurred whether by the failure to act or by the commission of action that was not professionally sufficient to fulfill that duty.

3. The aggrieved can demonstrate that an injury did occur.

4. The person with the duty owed to the aggrieved person committed the act (or lack of action) that was the proximate cause of the injury.

Emotional distress: Refers to acts that lead to emotional distress of the client.

Malicious prosecution: Occurs when a criminal accusation is made without probable cause and for improper reasons.

Negligence: Doing what a reasonably prudent person would not do in similar circumstances or failing to do what a reasonably prudent person would do in similar circumstances.

Liability Under Section 1983 Federal Lawsuits

This avenue for civil redress is one of the most frequently used by persons seeking damages from the government. Though there is substantial history associated with this particular form of liability, this text will delve right into the substantive issues associated with Section 1983 liabilities. Essentially, there are two simple requirements that must exist in order for liability to be imparted to a person so charged:

1. The person charged (the defendant) acted under color of state law.

2. The person charged violated a right secured by the constitution or by federal law.

Both of these requirements need some explanation. First, the term *color of state law* must be clarified. This term simply means that the actor committed his or her behavior while under the authority of some form of government. In this case, the term *state* is meant to imply government in general, regardless of the level of the government (local, state, or federal) that is associated with the agency. Thus, local jailers, state correctional officers, and community supervision officers all act under the color of state law when they are performing their duties in the employment of their respective agencies. This liability does not, however, apply during their off-time "normal" lives. In further clarifying this concept, del Carmen, Barnhill, Bonham, Hignite, and Jermstad (2001) note that anything correctional officers do in the performance of their regular duties and during the usual hours of employment is considered as falling under the color of state law. In contrast, whatever these same people do as private citizens during their off-hours falls outside and beyond the color of state law. In closing, it should be considered that not every violation of an inmate's rights rises to be a federal or constitutional issue. Rather, the violation must be severe enough and centered on a federal or constitutionally protected right to be considered as having sufficient merit.

Forms of Immunity and Types of Defenses

One key protection for correctional staff, as agents of the state, is their potential immunity from tort suits. *Official immunity* is a term that refers to being legally shielded from suit. Official immunity is granted to those professions that must be allowed to, at least in the majority of circumstances, actively pursue their duties without undue fear or intimidation. Otherwise, law enforcement, correctional, and judicial professionals could not fulfill their duties correctly. However, official immunity comes in a number of different forms, reflecting the different levels of responsibility and liability associated with different functions in the justice system. For purposes of this text, students need only to discern between *absolute immunity* and *qualified immunity*.

Absolute immunity, meaning that the individual is not able to be subject to a lawsuit when acting in his or her professional capacity, exists for those persons who work in positions that require unimpaired decision making. Judges and prosecutors have this type of immunity since their jobs require that they make very important decisions regarding the livelihood of persons in their courts; these decisions must be made free of intimidation or potential recrimination, and therefore these court actors enjoy absolute immunity from being sued when carrying out their responsibilities (del Carmen et al., 2001). Note, however, that this does not include an immunity from criminal charges, if such was applicable. For correctional officers and community supervision officers, protection through immunity is typically referred to as qualified immunity. **Qualified immunity** requires that the community supervision officer demonstrate three key criteria were met prior to invoking this form of defense against suit. These criteria, according to del Carmen et al. (2001), are as follows:

1. The community supervision officer must show that he or she was performing a discretionary act, not one that was mandatory by agency policy.

2. The correctional officer must have been acting in good faith—that is, holding the sincere belief that his or her action was correct under the circumstances.

3. The correctional officer must have acted within the scope of his or her designated authority. Thus, most correctional officers do not enjoy the same level of immunity as do their colleagues in the judicial arena since they have to demonstrate the grounds for their possession of immunity.

Beyond the initial forms of liability protection (i.e., qualified immunity and the public duty doctrine) afforded correctional officers, there are some defenses that officers can raise on their own behalf. First among these is the good faith defense. The **good faith defense** essentially buffers a correctional officer from liability in Section 1983 cases (not state tort cases) unless the officer violated some clearly established constitutional or federal statutory right that a reasonable person would have known to exist.

From the previous discussion of both state tort and Section 1983 forms of suit, it is clear that the issue of *good faith* is important. The notion that the officer acted in good faith (with the sincere belief that his or her action was appropriate) is important to establishing qualified immunity against state tort suits. The term *good faith* is the same, but its applications to both types of lawsuit (state tort and Section 1983) are different. Students are encouraged to examine Table 3.5 for further clarity in classifying each type of legal redress and its particular parameters.

Absolute immunity:
Protection for persons who work in positions that require unimpaired decision-making functions.

Qualified immunity:
Legal immunity that shields correctional officers from lawsuits, but first requires them to demonstrate the grounds for their possession of immunity.

Good faith defense:
The person acted in the honest belief that the action taken was appropriate under the circumstances.

Students should understand that even when the three criteria previously discussed are affirmatively answered, it is the good faith issue that is often the most relevant consideration, particularly in cases involving Constitutional rights. As noted, when considering Constitutional violations, the good faith defense means the officer did not violate a clearly established constitutional or federal right of which a reasonable person should have been aware. As the case of *Hope v. Pelzer*, 536 U.S. 730 (2002) demonstrated, qualified immunity did not apply to inmate challenges of use of the "hitching post" within the Alabama Department of Corrections. The hitching post was a punishment whereby inmates were cuffed to a metal bar and forced to stand, with hands held upward, out in the elements, for prolonged periods of time. This punishment caused muscle aches, sprains, chaffing of wrists, and sunburn when in the open sun. The Supreme Court did hold this practice to be a violation of the Eighth Amendment and also held that officers involved with this practice were not entitled to qualified immunity because according to the High Court, it was clear that this punishment did not comport with the Constitutional rights of inmates under the Eighth Amendment. Thus, qualified immunity does have its limits.

Indemnification and Representation

When correctional officers are faced with lawsuits (whether state or federal), the issue of legal representation is an automatic concern. The solutions to this issue as well as that of indemnification (payment for court costs) vary greatly from state to state. As a general rule, most states are willing to provide assistance in civil cases, but this is not nearly as true when the charges are criminal in nature. While it is typical for states to cover financial costs associated with civil proceedings, many do not necessarily do this automatically, and this means that officers from time to time face such instances without any financial assistance from their place of employment—a scary thought indeed.

Most states cover an officer's act or omission to act in civil cases, provided that it is determined that the incident occurred within the scope of the officer's employment (del Carmen et al., 2001). In some cases, this may also require a good faith element where it can be reasonably shown that the officer did act in good faith within the scope of his or her duty. In most cases where good faith is established, the officer will be represented by the state's attorney general. However, the attorneys general in all states have a wide degree of discretion in agreeing to defend an officer faced with a civil suit. If the AG (as the attorney general's office may be called from time to time) does not agree to defend the officer, that individual will have to retain private counsel at his or her own expense.

Once the issue of representation is resolved, the other issue of concern tends to revolve around the payment of legal costs. Most typically, when state tort cases are involved, both the plaintiff (in this case the offender) and the defendant (the community supervision officer) would pay their own attorney's fees, and this would remain the arrangement regardless of the case outcome. Thus, even if the correctional worker is found innocent of the allegations that the offender has claimed, the worker will still likely have to pay his or her own courts costs if the case is a state tort case. To remedy these concerns, officers may, in some cases, opt to purchase their own professional liability insurance, but they will typically be required to pay the premium themselves and, in several cases, may not even have the ability to purchase such insurance since companies may not be operating in the state to underwrite the policy.

Types of Damages

Lastly, most all civil cases (particularly tort cases) seek to obtain monetary damages. The amount can vary greatly, depending on the type of injury, the type of tort that is found (i.e., gross or willful negligence), and the severity of that injury. Most often, these financial awards come in the way of **compensatory damages**, which are payments for the actual losses suffered by a plaintiff (typically the offender). In some cases, **punitive damages** may also be awarded, but these monetary awards would be reserved for an offender who was harmed in a malicious or willful manner by agency staff; these damages are often added to emphasize the seriousness of the injury and/or to serve as a warning to other parties who might observe the case's outcome. The types of award for civil rights cases under Section 1983 suits can also vary greatly and often are similar to those under tort law.

Compensatory damages:
Payments for the actual losses suffered by a plaintiff.

Punitive damages:
Monetary awards reserved for the person harmed in a malicious or willful manner by the guilty party.

It is common for Section 1983 cases to also result in other types of remedies that go beyond financial awards. These remedies are typically geared toward agencies rather than individual officers, though both financial damages and additional awards can be made. One such nonmonetary award that is occasionally granted is the **declaratory judgment**, which is a judicial determination of the legal rights of the person bringing suit (Neubauer, 2002). An example might include a suit where an inmate sues for violation of certain due process safeguards in disciplinary proceedings. A court may award a declaratory judgment against the agency to ensure that future inmates in custody are afforded the appropriate safeguards established by prior Supreme Court case law.

Declaratory judgment: A judicial determination of the legal rights of the person bringing suit.

COMPLIANCE WITH JUDICIAL ORDERS

Inmates in state and federal prison have increasingly petitioned the courts for relief under a variety of statutes. Students will recall from Chapter 2 that some prison systems operated in a manner that was contrary to what progressive-minded prison advocates preferred. As such, the involvement of the courts in prison operations can be viewed as an example of how the U.S. judicial system is affected by the norms and mores of our society, which are reflected in case law and court rulings. These rulings ultimately extended into the prison world, which had, until that time, been separate and distinct from mainstream society.

Injunctions and Court-Imposed Remedies

In addition to the typical award of damages against a correctional agency or correctional staff, courts may also employ injunctions against an agency. An **injunction** is a court order that requires an agency to take some form of action(s) or to refrain from a particular action or set of actions (Neubauer, 2002). Though the PLRA has narrowed the scope of federal court interference and has limited the time frame in which judges can impose their edicts on a prison system or facility, injunctions remain an effective remedy, particularly when federal civil rights or federal statutes are violated.

Injunction: A court order that requires an agency to take some form of action(s) or to refrain from a particular action(s).

If the plaintiff inmate is successful in court (which is rare, but does happen on occasion), judges may issue injunctions that order correctional systems to correct a noted deficiency or set of deficiencies. But in these cases, prison officials must be given the time and opportunity to rectify conditions that are deemed unconstitutional. For the inmate population, this is essential because inmates will likely be living in the institution for years after the ruling has been handed down. Thus, ensuring that administrators do make the required changes and that they are given the necessary tools to do so is very important to the inmate population.

In the past, injunctions or court orders have required a variety of changes of institutions. First, they have mandated the abolition of certain prison regulations and state statutes relating to discipline, censorship, and court access. Second, these orders required improvements in institutions or in the services provided, including training for correctional staff, sanitation conditions, quality of food, and other aspects previously discussed in this chapter and in

PHOTO 3.6 Judges have the ability to issue injunctions that order correctional systems to correct a noted deficiency or set of deficiencies.

©iStockphoto.com/dcdebs

prior chapters. Third, jails, prisons, and/or sections of such facilities have been ordered closed when found to house inmates under conditions determined to be in violation of Eighth Amendment requirements. Thus, injunctions have had a widespread impact upon prison systems. A fairly recent example of this would be the Supreme Court case of *Brown v. Plata* (2011). In this case, the Court upheld a previous ruling by a three-judge panel from the U.S. District Courts of the Eastern and Northern Districts of

California in which California had been ordered to lower its inmate prison population to no more than 137.5% of its overall total capacity within 2 years. The Supreme Court agreed with the lower court, holding that the California Department of Corrections and Rehabilitation had violated the Eighth Amendment rights of inmates in its care. As a result, California was required to reduce its prison population within a very quick timeframe. This led to the passage of historic legislation in California, the Public Safety Realignment Act, which will be discussed in detail in later chapters.

Consent Decrees

As an alternative option to going to trial, inmate plaintiffs may have conditions remedied in a much more expedient fashion if they agree to settle with the agency in developing a suitable consent decree. Essentially, a **consent decree** is an injunction, but with the plaintiff and the agency both being involved (Carp & Stidham, 1990). This involves both parties agreeing to work out the terms of a stated settlement that is given official weight by the Court. The benefits to this remedy are that time and expense are spared and the uncertainty of a trial is eliminated. This also allows for a mutually agreeable solution rather than forcing the prison system into compliance.

Consent decree:
An injunction against both individual defendants and their agency.

CONCLUSION

This chapter demonstrates how during recent correctional history in America there has been a constant interplay between state-level correctional systems and the federal courts. This interplay has been marked by controversy, challenges, and conflict between state priorities regarding crime and prison operations and the need to adhere to constitutional standards. In addition, the legal issues related to the death penalty have been examined. Offenders who are given the death penalty, because of the seriousness of the sanction, must have their civil liberties protected and are given a full range of appeals guaranteed by law. The central feature of the evolution of prison operations in the country has been the interpretation of those constitutional standards as well as the Supreme Court's interpretation of its own role in ensuring that those standards, as interpreted, are met.

The legal history of corrections in the United States reflects the norms and mores of our society in relation to crime, punishment, and prison operations. The hands-off era mirrored common public understanding regarding the incarcerated, which claimed inmates were slaves of the state and entitled to nothing in particular. Thus, the need for intervention, whether by the mainstream public or the Supreme Court, seemed to be unnecessary to all but the most ardent of reform-minded persons.

Conversely, the hands-on era of sweeping prison reform that took place during the civil rights movement reflected the sentiments of a court that was in tune with social changes occurring throughout society. The hands-on era was tumultuous for many state prison systems, and its results created larger expenses for the taxpaying public, but the ethics, integrity, and general social conscience of the corrections field were improved. It is due to these changes that the corrections field became professionalized.

Legal issues that face both administrators and line staff are important to know and understand. The distinction between federal suits and state suits is important since they have different standards that must be met. Staff must understand where their liability begins and where it ends when working in the field of corrections. Individual knowledge of the parameters associated with liability aids correctional agencies as a whole since such knowledge can act as a preventative mechanism against inmate suits being filed against correctional staff.

Lastly, a brief overview of injunctions and other forms of court-oriented remediation was presented. These actions are what ultimately led to the significant changes that we have seen in this and the preceding chapters. However, the courts have now adopted a more neutral stance in intervening in prison operations. The passage of the PLRA has been instrumental in limiting the role of the courts and reflects a growing sentiment toward a "middle ground" approach to handling inmate claims and standards in prison operations. It appears that the field of corrections has matured into a professionalized and widely recognized discipline that has adopted a new, more balanced means of operation.

Want a Better Grade?

Get the tools you need to sharpen your study skills. Access practice quizzes, eFlashcards, video, and multimedia at **edge.sagepub.com/hanserbrief**

Interactive eBook

Visit the interactive eBook to watch SAGE premium videos. Learn more at **edge.sagepub.com/hanserbrief/access**.

 Career Video 3.1: Corrections Sergeant

 Criminal Justice in Practice Video 3.1: Deprivation of Rights

 Prison Tour Video 3.1: Laws

 Prison Tour Video 3.2: Legal Research

 Prison Tour Video 3.3: Religion in Prison

DISCUSSION QUESTIONS

Test your understanding of chapter content. Take the practice quiz at edge.sagepub.com/hanserbrief.

1. What was the hands-off doctrine, and how were prisons run during this era?

2. What was the importance of *Ex parte Hull?* How did it change corrections?

3. Give a brief overview of *Turner v. Safley.* How did this case define a variety of aspects related to correctional decision making?

4. Identify and discuss legal liabilities associated with correctional staff.

5. How would you describe the current relationship between corrections and judicial oversight? Is it now balanced, or are prisons systems still under close scrutiny?

6. Compare and contrast state tort cases and federal Section 1983 cases.

7. Discuss injunctions and consent decrees. What is the purpose of each when leveled against a correctional system?

KEY TERMS

Review key terms with eFlashcards at edge.sagepub.com/hanserbrief.

Absolute immunity, 72

Compensatory damages, 73

Consent decree, 75

Declaratory judgment, 74

Defamation, 70

Emotional distress, 71

Good faith defense, 72

Hands-off doctrine, 54

Injunction, 74

Intentional tort, 69

Libel, 70

Malicious prosecution, 71

Negligence, 71

One hand on, one hand off doctrine, 68

Prison Litigation Reform Act (PLRA), 68

Punitive damages, 73

Qualified immunity, 72

Rational basis test, 56

Religious Land Use and Institutionalized Persons Act of 2000, 59

Slander, 70

Tort, 69

Totality of the conditions, 62

Writ writer, 53

APPLIED EXERCISE 3.1

For this exercise, the instructor should organize several students (perhaps up to 12 students per group) into a mock jury, with one student designated as the foreman. Students must consider the following case, which is the same case that will be introduced in the "What Would You Do?" exercise that follows (but that version of the case will have a twist not encountered in this applied exercise). Students must identify specific legal criteria to determine liability involved with Officer Jimmy Joe, Officer Hackworth, Officer Guillory, Officer Killroy, Officer Ortega, Sergeant Smith, and Nurse Hatchett. Use information from this chapter regarding personal liability issues among correctional workers as the basis in providing your response.

The defendant, Jimmy Joe, a correctional officer at the Charles C. Broderick Maximum Security Facility, has claimed that he was unable to observe any of the events that are alleged by inmate Baker. Jimmy Joe's incident report notes that he was upstairs conducting a security check of the cell block; his documentation for that day does corroborate his story, according to date and time of the activity.

Inmate Baker alleges that Jimmy Joe watched as he was beaten in the foyer of the dayroom by Officers Hackworth, Guillory, and Killroy. Baker also insists that when he was removed from the cell block and taken to the infirmary, Ortega operated the video camera that recorded his escort. Baker contends that while Ortega was recording, he deliberately moved the angle of the camera during key times when the officers "accidentally" bumped Baker into infirmary doors and wall corners. As this was occurring, Nurse Hatchett, who Baker alleges was working in tandem with the officers, would ask the officers to have Baker moved to another room, thereby repeating the haphazard escorting process. Throughout his visit to the infirmary, Baker was handcuffed behind his back as officers Guillory and Killroy escorted him from room to room.

Baker also contends that Hatchett was deliberately rough with him when applying bandages. This process was not captured on video camera, and, as Baker contends, there are some periods where the camera appears to be pointed away from the inmate for a few seconds.

Officer Ortega claims that he has not been trained with the video camera and that this was his first time recording an escort after a use-of-force incident. An investigation of official prison records indicates that what Ortega says is true.

Sergeant John Smith had arrived on the scene with Ortega, and it was Smith who instructed Ortega to grab the camera and make his best effort to record the incident. Institutional policy precludes sergeants from using the camera when supervising infirmary escorts. According to Smith's incident report, Ortega was the only officer available at the time, and Smith had to make a quick discretionary call. He chose to not delay in responding to the situation and to aid Ortega with verbal instruction on how to operate the camera while they were on the scene and as the circumstances permitted.

When Ortega and Smith arrived on the scene with the camera, they saw Baker in cuffs and restrained on the floor by Guillory and Killroy. Hackworth had placed the cuffs and appeared to be on standby, waiting to aid in the escort once the supervisor and the camera appeared. All of these actions were consistent with institutional policy, including the restraining techniques used by Guillory and Killroy.

Baker contends that prior to the arrival of Ortega and Smith, Hackworth, Guillory, and Killroy had trapped him in a nonvisible portion of the dayroom and had repeatedly assaulted him. He also noted that Officer Jimmy Joe was near the scene the entire time and that he refused to render aid. Rather, he watched from a corner as Baker was beaten for 10 minutes by Hackworth.

Guillory, Killroy, and Hackworth contend that Baker had been acting aggressively when they tried to allow him off the cell block to go to commissary. Hackworth, noting that the cell door was closed and that the dayroom was clear of other inmates, instructed the junior officers to temporarily place Baker in the dayroom to contain the situation. Hackworth claims that his intent was to either defuse the situation or, if that did not work, escort Baker to his cell once staff in the picket (an area from which the cell doors are opened) were able to get his cell back open.

Hackworth's incident report claims that Baker began to yell and took a swing with his right arm at Guillory, whom he had a strong dislike for. When Baker did this, all three officers moved to restrain him. Baker lifts weights in the prison and has a past history of violence. Hackworth, Guillory, and Killroy all contend that Baker resisted with great force and fury and they found it difficult to restrain him. In the process, Baker received several welts and bruises from bumping into walls (while trying to get distance from the officers) and from fighting after being restrained on the concrete. He also had three chipped teeth from landing face first on the floor when placed in an arm-restraining hold. No other serious injuries were sustained.

Once the incident was over, Hackworth said to Jimmy Joe, "Quick, call a supervisor; we have an inmate down, and we need him escorted to the infirmary."

Now, as you sit on the jury, you must decide whose version of the story is true. Should any or all of these officers be found guilty of cruel and unusual punishment, per the Eighth Amendment? Are there any areas of potential liability for the three officers in the altercation, Officer Jimmy Joe, Nurse Hatchett, Officer Ortega, Sergeant Smith, or the correctional agency? Be sure to provide specific details in regard to your response.

WHAT WOULD YOU DO?

Jimmy Joe, a corrections officer at the Charles C. Broderick Maximum Security Facility, remembers the situation all too clearly. Baker had always been a difficult inmate, and he was a tough number. While serving time at Broderick Maximum, Baker had shanked two inmates on one occasion, and he had done the same with a correctional officer on another occasion. So everyone knew that Baker was dangerous.

Even though Baker was dangerous and had been a serious problem for many of the guards working at Broderick Maximum, Jimmy Joe was not really interested in getting "revenge" on Baker. However, Officer Hackworth could not get it out of his mind. For some reason, he had a constant fixation on Baker and would often mention that no inmate should be allowed to assault an officer and live to talk about it. For a long time, Jimmy Joe did not think much of Hackworth's comments—he was just another frustrated prison guard blowing off steam about the hardhead inmates on the cell block.

But one day, a hot summer day, Baker was hollering because he wanted off the cell block to go to commissary. Two officers told him no, not until they were ready to allow inmates to leave for the commissary line. On that particular day, the commissary was running late, but other inmates had told Baker that this was not true; instead he thought that the officers were lying to him as a means of "messing with his head." In fact, one inmate explained to Baker, "Man, it's like they're tryin' that guard crap again, you know, mentally punkin' you out." The inmate then added, "Ya know, they're always doing that crap with only you, all because they know you won't take it. They want you to stud up, man—what you gonna do?"

The excessive summer heat of the prison, located in the Deep South of the United States, was working on Baker, and his overall mood and demeanor had become quite sour. As the officers stood there, Baker said, "Look, you two little girls better get out of my way, and you'd better quit runnin' game on me." At this moment, Hackworth walked onto the cell block. He had arrived to assist in moving an inmate to another cell block but overheard Baker's comment and instead decided to walk over to Baker and the other two officers. Jimmy Joe was the officer in charge of the cell block and, seeing Hackworth head in that direction, kept an eye on everyone.

The dayroom was nearly empty because most of the inmates were either at work or out on the recreation yard, but a few were watching the television bolted near the ceiling. Hackworth told the officers near Baker that he would be right back. Hackworth then walked up to Jimmy Joe, who was a very new officer still on his 6-month probationary period, and said, "Look, you need to get them inmates out of that dayroom. If this gets sloppy, we don't want them hangin' around here making up crap that they think they've seen. Tell them that TV time is over and they gotta get on the rec yard." He followed this by adding, "If you ain't man enough to run them out, then let me know, and I will get their butts out of here myself."

Jimmy Joe ordered the inmates to go to the rec yard, and they went without much grumbling. They did look at Hackworth in a strange manner, almost as if they knew what was going on, and they aimed sympathetic gazes at Baker. Hackworth then told Jimmy Joe, "Look, Bub, you stay here and just man your post. If someone is about to come on the cell block, let us know."

Then "the Hack," as everyone referred to him, went back down the cell block and told the other two officers that maybe they should go ahead and let Baker approach the outer door of the cell block, just so he would be ready for the commissary call when it happened. Baker said, "Yeah, man, let me up front so I'll be ready. You know it ain't even closed—y'all are just making this difficult." Hackworth walked with the two officers and Baker. As they approached the front of the cell block, the door to the dayroom was on their right. Hackworth said, "Hey Baker, look, we don't want no problems, so if you want, feel free to go into the dayroom and watch the TV until you hear the commissary call. We'll leave the door open so you can come as soon as the call is sounded."

Baker began to think that perhaps the commissary really was running late and, since he had no TV in his own cell, he liked the idea of watching one of the channels while waiting for the commissary line. He went in, sat down on a bench facing the TV, and was switching the channels when Hackworth entered with the two other officers. From his location, Jimmy Joe could barely make out their position. Hackworth then said, "Hey, Baker, next time you pick up that channel changer, you need to ask to do so—that is a prison regulation, and you just broke that regulation." Before Baker even responded, he continued. "And it looks like you are being resistant and assaultive, so I am going to have to put you on the ground."

What happened next was a blur. Hackworth and the two other officers tackled Baker and began punching him repeatedly. They made a point to do this in an area of the dayroom foyer that was not clearly visible to Jimmy Joe. Baker tried to scream, but Hackworth shoved a bandanna in his mouth, cursed at him, and continued with the pounding. Once the incident was over, the officers said to Jimmy Joe, "Quick, call a supervisor; we have an inmate down, and we need him escorted to the infirmary."

Fifteen months later, Jimmy Joe, Hackworth, the two other officers, and Baker are all together in the federal district court. Baker filed a Section 1983 lawsuit that has made it to court. A jury sits across from Jimmy Joe—a jury of real people who do not seem to give any clue as to how they feel about the case in front of them.

At the time of the incident, Hackworth and the other two officers wrote in their paperwork that the incident started because Baker had tried to assault one of the officers. Hackworth claimed that the use of force was necessary to contain the inmate and to ensure officer safety. It is now Jimmy Joe on the stand, and he is being asked whether Baker initiated the assault. In his paperwork, Jimmy Joe had indicated that he did not see or hear anything during the incident but had been doing his rounds on the cell block. If Jimmy Joe changes his testimony, he will of course end up in trouble and will be labeled a "snitch" among his coworkers. In essence, he would be better off quitting his job. If he sticks with his story, he knows that he is lying and runs the risk of the jury not believing him.

If you were Jimmy Joe, **what would you do?**

PRACTICE AND APPLY WHAT YOU'VE LEARNED

▶ edge.sagepub.com/hanserbrief

SAUL LOEB/AFP/Getty Images

4 Jail and Detention Facilities

Learning Objectives

1. Describe the evolution of jails.
2. Compare issues for large metropolitan jails to those for small rural jails.
3. Compare the differences between jails as short-term institutions and jails as long-term institutions.
4. Discuss the state and challenges of health care in jails.
5. Discuss challenges with jail staff, staff motivation, and training.
6. Identify special types of sentencing options in jails.
7. Identify special needs inmates within the jail environment.

Learning the Ropes

As Jeff walked toward the visitation room with the guard, he remembered something he had read in a guide titled *County Jail: A Survival Guide for Inmates, Friends and Families*: "Anybody can survive a little time in county jail." He had read this back when he started worrying about the likelihood of doing time. He had been working at a department store and, on a dare from some other coworkers, began stealing merchandise by sneaking it out the back in the receiving warehouse area. He kept doing this sporadically, as much for sport as for profit. Eventually, however, the company began to suspect Jeff, and he knew it. He stopped stealing, but they had him on camera. He remembered other words from the guide: "It may not always be pleasant, but you just have to take things each day at a time. The time will pass and your life will continue. The first week or so is a major adjustment." Jeff thought to himself, *Boy, was that the truth.*

Jeff replayed in his mind some of the rules that he read in the guide:

1. Avoid talking about your sentence. The details let others know how seasoned you are and whether you might be easy to exploit.

2. Avoid gambling with anybody who has been to prison because they are probably a lot better at cards than you are and more prone to violence.

3. Try to get a job while in jail; it is the best way to pass the time.

On his way to the visitation room, he thought to himself that reading that survival guide was the smartest thing he ever did. At that moment, the guard told him to stop and "shook him down," a term used in jails and prisons for when officers check if inmates have contraband items upon their person. All went well; the officer found nothing.

"Go ahead," said the officer, pointing to the visitation room.

Jeff walked in and saw Cindy on the other side of the glass. She was smiling and waving. Jeff waved back and took his seat across from her. They talked through the glass on the phones that were provided at each visitation booth.

Cindy said, "Oh, Jeff, I'm counting the days. Just 2 weeks until you're out!"

Jeff smiled and said, "Yeah, I know. This part of the countdown is going to seem like forever."

"Yeah, but remember when you had almost a year?"

"Yep, it has gone by faster than I thought it would; first week was the worst."

Cindy nodded. "Yeah, I know, but nobody really messed with you, which surprised us."

"Right, unlike all that TV crap, most guys don't get raped and stuff, in jails, at least not that often, but. . . ." Jeff trailed off.

"What?"

"Well, some guys do get beat up; they just kind of either don't know the ropes or they basically ask for it. Take this one guy, he was booked 2 days ago and he went to take a crap and was on the *wrong* toilet."

Cindy looked at him askance. "The *wrong* toilet?"

Jeff explained to her that among the inmates, certain toilets were reserved for each race and some for gang members. Each racial group had its own designated toilets that looked the same but were understood by everyone else to be off-limits. In addition, nobody but the correct gang members could use "gang-owned" toilets. He explained that the guards were aware that this went on but usually did little to prevent these inmate rules.

"He got beat up pretty bad; they hit him while he was there, sitting and defenseless, taking a crap," said Jeff.

Cindy's eyes were huge. "Jeff, that is crazy!"

"Yeah, I know. I'm lucky that I was able to watch how things worked for a while and that my neighbor showed me the ropes." Jeff was referring to Jim Hammond, who, over time, had befriended Jeff. Hammond slept in the bunk across from Jeff.

Jeff paused, then said, "Cindy?"

"Yes?"

"One thing that I have learned from all of this is that you have to be observant of your surroundings and very careful when selecting friends."

The guard moved behind Jeff, glanced from him to Cindy, and declared, "Five more minutes."

"That sure did go by fast," Cindy lamented.

Jeff sighed. "Yeah, all the good moments in life do."

INTRODUCTION

Jails are a unique aspect of the correctional system because they fulfill many different functions. This multiplicity of functions and the rapid processing of offenders within their confines separate jails from prisons. Indeed, it is common for the population in any typical jail facility to change in composition quite quickly throughout the year, whereas a prison population will remain more stable. Jails have a very long history within the field of corrections, though their original purpose was not to correct at all. Nowadays, jails perform a number of functions that go well beyond the mere housing of inmates.

JAILS IN THE PAST

As with many aspects of the American criminal justice system, the use and operation of jail facilities owe their origin to England. During early American history, jails were informal and were used to detain persons awaiting trial. Eventually, as corrections officials came to rely less on corporal punishments, jails began to be used more frequently to house offenders (Giever, 2006). Jails eventually became a form of punishment in and of themselves, considered suitable for minor offenders.

Gaol: A term used in England during the Middle Ages that was synonymous with today's jail.

Tower of London: One of the earliest examples of a jail used for confinement purposes.

Jails in England were referred to as **gaols** during the Middle Ages. One of the earliest examples of a gaol used for confinement purposes was the **Tower of London**, which was constructed after 1066 CE during the reign of King William I of England (Giever, 2006). Following this time period, various other gaols were constructed and used to house a variety of offenders. Eventually, gaols were constructed in every shire (the equivalent of a modern-day county), and each was placed under the care of the shire reeve, the early version of today's sheriff (Giever, 2006).

The first jail built in the United States was constructed in Jamestown, the first true settlement in the country. Inmates were housed in the jail as early as 1608, according to official records (Johnson, Wolfe, & Jones, 2008; Roth, 2011). Early jails had no specific architecture or design. However, most were located within close proximity of the stocks and pillory. These facilities also did not usually utilize individual cells. Rather, inmates were housed in groups, often with a mix of serious and minor offenders in each room (Carlson & Garrett, 2008; Giever, 2006).

During this time, jail inmates were required to provide their own amenities. This could be done by having family and friends provide needed items or by purchasing them from jailers (Giever, 2006). Obviously, this system led to a high level of corruption as jailers tended to exploit inmates who were in need. The practice of collecting monetary compensation from inmates for services became known as the fee system, and there is even record that some jails charged inmates for their very room and board (Giever, 2006). This system was generally successful in supplementing the jailer's meager salary, and offenders were often able to afford it since most were kept only for a short time while awaiting trial or corporal punishment (Giever, 2006). Nevertheless, some were kept in jail for exceedingly lengthy amounts of time, during which these inmates would have little recourse if family or friends were not available to provide aid. In such cases, survival was bleak and meager, with these inmates completely at the mercy of their jail caretakers.

PHOTO 4.1 A jailhouse visitation room, similar to the one where Cindy visited Jeff (see chapter vignette). Visitors sit on the side where the stools can be seen while inmates sit on the other side. They talk to each other through the glass by using the phones that can be seen in this picture.

Robert Hanser

THE MODERN JAIL

In today's justice system, jail facilities are typically the first point at which an offender is officially classified as being in the correctional component of the criminal justice system. However, this is a bit deceptive since most persons are only being detained after they have been arrested. This detainment, or detention, occurs at a local detention facility that is typically administered by the county and operated by the sheriff's office. This detention facility is typically what is thought of when we use the term *jail*. In simple terms, a **jail** is a confinement facility, usually operated and controlled by county-level law enforcement, that is designed to hold persons charged with a crime who are either awaiting adjudication or serving a short sentence of 1 year or less after the point of adjudication, similar to Jeff's experience in the chapter vignette. Similarly, the Bureau of Justice Statistics defines jails as "locally-operated correctional facilities that confine persons before or after adjudication. Inmates sentenced to jail usually have a sentence of a year or less, but jails also incarcerate persons in a wide variety of other categories" (Sabol & Minton, 2008). This is important because this means that there is quite a bit of flow in and out of a jail facility. This is the case for two reasons. First, persons who are arrested are automatically held within a jail facility, but many are released within 2 to 3 days due to the setting of bond and/or a judge releasing them on their own recognizance. Second, offenders who serve jail terms do so for 1 year or less, as longer sentences are most often reserved for persons serving true prison sentences. Thus, even among those serving a jail sentence, the turnover tends to be rapid because most sentences are only for a few months to a year.

Recent findings by Zeng (2018) regarding jail operations and jail populations in the United States indicate the following:

1. After a peak in the number of inmates confined in county and city jails at midyear 2008 (785,500), the jail population was significantly lower by midyear 2016 (740,700).

Jail: A confinement facility, usually operated and controlled by county-level law enforcement, designed to hold persons who are awaiting adjudication or serving a short sentence of 1 year or less.

2. At midyear 2016, only 80% of jail bed space was occupied; this is a significant decline from 95% occupancy in 2007. The jail incarceration rate—the confined population per 100,000 U.S. residents—declined slightly between midyear 2007 (259 persons per 100,000) and midyear 2016 (229 per 100,000). This decline continues a downward trend that began in 2007. On average, the expected length of stay in jail was 25 days in 2016. Further, admissions were under 10.6 million around the nation. This is a decline from the previous year and represents a continued decline since 2008.

Table 4.1 demonstrates that jail jurisdictions around the nation tend to hold lower numbers of inmates. Indeed, the largest category jails (those holding 1,000 or more inmates) held the largest portion of the overall jail population (351,239 out of 744,592) in 2014, and this same category dropped at a larger rate than all of the other smaller categories of jail facilities resulting in 7.9% less inmates housed in these larger jail facilities in 2017. All combined, the jail population increase by 1,800 (0.13%) inmates between 2014 and 2017. As Table 4.1 demonstrates, the largest percentage growth is in the category of jails housing 100–249 inmates, with the second largest percentage growth occurring among jails with 500–999 or fewer inmates.

According to data provided by the Bureau of Justice Statistics, roughly 745,600 persons are held in jails throughout a given year (Zeng, 2019). However, this number does not truly reflect how integral the jail is to the American criminal justice system as a whole. In terms of the total number of persons processed within jails throughout a year, it is estimated that anywhere from 10 million to 14 million persons enter and exit jail facilities in the United States. This figure includes persons who are briefly held in jail until their bond is secured and/or are detained temporarily for minor offenses. When compared with an overall prison population of approximately 1.4 million, it seems that jails process approximately 10 times the amount of inmates than do prisons throughout a given year, due in part to the transient, in-and-out nature of the jail population when compared to the more stable and long-term prison population.

Most individuals who are booked in jails remain for short periods of time that range from just a few hours to several months. The majority of infractions that entail

TABLE 4.1

Inmates Confined in Local Jails

INMATES CONFINED IN LOCAL JAILS AT MIDYEAR, BY SIZE OF JURISDICTION, 2014 AND 2017					
JURISDICTION SIZE	**INMATES CONFINED AT MIDYEAR**				**PERCENTAGE OF JAIL CAPACITY OCCUPIED AT MIDYEAR**
	2014	**2017**	**DIFFERENCE**	**PERCENT CHANGE**	**2017**
Total	744,592	745,600	+1,008	+0.13%	81.4%
49 or fewer	25,058	21,600	−3,458	−13.8%	58.7%
50–99	42,172	35,500	−6,672	−15.8%	72.8%
100–249	96,443	111,300	+14,857	+15.4%	85.1%
250–499	101,609	109,200	+7,591	+7.5%	87.3%
500–999	128,070	144,500	+16,430	+12.8%	80.9%
1,000 or more	351,239	323,600	−27,639	−7.9%	81.7%

Sources: Zeng, Z. (2019). *Jail inmates at midyear 2017.* Washington, DC: Bureau of Justice Statistics; Minton, T. D., & Zeng, Z. *Jail inmates at midyear 2014—statistical tables.* Washington, DC: Bureau of Justice Statistics.

FIGURE 4.1

Characteristics of Jail Inmates 2017

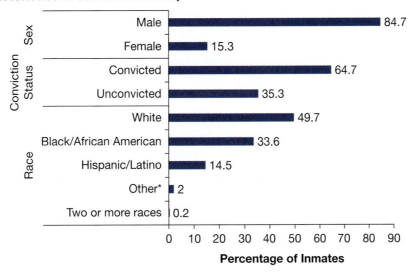

*Other includes American Indian or Alaska Native, Asian, Native Hawaiian, or other Pacific Islander.

Source: Zeng. Z. (2019). *Jail inmates in 2017.* Washington. DC: Bureau of Justice Statistics.

contact with jails include misdemeanor crimes, substance abuse issues, domestic crimes, and public safety issues. In addition, it is common for persons involved in the illicit sex industry (i.e., prostitutes and their customers) and persons who fail to appear for court hearings to routinely land in jail. Thus, jails consist of a changing population of offenders who have committed a variety of legal infractions, leading to a diverse array, legally speaking. Students are encouraged to examine Figure 4.1 for additional information regarding offenders in jail facilities.

Rural Jails

Rural jails are often challenged by tight budgets and limited training for staff. The author of this text has himself visited numerous rural jails and trained their personnel. It is clear that county-level governments tasked with operating jails do so amid a number of problems, both financial and political. Unlike most urban area jail staffs, Wallenstein and Kerle (2008) note that few jail staffs in rural areas receive official police or corrections training, and, in jurisdictions where such training *is* offered, the sheriff is hard-pressed to allow officers to attend the training due to limited numbers of security staff (Kerle, 1982).

As a result, the regional training facility frequently has instructors travel to the jail facility itself to train persons, though this still impacts county budgets since staff members must be paid while attending the training (an additional expenditure) and they must stay late after work or arrive prior to their work shift. Thus, unless training providers with great schedule flexibility are available, training issues for rural jails are a serious concern that provides a daunting challenge to sheriff agencies and county governments.

In addition, the conditions in rural jails are often substandard, and this is again exacerbated by the lack of funds that tends to be common among most such jurisdictions. Ruddell and Mays (2007) note that rural jails face most of the same challenges of larger jails but must work with additional disadvantages related to the far-flung locations of their facilities and the small tax bases generated by rural counties. Further, many rural jails in the United States have no physician services, and less than half have regular nonemergency medical services (Ruddell & Mays, 2007). Dental services are often limited to simple tooth extractions for jail inmates. In addition, research by Applegate and

Rural jail: Usually small jails in rural county jurisdictions that are often challenged by tight budgets and limited training for staff.

Sitren (2008) found that rural jails provided less rehabilitative programming than did jails in other categories. This included fewer opportunities for work release programs and educational achievement as well as treatment programs for substance abuse and mental health interventions. Psychoeducational programming for life skills development, parenting, and/or money management were also lacking. Simply put, rural jails cannot provide the services that larger and better-funded jails can.

Because of these challenges, there is incentive to use alternatives to jail incarceration in many jurisdictions. Some proponents have advocated for placing jail operations under state-level government, to be managed by the state's correctional department (Wallenstein & Kerle, 2008). This may be one means of streamlining jail operations, but, as we will see with state prison systems, this is no panacea. In fact, many state prison systems are overcrowded, making it unlikely that states will be any better at managing jail facilities than are persons at the county level of government.

Even more interesting is that in some states there is incentive for rural departments to embrace the jail operation industry. In these cases, county facilities tend to house state-level inmates due to the fact that many states have burgeoning prison populations and there is simply no room to warehouse offenders. Because of this, the state will pay county-level jails to house state-level inmates. This can generate substantial revenue for the jail, and, as a result, sheriffs may actively solicit the state for additional inmates (Giever, 2006; Wallenstein & Kerle, 2008). This leads to a bit of an unusual situation for jails where instead of seeking to limit the inmate population, the jail administration will instead attempt to remain full to capacity as much as possible. As will be discussed in subsequent sections of this chapter, this has been the case in California, where the state prison system is under pressure to reduce its count and, as a result, relies more on its jail system to house excess inmates.

Metropolitan Jail Systems

Jails in large metropolitan areas face numerous challenges that impact their operation. First off, these facilities tend to have sizeable populations that are diverse and that present with a variety of issues. Second, there is a tendency for offenders who live rough-and-tumble lifestyles to process in and out of the jail facility. Indeed, the average daily population data used to mark the number of persons in jail do not adequately reflect the important role of the jail to the correctional and judicial arms of the justice system.

In 2016, the midyear jail population throughout the United States was 731,300 inmates, with 10.6 million admissions in jail systems during that same year (Zeng, 2018). This statistic suggests that jail facilities around the nation process roughly 14 times the number of persons than is reflected in a count taken on any given day of the year (Zeng, 2018). In past years, many of the larger jail systems have operated with capacities over 100% of what they are intended to hold (Zeng, 2018). Even with the declines that have been observed among larger jails systems, many of these systems in large metropolitan areas still tend to have large offender populations that push the limit of what dedicated facilities can handle. In the state of California, rising jail populations have become a problem due to, as we saw in Chapter 3, the ruling of *Brown v. Plata* (2011), which has required the "realignment" of California's prison population, resulting in the reduction of state prison inmates at an unprecedented rate. Many of those inmates have simply been transferred to jail systems in California, leading to an increase of inmates in jails throughout the state (Lofstrom & Martin, 2017). Indeed, since the realignment initiative began in October 2011, county jails in California have had oversight over most nonserious, nonviolent, nonsexual felons and parolees who violate their parole (Lofstrom & Martin, 2017). Before the realignment process, the maximum sentence in county jails was one year. The realignment has required that lower-level felony offenders serve sentences in county jail and, just as important, these offenders now often stay for more than one year in these facilities (Lofstrom & Martin, 2017).

The large number of offenders in metropolitan areas, coupled with the various issues prevalent among the offender population (e.g., substance abuse, mental health issues, medical issues), requires that these facilities provide more comprehensive services. While larger jail systems may have more staff and the ability to provide more extensive services than systems located in rural regions of the United States, it seems that these services are just enough to meet the demand; more money from city tax dollars simply corresponds with the need for more extensive services. Thus, bigger jail systems often find themselves strapped for cash despite their larger revenue base in the community because of the greater needs of the offenders in these areas.

In many cases, particularly in urban jail facilities, a similar group of offenders may cycle in and out of the facility, perhaps going through intake and exit several times throughout the year. This provides a number of challenges and difficulties for jail staff who must contend with this constantly changing offender population. This also means that jail

PHOTO 4.2 This central area serves as an area where staff can observe the movement of inmates and also have quick access to numerous adjacent areas of the prison.

SAUL LOEB/AFP/Getty Images

facilities have a substantial impact on the public safety of the communities that surround them. It would then appear that jail administrators have a very big responsibility, both to the jail staff and to the community at large. The jail agency, therefore, is pushed and pulled by the ingress and egress of inmates as well as the demands of and concern for the community.

Urban jails also book a large number of persons with mental disturbances, this often being comorbid with drug and alcohol problems. It is for this reason that large jail facilities tend to have mental health personnel and substance abuse specialists on staff and available 24 hours a day to diagnose and manage the array of problems that these offenders may present (James & Glaze, 2006; Treatment Advocacy Center, 2014). Smaller jails in rural areas may have no such staff at all, however. And even with larger jail facilities, these staff may be so overworked as to hardly be available during times not considered peak hours for intake. Further, the risk of suicide is greater in jail facilities than in prisons, particularly during the first 48 hours and especially if the person is under the influence of alcohol or drugs. The booking officer and other staff must be quick to screen for potential suicide in all circumstances, noting mental health, substance abuse, or other factors that might exacerbate its likelihood.

As noted earlier, jails may commonly house persons who cycle in and out of their confines. The reason for this is that the majority of criminal activity is committed by a small group in a community. These offenders, who are a small segment of the total offender population, commit well over half of all the crime in a local jurisdiction. While much of this crime may be petty, these repeat offenders tend to cycle in and out of jail between charges, with long-term prison terms not occurring due to the low priority of the criminal activity. Further, these offenders tend to know each other (Giever, 2006). Indeed, many are drug users who may sell, share, and/or use drugs with one another. Others may be partners in criminal activity, and, even more disturbing, some may be mutual members of a street gang. The point is that interconnections between members of the criminogenic population tend to occur due to chance meetings that happen on the streets or during their periodic contact while in jail. Thus, in many larger jurisdictions, this offender population tends to maintain contact, both in and out of jail, revolving back and forth from the community to the jail and back again.

A number of these petty and small-time offenders may be homeless; this is especially true in urban areas. The homeless are particularly a problem for larger jurisdictions, where most beat cops know these individuals by name because the contact between police and the homeless is so frequent. Many homeless may have substance abuse issues, problems with trauma and anxiety, or other mental health disturbances. All of these factors are further worsened by an unstable lifestyle that consists of poor nutrition, inadequate health maintenance, and often substance abuse. Further still, communicable diseases may be more common among these individuals due to poor personal maintenance and risky lifestyle choices. This is particularly true for female offenders who may resort to prostitution to pay for either their drug habit or their basic needs. In such cases, these offenders are likely to be "regulars" for police officers in those jurisdictions and for jail staff, who will book them multiple times throughout the course of a year. It is even common among the homeless population for offenses to coincide with colder months of the year, with such persons committing petty crimes so that they may spend the winter indoors within the jail facility rather than outside on the cold streets.

Podular Direct-Supervision Jails

Podular jail: Includes rounded architecture for living units and allows for direct supervision of inmates by security staff.

During the 1980s, a new kind of jail came under construction in the United States. The two key components of this jail, the **podular jail**, are rounded or "podular" architecture for living units and a "direct," as opposed to indirect, form of supervision of inmates by security staff. In short, security staff are present in the living units at all times. It was thought that this architectural design would complement the staff's ability to supervise the inmate population while negating the ability of predatory inmates to control a cell block or dormitory. Two other important facets of this type of jail are the provision of more goods and services in these units (e.g., more access to telephone privileges, visiting booths, and library facilities) and a more enriched dialogue between staff and inmates. This, in turn, enhances the staff's knowledge of individual inmates—their personalities, issues, and challenges. Such knowledge improves both security and the day-to-day operational aspects of the jail.

Though podular direct-supervision jails or prisons are not necessarily a panacea for all the challenges associated with jail operation, they are a significant improvement over more traditional jail designs. When operated effectively and when they have the most critical design components, they tend to be less costly, primarily due to the fact that fewer lawsuits occur and fewer prevail than at jail facilities using other designs (Stohr, Walsh, & Hemmens, 2009). This is due to the fact that the open nature of the design prevents most incidents from escalating into grounds for a lawsuit.

Innovations in Jail Operations

Other innovations in jail operations include the development of community jails and reentry programs (Hanser, 2018). Community jails are devised so that programming provided to offenders does not end when those offenders are released back into the community. Rather, the programming is a continuum that extends beyond incarceration and well into the community. This is an important concept because it addresses the fact that many jail inmates return to the jail facility, primarily because many of their needs are not met in the community (Hanser, 2018). Once in the community, inmates with special needs and/or a lack of resources simply resort to criminal behavior, and the already full jail system once again must absorb the offenders into the burgeoning population. If programming occurs in both settings, regardless of whether the inmates are in or out of the facility, their needs are addressed and services are provided so that they can reintegrate into the community with greater success (Hanser, 2018, 2019). This then lowers crime in the long term since recidivism is lowered.

Further, a wide range of intermediate sanctions can be used with offenders, including jail offenders (see Table 4.2). These intermediate sanctions will be covered in much more depth in Chapter 5, but for now students should know that some jails include weekend programs, electronic monitoring, home detention, day reporting, and community service, among others. This means that a stint in jail for some offenders may actually be a stint that is served partly in jail and partly in the community. This blurring of the sentence and the type of confinement serves many purposes, such as easing overcrowding and providing a means for jail offenders to keep their employment so they can afford fines and fees while they are in jail.

Jail reentry programs: Programs usually interlaced with probation and parole agencies as a means of integrating the supervisory functions of both the jail and community supervision agencies.

Jail reentry programs are somewhat similar in nature but focus more on the fact that ultimately the inmate is likely to be on some form of community supervision. Thus, jail reentry programs tend to be interlaced with probation and parole agencies as a means of integrating the supervisory functions of both the jail and the community

TABLE 4.2

Confinement Status of Jail Inmates

PERSONS UNDER JAIL SUPERVISION, BY CONFINEMENT STATUS AND TYPE OF PROGRAM, MIDYEAR 2000 AND 2009–2017										
CONFINEMENT STATUS AND TYPE OF PROGRAM	NUMBER OF PERSONS UNDER JAIL SUPERVISION									
	2000	2009	2010	2011	2012	2013	2014	2015	2016	2017
Total[a]	687,033	837,647	809,360	798,417	808,622	790,649	808,070	782,300	794,900	801,100
Held in jail[a]	621,149	767,434	748,728	735,601	744,524	731,208	744,592	727,400	740,700	745,200
Supervised outside of a jail facility[b]	65,884	70,213	60,632	62,816	64,098	59,441	63,478	54,900	54,200	55,900

Source: Minton, T. D., & Zeng, Z. (2018). *Jail inmates at midyear 2017—statistical tables.* Washington, DC: Bureau of Justice Statistics.

Note: Most recent data available.

[a]Number of inmates held on the last weekday in June 2017.

[b]Number of persons under jail supervision but not confined on the last weekday in June 2017. Excludes persons supervised by a probation or parole agency.

supervision agencies (Hanser, 2018, 2019). Amid this, the continued care of the offender's specific needs is addressed (similar to community jails). One study found that effective interventions to improve reentry include everything from referral to counseling to drug treatment, depending on the needs of the offender. However, community safety must remain paramount, especially if agencies hope to have support from community members in the surrounding areas. Because reentry is a complicated process that requires a team approach among many different agencies and personnel, it requires that jail staff and administrators prioritize the needs that they target and the interventions that will be applied while also considering the network of community agencies that will provide these services. In fact, Hanser (2019) notes that this requires a collaborative process between jail security and treatment staff and members of surrounding agencies, while at the same time including members of the community in the process. Such collaborative efforts can educate the community on the process of reentry, build rapport between the community and the jail facility, and enhance the supervision of offenders since more eyes will be upon them within the community. Incidentally, this also provides the offenders with enriched support, thereby increasing the likelihood for motivated offenders to rebuild their future lives.

JAILS AS SHORT-TERM INSTITUTIONS

According to Kerle (1999), **short-term jails** are facilities that hold sentenced inmates for no more than 1 year. A jail, particularly a short-term jail, is an institution where both pretrial and sentenced inmates are confined. A lockup is often a police-operated facility where individuals who have been arrested are held for 24 to 72 hours, depending on the circumstances and the jurisdiction. Once a maximum of 72 hours has elapsed, the inmate held in the lockup is transferred to the local jail, where he or she is admitted. Lockups exist in most all larger cities because they make the arrest and booking process more convenient and less complicated. Once inmates are in the lockup, other personnel transport them to the jail. Lockups are often housed at the police station itself, and they are the true "holding cell" type of facility, used for no other purpose.

Short-term jail: A facility that holds sentenced inmates for no more than 1 year.

Kerle (1999) notes that lockups should actually be used sparingly, if at all. Much of this has to do with the fact that most police agencies cannot sufficiently staff and operate such a facility. Also, it is during the first 24 to 48 hours that inmates are most vulnerable to mental health issues, potential medical complications, and suicide. Due to these issues, Kerle points to the International Association of Chiefs of Police (IACP), which encourages police agencies to refrain from using lockups as much as is feasible.

The Booking Area

The most important area in short-term and long-term jail facilities is where intakes occur and is usually called the booking area (Kerle, 1999). There are greater risks in the booking area than in other areas of the jail due to the fact that so many offenders and suspected offenders enter and exit the jail from this point. It is important that staff in booking areas keep very good records related to the intake of an inmate. Indeed, if the appropriate arrest or commitment papers are not possessed, then the suspected offender cannot be legally confined. Because it is the responsibility of booking officers to ensure that these records are maintained, their job as the entry point in the jail custodial process becomes even more essential (Kerle, 1999).

Sally port: Entry design that allows security staff to bring vehicles close to the admissions area in a secure fashion.

Most all booking areas have what is commonly referred to as a sally port, which is adjacent to the booking area. The **sally port** is a secure area where vehicles can enter, with the area being closed after the transport vehicle has entered the compound. A sally port allows security staff to bring vehicles close to the admissions area in a secure fashion. In many small jails, a sally port may not exist, and, in such cases, additional care in the processing of inmates is required in order to make sure that security is provided (Kerle, 1999).

Most offenders who are arrested and brought into the booking area are under the influence of drugs or alcohol, or they may have some other factor that impairs their functioning. The stress level of those being booked is likely to be heightened, and this means that they are likely to be more problematic. Some of these individuals will present aggressive responses and may be assaultive. It is commonly observed that jail altercations occur in the booking area more often than in other areas of the jail.

Jails also book a large number of persons with mental disturbances, which often occur in tandem with drug and alcohol problems. Because of this, the booking officer must be able to identify unusual behavior; it is useful if he or she is trained through in-service processes to observe sudden shifts in mood or personality, hallucinations, intense anxiety, paranoia, delusion, and loss of memory (Wallenstein, 2014). Further, as noted above, the risk of suicide is greater in jail facilities than in prisons, particularly during the first 48 hours and especially if the person is under the influence of alcohol or drugs.

Due to the challenges associated with the booking process and the types of offenders that will be seen, it is important to have some kind of holding cell located near the booking area. Holding cells can be useful for detoxifying short-term inmates and can allow staff to book inmates without having distractions from other inmates while completing the admission process (Ruddell & Mays, 2007; Wallenstein, 2014). In short, holding cells tend to alleviate much of the chaos associated with booking. While the use of a holding cell may seem to be common sense, many smaller jails cannot afford the luxury of jail space that is not specifically used to house an inmate for a period longer than 72 hours—an indication of how cash-strapped many of these agencies tend to be (Ruddell & Mays, 2007).

Issues With Booking Female Inmates

In the past few years, a rise in the number (and proportion) of females who are housed in jails has occurred. Because female offenders, just like male offenders, may stay anywhere from a day to a year in the jail facility, it is important that staff are trained on issues related to female confinement. In this instance, it is certainly a benefit to have female employees available to assist in processing these offenders. In recent years, there has been an emphasis on recruiting women into the ranks of corrections officers, which has resulted in an increased representation of women working in jails and prisons. Nevertheless, the job of "jailer" still tends to be male dominated, and this means that in some jails, especially rural ones, female officers may not be present during one or several shifts (Morton, 2005).

Male officers who work in the booking area should be provided with additional training to ensure sensitivity to female issues. This also has advantages from a legal perspective, since male officers should use good professional discretion when processing female offenders. Given that inappropriate staff-on-inmate relations have been identified as an area of concern in recent years, it is important that male and female staff make a point to safeguard themselves from allegations.

Lastly, Kerle (1999) notes that booking forms and documents should be inclusive enough to contain questions about physical and sexual abuse, thus signifying that jail administrators recognize these issues as serious ones. This provides documentation that can be important should later allegations result in grievances or courtroom lawsuits. Aside from legal concerns, it is also useful for jail staff to develop an awareness of their own impact upon the booking process and realize how they may serve to heighten stress and generate hostility among entering inmates.

PRACTITIONER'S PERSPECTIVE

"Now there are digital search warrants . . . technology is growing leaps and bounds from all aspects."

Go to the IEB to watch Michael Verro's video on his career as a former drug investigator.

Information Technology and Integration

Unfortunately, there has been very little integration of systems between agencies and even within the police agency in regard to jail data. Often, police officers and jailers of the same agency do not engage in routine information exchange. This is not because high-quality information technology products do not exist. For instance, the SmartJAIL system, developed by CTS America, provides all the information needed to manage data and processes for jails of any size or complexity. SmartJAIL tracks and manages all aspects of an inmate's stay in a correctional facility. The system is integrated with a name index and initial arrest reports (within the records management system) to help speed up the booking process by using information already collected in these other programs. SmartJAIL is equipped to do the following:

- **Jail booking:** Tracks inmate booking, including mug shots, medical screening information, inmate property items, and all events involving the inmate

- **Inmate visitation:** Tracks the date and time of each visit made to an inmate

- **Inmate transportation and movement:** Tracks the movement of inmates by displaying all scheduled inmate movements, allowing time for the arrangement of transportation and scheduling

- **Jail management configuration:** Maintains lists for use by other jail management modules, including bed and cell locations, vehicles for transport, destinations, judges, and gain-time calculations

- **Jail log:** Enables officers to keep an electronic duty log to report activities or information to other officers
- **Jail incidents:** Records incidents requiring use of force, disciplinary action, or institutional charges against an inmate
- **Jail medical:** Records medical visits made by an inmate, including the initial medical screening; historical records are maintained for easy review of an inmate's health
- **Jail commissary:** Tracks inmate banking and purchasing at the commissary
- **Jail search:** Provides administrators an easy, fast way to locate inmate records based on extensive search criteria
- **Jail administrative reports:** Provides administrators easy access to dozens of the most commonly used reports in the system, including billing reports and statistical reports required by the federal government
- **Mobile inmate tracking:** Designed for handheld devices, this module checks inmates in and out of a facility by scanning inmate ID tags and providing accurate historical tracking records

From the above description, it is clear that automated systems exist that hold a large amount of information. Whether this information is utilized to its fullest potential is debatable. The lack of communication between agencies and even within agencies limits the use of much of the information that is stored. Over time, it is expected that jail systems will increasingly share data, both with other jail facilities and with police agencies whose officers may come into contact with jailed offenders after their release. This last application of information exchange can be particularly valuable when combating gang offenders who tend to cycle in and out from the community to the jail and back again.

JAILS AS LONG-TERM FACILITIES

Since the 1990s, an increasing number of larger jails have been required to house inmates who serve sentences that exceed the typical 1-year term. Understandably, jail facilities and the staff at those facilities have had difficulty maintaining these types of populations. Most jails are simply not designed or constructed to house inmates on a long-term basis. These facilities often lack the space for programming as well as the full recreational facilities that would be necessary for long-term populations. The use of jails for long-term inmates can be attributed to the following three broad reasons:

1. The use of longer jail sentences has resulted in the more extensive use of jails in many jurisdictions.

2. Many local and regional jurisdictions have been required to house inmates who have been sentenced to state prisons. This is because many state prisons have been so crowded that they have been unable to accept the newly sentenced offenders.

3. Some local jurisdictions have leased out beds to other jurisdictions, such as state correctional systems. These contractual agreements have created income for local sheriffs who oversee these jail facilities. This has created a situation where some sheriffs actively solicit state correctional systems for inmates to fill their beds—sometimes even soliciting systems outside of their own state, when feasible.

As crowding in state prisons became a longer-term problem, many jails began to hold sentenced inmates for anywhere from 2 to 5 years (Wallenstein & Kerle, 2008). Table 4.3 contains data regarding jail facilities that house long-term inmates. Note that the states of Louisiana, Tennessee, and Kentucky utilize this approach with great frequency. Indeed, over half of all state inmates are in local jail facilities in Louisiana.

TABLE 4.3

State and Federal Inmates Held in Local Jails, End of Year, 2017

TOTAL POPULATION OF STATE AND FEDERAL INMATES HELD IN LOCAL JAILS	
2015	81,196
2016	83,700
2017	80,900
PERCENTAGE OF STATE INMATE POPULATION HOUSED IN LOCAL JAILS (THOSE HAVING 20% OR MORE)	
Louisiana	55%
Kentucky	49%
Mississippi	27%
Tennessee	24%
Utah	22%
Virginia	20%

Source: Carson, E. A. (2018). *Prisoners, 2017.* Washington, DC: Bureau of Justice Statistics.

Local Jails That House State Inmates

Many of the excess jail inmates in recent years have been the legal responsibility of various state correctional facilities. However, many state correctional administrators allowed these state inmates to be held in jails to relieve prison overcrowding. Some states have had more problems with overcrowding than others. We now discuss two states that exemplify differences in overcrowding and the means by which jails provided relief to these burgeoning systems. We will focus on the prison systems of Texas and California.

In Texas, thousands of state inmates were placed into county-level jails. Much of the reason for this had to do with a significant legal development related to the Supreme Court ruling in *Ruiz v. Estelle* (1980). In this ruling (which students should recall from Chapter 3), the Texas prison system was found to be in violation of the Eighth Amendment prohibition against cruel and unusual punishments due to inadequate, unhealthy, and abusive prison practices in operation. One of the key issues in this case addressed the overcrowded facilities that existed in that system. Ultimately, this case resulted in population capacity limitations being imposed on various facilities throughout the system. This resulted in approximately 21,000 state offenders being detained in jails since the state prison system was unable to accommodate these inmates.

Later lawsuits between county jails and the state prison system resulted in the state being forced to meet its statutory obligation (see *County of Nueces, Texas v. Texas Board of Corrections,* 1989) to house inmates. If the state was unable to meet this obligation, the plaintiffs for the county jails contended that Texas would need to, at a minimum, reimburse county governments for the expense of housing such inmates. The Fifth Circuit federal court agreed, and the state was forced to pay counties a specified daily amount for each state-level inmate held within county jail facilities. This total amount ultimately added up to well over $100 million in compensation, a hefty sum by any account (Wallenstein & Kerle, 2008).

More recently, and as noted earlier in this chapter, the Supreme Court in 2011 upheld the ruling by a lower three-judge court that California must reduce its prison population to 137.5% of design capacity, which amounted to approximately 110,000 inmates being kept in a state system designed to hold about 80,000 inmates. As a result, the California State Legislature and governor enacted two laws—AB 109 and AB 117—to reduce the number of inmates housed in state prisons starting October 1, 2011 (Minton & Golinelli, 2014). These two laws are often referred to as the **Public Safety Realignment (PSR)** policy, and they are designed to reduce the prison population, in part, by placing new nonviolent, nonserious, nonsex offenders under county jurisdiction

Public Safety Realignment (PSR): A California state policy designed to reduce the number of offenders in that state's prison system to 110,000.

FIGURE 4.2

California's Confined Jail Population, 2010 to 2013

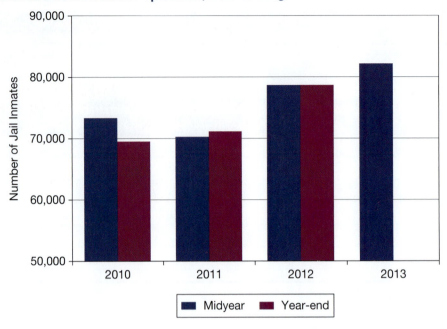

Source: Minton, T. D., & Golinelli, D. (2014). *Jail inmates at midyear 2013—statistical tables.* Washington, DC: Bureau of Justice Statistics.

Note: Most recent data available.

for incarceration in local jail facilities (Minton & Golinelli, 2014). Inmates released from local jails are placed under a county-directed post-release community supervision program instead of the state's parole system. In addition, California has given additional funding to the 58 counties throughout the state to deal with the increased inmate population; amid this, each county is left to develop its own plan for custody and post-custody that best serves its needs. As can be seen in Figure 4.2, California had a low jail population between 2010 and 2011. However, as a result of the PSR, the California jail population increased by an estimated 7,600 inmates between year-end 2011 and midyear 2012 and by an estimated 3,500 inmates between midyear 2012 and midyear 2013 (Minton & Golinelli, 2014).

Because of this atypical influx of jail inmates from the state prison system, by 2013 it was determined that characteristics of inmates held in California jails, including inmates' race and conviction status, differed substantially from the rest of the national jail population. For example, in California jails, Hispanics accounted for 45% of the total inmate population, Caucasians represented 32%, and African Americans represented 20% (not shown). In comparison, in the rest of the country, Hispanics accounted for 11% of the national inmate population, Caucasians represented 49%, and African Americans represented 38% (Minton & Golinelli, 2014). There was also a slight difference in the inmate conviction status. At midyear 2013, 43% of inmates held in California jails were convicted, compared to 37% confined in non-California jails (Minton & Golinelli, 2014).

Jails as Overflow Facilities: For a Fee

In many states, sheriffs have begun to see the operation of jail facilities as a potential money-making endeavor. This is true in states like Louisiana, where inmates from the Department of Corrections are routinely assigned to jail facilities operated by local sheriffs for a set fee that is paid to the agency of the presiding sheriff. For rural regions, this may be a primary source of revenue and may also provide several jobs in the area that would otherwise not exist. Some critics of this type of operation note that this creates a sort of dependency on the state inmate population. On the other hand, proponents note that this type of system helps to keep costs down, eliminates the need for reduced sentences among offenders who should be kept locked up, and also disseminates revenue to areas of the state that are cash-strapped. So long as the local population does not object, it would seem that such forms of operation are win–win for both local and state government.

Some jail systems have even taken inmates from other states. For instance, the Spokane County Jail in Washington State agreed to take inmates for a fee from the District of Columbia Department of Corrections (Washington, D.C.). In another instance, Texas jails in Denton County contracted with the state of Oregon's prison system to house inmates. In this case, Denton County added stipulations that Oregon inmates would have to return to Oregon for their release and also that all accepted inmates would need to have at least 2 years remaining on their sentence.

Denton County also later set up similar contracts with the U.S. Immigration and Naturalization Service (since 2003 called U.S. Citizenship and Immigration Services) to hold federal detainees. Thus, it is clear that jail systems have been used as money-making ventures with such partnerships often involving very different regions of the nation. It is important to understand that immigration law is civil in nature, not criminal. Thus, when counties such as Denton agree to hold immigrants by contract on behalf of the federal government, they must be careful not to mix individuals who have committed no criminal offense—for example, those persons being detained only for immigration law violations, who are often referred to as irregular immigrants—with their criminal population. There is an exception to this, however, which is when agencies arrest and jail an individual who does commit a criminal offense and it is then discovered that the individual is also inside the borders of the United States illegally.

Jails and Immigration Detention

The housing of irregular immigrants in local jails has also occurred due to agreements with local law enforcement agencies and the federal government derived from Section 287(g) of the Immigration and Nationality Act. According to Section 287(g), state and local police possess an inherent authority to arrest illegal aliens who have violated *criminal statutes* (Hanser, 2015). This section also broadens police powers for agencies that sign formal agreements (memorandums of assistance or MOAs) with Immigration and Customs Enforcement (ICE). Section 287(g) was established in 1996 to aid federal agencies in enforcing immigration issues within the nation's interior (Hanser, 2015). These agreements are sometimes referred to as jail enforcement models. The jail enforcement agreements permit approved jail custodial staff to enforce immigration laws once an irregular immigrant is brought in for booking. These agreements exist throughout the nation but are especially prevalent in areas of the nation where immigration issues are seen as a major concern (Hanser, 2015).

As a result of these agreements and due to the overflow experienced by ICE, the present immigration detention system is sprawling and has gone through continued changes and restructuring to improve federal oversight and management. While ICE has over 32,000 detention beds at any given time, the beds are spread out over as many as 350 different facilities largely designed for penal, not civil, detention (Immigration and Customs Enforcement, 2011). ICE employees do not run most of these. The facilities are either jails operated by county authorities or detention centers operated by private contractors. Table 4.4 provides an annual tally of the number of undocumented immigrants who were held by local jail jurisdictions. From these data, it can be seen that in 2014 there were 16,384 persons held in jail facilities for ICE.

Civil Rights Violations in Immigration Detention Facilities

In late 2015, the U.S. Commission on Civil Rights, established by Congress in 1957, conducted an investigation into the actual types of response and conditions of confinement that existed within federal detention facilities holding immigrants. They examined a variety of facilities administered through the Office of Refugee Resettlement (ORR), which is required to maintain certain standards of care and custody of immigrant children.

TABLE 4.4

Jail Jurisdictions Reporting Confinement of Persons Held for U.S. Immigration and Customs Enforcement (ICE), Midyear 2005–2014

	NUMBER OF JAIL SYSTEMS HOLDING ICE DETAINEES	TOTAL PERSONS CONFINED FOR ICE	PERCENTAGE OF JAIL INMATE POPULATION
2005	2,824	11,919	1.7
2006	2,784	13,598	1.9
2007	2,713	15,063	2.2
2008	2,699	20,785	3.0
2009	2,643	24,278	3.5
2010	2,531	21,607	3.5
2011	2,758	22,049	3.3
2012	2,716	22,870	3.3
2013	2,685	17,241	2.6
2014	2,634	16,384	2.5

Source: Minton, T. D., & Zeng, Z. (2015). *Jail inmates at midyear 2014—statistical tables.* Washington, DC: Bureau of Justice Statistics.

Note: Most recent data available. Data are based on the reported data and were not estimated for survey item nonresponse. Comparisons were not tested due to changing coverage each year. See appendix Table 9 for standard errors.

These standards of care include medical and mental health, education, family reunification efforts, and the provision of recreational activities.

This investigation resulted in the generation of a corrective action plan outlining the needs for providing better medical and mental health services. Further, this plan recommended that programs provide youth care workers with additional communications training when working with unaccompanied immigrant children and to foster nurturing and positive interactions (U.S. Commission on Civil Rights, 2015). Keep in mind again that these are youth who are held without parents, legal guardians, or other adult figures. They are also detained under civil law, not criminal law, themselves not being suspected of criminal activity and, in the United States legal system, being considered juveniles not adults, which implies that they should be provided with added protections due to their age.

Detention centers that house children are required to provide classroom education taught by teachers with a minimum 4-year college degree. Upon investigation, the Commission found that numerous educational workers did not meet minimal standards for hire. The Commission, in its report, pointed out that educational programming is vital to a child's development and that education for unaccompanied immigrant children gives them better odds of integrating in the American society, as a whole, and into the American job sector, specifically, should they remain in the United States. A failure to provide adequate educational programming essentially ensures that these youth remain in an unstable and disadvantaged status

PHOTO 4.3 Children in ICE Detention Center.
ASSOCIATED PRESS/Charles Reed

and increases the likelihood that they will not be productive within mainstream American society. This again is counterproductive to the welfare of these youth and to the United States economy and social dignity.

Likewise when considering private facilities involved with immigration detention, it would seem that human rights abuses have been reported more often in those facilities than in those managed by government agencies. Indeed, among the 179 detainees in ICE custody who died between October 1, 2003, and February 19, 2018, 15 were housed in the CoreCivic-operated Eloy Federal Contract Facility in Arizona (Luan, 2018, p. 1). Further, in a 2014 investigation of 5 of 13 Criminal Alien Requirement prisons in the United States, all of which are privately managed, it was found that these companies not only placed excessive numbers of detainees in isolation, but they also overcrowded the prisons, reduced medical staff, and denied medical treatment (Luan, 2018). That same year, a lawsuit was filed on behalf of nine detainees at Aurora Detention Facility in Colorado against GEO Group, accusing the company of forcing detainees to work without pay and threatening them with solitary confinement if they refused (Luan, 2018). The suit was ultimately settled between the two parties.

Detention Facility Housing Conditions

During the past few years, there has been growing concern that the U.S. Customs and Border Protection (CBP) agency has maintained facilities that are no different than prison facilities designed for common criminals. This includes facilities that house entire families (men, women, and children) who are being detained civilly, not for criminal offenses committed in the United States. It should be emphasized that these concerns are not simply liberal or antiestablishment rhetoric. Indeed, even members of Congress have voiced concerns about the punitive nature of immigration detention facilities. In fact, 136 members of Congress signed and endorsed a letter to the Department of Homeland Security's chief administrator, indicating the following:

> We are disturbed by the fact that many mothers and children remain in family detention despite serious medical needs. In the past year, we have learned of the detention of children with intellectual disabilities, a child with brain cancer, a mother with a congenital heart disorder, a 14-day-old baby, and a 12-year-old child who has not eaten solid food for two months, among many others. Recently, we learned of a three-year-old child at the Berks County Residential Center who was throwing up for three days and was apparently offered water as a form of medical treatment. It was only after the child began throwing up blood on the fourth day that the facility finally transferred her to a hospital. This is simply unacceptable. (U.S. Commission on Civil Rights, 2015, p. 105)

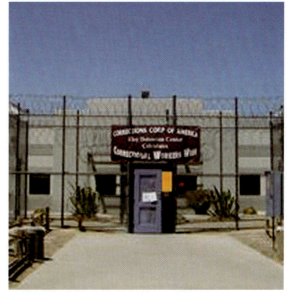

PHOTO 4.4 Eloy Detention Center in Eloy, Arizona. Operated by Corrections Corporation of America, a private prison company.

Photo courtesy of Immigrations and Customs Enforcement

Naturally, the U.S. Sentencing Commission supported these Congressional concerns, adding that additional concerns included not just family detention centers, but reports of unsuitable conditions in border patrol facilities, and adult-only detention facilities.

In particular, the Commission found conditions of extreme cold, overcrowding, and inadequate food to be a common problem throughout numerous CBP facilities. Again, this should not be interpreted as isolated incidents but was, instead, the norm throughout these facilities. Some facilities resembled the conditions of a prison rather than detention. Naturally, while some locations house immigrants convicted of serious crimes, lesser crimes, and those that have failed to appear for immigration hearings, it is important that others held on civil immigration issues not be comingled with the criminal population and that their treatment be substantially different. Altogether, the

Commission found evidence to conclude that the Department of Homeland Security (DHS) in general and the Customs Border Protection in particular detained undocumented immigrants in a manner more akin to prison rather than civil detention, which is a violation of the Fifth Amendment. Prior to this and since this time, other various levels of federal oversight, including federal district courts, have had similar decisions and left similar rulings as precedent.

Jail Overcrowding and the Matrix Classification System

In many of the larger jail systems, issues related to overcrowding have led to lawsuits and legal concerns that plague jail administrators. This has prompted officials in various jurisdictions to build even more jail facilities and to add onto those that already exist. In many cases, this results in the need to generate revenue to finance this construction, which often is obtained through the use of bonds. In many cases, bonds must be passed by the voting community, and this means that in addition to everything else, the jail becomes a political issue whereby sheriff's offices attempt to get bond referendums passed for the construction of more jail space. In some instances, voters are not receptive to the needs of the sheriff's agency and fail to vote for such bonds.

Multnomah County in Oregon developed what is now known as the matrix system, which is designed to release inmates early from the jail when the facility's population exceeds the capacity of the structure. This system is designed to release the least dangerous offenders first; this is determined via a computerized scoring system. An individual booked into the jail is scored on the basis of the charges and, if applicable, failure to appear after being served a warrant. The matrix system is used to control the population in custody without placing the community at additional risk of harm. Specifically, this system is designed to meet the following goals: (1) remain an objective tool of assessment; (2) allow for the use of additional information related to potential danger, if it can be measured objectively; (3) have all functions fully automated; and (4) identify dangerous inmates and prevent their release.

Wallenstein and Kerle (2008) note that since Multnomah County began to use the matrix, over 10,000 inmates have been released to reduce crowding. Assessment and classification programs such as the matrix have become more and more important in many jail and prison facilities. When combined with the housing of long-term inmates, the overcrowding issue in jails becomes much more difficult in prisons, particularly since jails hold both convicted offenders and persons being detained until they see a judge.

HEALTH CARE IN JAILS

One of the largest health care systems that exists within the United States can be found in jails. Nearly every person who enters a jail receives some sort of basic medical screening and evaluation. This is necessary for risk avoidance, maintaining constitutional requirements, and institutional safety concerns. Indeed, offenders in jails often have a number of health care problems. According to a 2002 study of jail inmates conducted by the Bureau of Justice Statistics, more than a third of all jail inmates, or approximately 229,000 inmates, reported some sort of medical problem that was more serious than a cold or the flu (Maruschak, 2006). Most medical problems tend to precede the offender's placement into the jail system and include HIV/AIDS, hepatitis, sexually transmitted diseases (STDs), tuberculosis, heart disease and diabetes, women's issues, and disorders related to aging. And of course substance abuse issues affect many of the inmates encountered in jails.

As one might expect, the elderly are much more prone to certain medical problems than young inmates. Older jail inmates have been found to be more likely than younger inmates to report ever having a chronic condition or infectious disease. Indeed, jail inmates age 50 or older were twice as likely as those ages 18 to 24 to report ever having a chronic condition. In addition, jail inmates 50 years of age or older were about three times more likely to report having an infectious disease than younger jail inmates (Berzofsky, Maruschak, & Unangst, 2015).

Thus, when and where feasible, it is often in the best interest of the jail administrator to have nondangerous elderly offenders released as soon as possible due to their exorbitant medical costs. Naturally, this should not be done at the expense of public safety, but in those cases where elderly offenders are incarcerated for nuisance crimes and other such petty issues, it is likely that jail facilities will process these offenders out as soon as is reasonable.

Like older inmates, women in the study were also much more likely to report medical problems (53% for female inmates as opposed to 35% for male inmates). Female inmates indicated a cancer rate that was nearly eight times that of male inmates, with the most common form of cancer for women being cervical. Indeed, for every medical problem documented in the study, female inmates reported more prevalence than male offenders, except for paralysis and tuberculosis (Stohr et al., 2008). Since female inmates tend to have more health care concerns than do male inmates, housing female inmates presents a plethora of issues regarding security procedures, specialized needs, and health care considerations.

Clinics, Sick Call, and Standards of Care

In many facilities, sick call often requires that inmates fill out a form to request a visit to the unit infirmary. Inmates should be appraised of the procedures used to obtain medical services since, unlike in the free world, they cannot simply decide to see a doctor whenever they wish. Perhaps the best time to ensure that inmates are informed of these procedures is during orientation, when written materials are handed out to inmates and they are expected to ask questions.

In order to avoid the abuse of sick call and the likelihood of malingering (faking an illness), many facilities have instituted a modest inmate co-pay system. In essence, inmates must pay a small fee to see a medical professional, and this tends to discourage most inmates from taking trips to the prison infirmary unless the request is legitimate. Administrators must be careful that these types of programs are not seen as punitive. The key is for administrators to lend assistance to inmates who are truly ill while holding inmates partially accountable for medical services when the illness is not severe.

Some issues are unique to jails and prisons regarding the standard of care to which inmates are afforded. First, there can sometimes be problems with confidentiality due to the close quarters and constant traffic in and out of spaces, including prison infirmaries. It is important to remember that inmates have a right to medical privacy, which means that medical personnel may not share details about an inmate's health or medical status with other correctional staff. This is particularly true if those persons are not medical staff. However, this does not prevent security staff and other personnel from gaining access to protected medical information in the process of completing work assignments or through conversations with inmates. There are several exceptions to privacy, which are delineated in the **Health Insurance Portability and Accountability Act (HIPAA)**. HIPAA guidelines are well known among medical and mental health personnel, and, for the most part, this act guides professionals on matters regarding the confidentiality of medical information.

Health Insurance Portability and Accountability Act (HIPAA): Guides professionals on matters regarding the confidentiality of medical information.

The Lifestyle of Offenders Inside and Outside

The inmate population tends to cycle in and out of prison. This has been discussed in prior chapters but is again important in the current discussion. When inmates enter prison, they may bring with them conditions and illnesses acquired on the streets due to their unhealthy lifestyles. Likewise, when inmates leave prison and go back to their communities, they bring any illnesses they may have acquired while incarcerated.

The use of drugs and participation in risky life choices deteriorate the health of offenders, and any negative effects from this are exacerbated by the prison environment. Thus, all individuals (including both inmates and correctional staff) within the prison should be educated on communicable diseases and take precautions to avoid contracting germ-based illnesses.

PHOTO 4.5 Correctional health care is an important area of service delivery in institutions. These professionals deal with a variety of issues that are complicated by the criminogenic lifestyles of most inmates who are their patients.

Guido Kirchner/picture alliance/Newscom

National Commission on Correctional Health Care (NCCHC): Sets the tone for standards of care in correctional settings.

Communicable Diseases in Jails

Blood-borne and airborne pathogens are now a serious consideration within most jail facilities (Berzofsky et al., 2015). Concerns regarding jail security and the prevention of communicable diseases are integral to most jail facilities as these diseases affect inmates, staff, and the outside community simultaneously. Communicable diseases found in jails include hepatitis A, hepatitis B, hepatitis C, human immunodeficiency virus/acquired immunodeficiency syndrome (HIV/AIDS), tuberculosis (TB), measles, and rubella (Berzofsky et al., 2015). To ensure that both staff and inmates are sufficiently protected, jails must have a written exposure plan that includes the engineering of physical facilities that help to isolate identified pathogens, work practices that control the spread of pathogens, the availability of personal protective equipment, routine staff training, the availability of appropriate vaccines, and detailed records of training provided and vaccines available (Berzofsky et al., 2015).

Further, most guidelines require that written protocols for responding to spills, occupational exposures, and other possible means of pathogen transmission be kept and disseminated among staff. In addition, confidential counseling for employees due to occupational exposure to a communicable disease must be made available. Though these procedures are becoming more commonplace, there are still some areas of operation regarding communicable diseases that need improvement in many jail systems.

Data from the **National Commission on Correctional Health Care** (NCCHC, 2002) suggest that many jails are not adequately addressing three communicable diseases: HIV/AIDS, syphilis, and TB. Although rudimentary HIV/AIDS education programs are becoming more widespread in jails, few jail systems have implemented comprehensive HIV/AIDS prevention programs in all of their facilities. Most jail systems provide HIV/AIDS antibody testing only when inmates ask to be tested or have signs and symptoms of HIV/AIDS. Testing is not aggressively "marketed" in most jail systems (Berzofsky et al., 2015).

Despite the availability of fairly inexpensive diagnostic and treatment modalities for STDs such as chlamydia and syphilis, testing is not thought to be automatic and/or widespread among jail systems. Research by the Centers for Disease Control and Prevention (2011) has concluded that often testing is only conducted when the presence of symptoms is observed by staff or when an inmate specifically requests the screening—and even when symptoms are observed, testing does not always occur. Even jails that report aggressive screening policies actually screen less than half (48%) of inmates (Centers for Disease Control and Prevention, 2011). As a result, on average, less than one third of jail inmates undergo laboratory testing for chlamydia or syphilis while incarcerated. Continuity of care for inmates released with chlamydia, syphilis, and other STDs is also inadequate (Centers for Disease Control and Prevention, 2011).

Although more jails screen for TB than for STDs, this number should still be higher. According to the Centers for Disease Control and Prevention (2006), at least three factors have contributed to the high rates of TB in detention and jail facilities. First is the disparate number of individuals who are at high risk for TB, such as substance abusers, individuals of low socioeconomic status, and individuals who are HIV positive. Second, the physical structure of many facilities enables transmission of the disease, given the close living quarters and inadequate ventilation that may exist. Third, the constant flow of persons in and out of these facilities (keep in mind this is approximately 10 to 14 million people per year), when combined with the first two factors, creates an elevated risk for contracting tuberculosis. In the United States, tuberculosis is often concentrated among disadvantaged populations, particularly immigrant populations. Immigrants in detention or in jail often come from countries with a high prevalence of TB. Due to social and legal circumstances, these individuals may not have access to testing and treatment for TB, making the jail or detention facility their first point of access to some type of medical service for the infection (Centers for Disease Control and Prevention, 2006).

JAIL TRAINING STANDARDS

In many respects, jails have not been given the attention that they deserve. This is particularly true with regard to the training of staff. The National Institute of Corrections has in recent years published numerous manuals and videos and has also provided a variety of training programs for jail staff and leadership. This has been in response to the realization that jails have traditionally been overlooked until the last decade. Indeed, the training of jail staff has a fairly short history, having started in 1948 with the Federal Bureau of Prisons (BOP). Formal training began in the BOP when federal agencies became concerned that states were not adequately training their jail staff. In response to this dearth of training, the BOP established a training school for sheriffs and jailers to remedy this problem.

PHOTO 4.6 The author is a trainer at a regional academy for jailers and police officers. Hanser is seen here providing instruction at North Delta Regional Training Academy.

Robert Hanser

The type of formal training standards used varies quite considerably from state to state. Most of the short-term jails tend to use on-the-job training. This can create many problems since such training tends to be unorganized, and often there are gaps in the required knowledge that staff must have to be truly competent. Since many new staff must learn through a process of trial and error, the incidence of lawsuits tends to rise, and, even worse, the likelihood that inmate plaintiffs prevail in court against the agency also correspondingly increases. This is one of the primary reasons that many police agencies choose to refrain from operating a jail facility; the legal liability and responsibility offset the value of the convenience.

Language, Ethnic Diversity, and the Selection of Staff

Inmates processed in jail intake units reflect the cultural and linguistic diversity that exists throughout the United States. Because of this, it is important that staff have linguistic skills and are knowledgeable about different cultures, particularly those likely to be in the jail facility in their area. Due to this, jails need to have a diverse staff. This can even extend into the multinational realm as well as the multicultural realm. Consider, for example, that in California jails may house a substantial number of Asian offenders and Latino offenders. In some areas of the northeastern United States, the need for Jamaican, Russian, or Ukrainian staff may exist. Thus, jail administrators face the need to develop a workforce that reflects the composition of the offender population in their area.

Challenges Faced by Female Staff

Female staff face a number of challenges that are encountered with less frequency by male staff. In many jails, women are not afforded the same respect as men, both by the inmates and by their fellow officers. Much of this may have to do with the inmate subculture, the organizational culture of the agency, and other factors that impact a female officer's professional development in law enforcement. Likewise, since so many of the inmates are male, female officers find themselves a minority among both the inmates and the officers. Thus, many female officers are utilized with female inmates when and where possible to alleviate concerns over searches and maintenance security as well as medical issues common to female offenders.

Other Employee Issues

There are approximately 300,000 jail staff around the country. Typically, these staff are underpaid in relation to other law enforcement officers and/or correctional officers in state (rather than local) facilities. Many of these employees take these jobs on a temporary basis while waiting for openings in other areas of their sheriff's agency. However, often these employees have only limited education and/or skill development and do not fare as well in the competition for better positions. Therefore, they often find themselves on the jail staff for longer periods of time. Some truly do find the work rewarding, though, and/or develop a set of specialized skills within the jail that make them invaluable to the jail's operation. For these individuals, working in the jail becomes as lucrative as (and perhaps even more lucrative than) working in other positions in the agency. One example is training officers in jail settings. Over time, these persons can become quite sought out among other agencies and can command a high wage for their knowledge and expertise.

For the most part, jailers in many jurisdictions do not seem to have an effective career track that is sufficiently rewarding. Turnover is very high, with many jails reporting a complete change in staff every few years. This has generated a serious concern regarding the quality of employees in many jails as well as the quality of their training. To help improve the quality of training for jailers throughout the nation, the National Institute of Corrections Jails Division has implemented programs to increase the number of trainers for jail staff to ensure that there are enough opportunities for training among employees of various agencies. While this is an admirable approach, the true long-term answer to this problem is an increase in pay and career incentives for these employees.

Evolving Professionalism

One of the most direct means of increasing professionalism in the jailer's career track is to utilize what Stohr, Walsh, and Hemmens (2008) refer to as coequal staffing. In this type of staffing, programs provide comparable pay and benefits to those who work in the jail and those who work in routine law enforcement positions. In some jurisdictions, agencies have created two career tracks—one for law enforcement and the other for jailers—with equal pay. There is some evidence that this approach has had a phenomenal effect on the professional operation of jails (they are better staffed) and on the morale of those assigned to them (Stohr et al., 2008).

As our jails become more professionalized, the requirements for training and higher entry standards become more important. In fact, some states now require that jail officers be certified through some type of official academy training. Even in many places that require such training, however, that training still tends to fall short of the number of hours required in traditional law enforcement positions. Silverman (2001) notes that the reluctance of local jurisdictions to meet their training obligations has served as the impetus behind the American Jail Association's monthly training bulletins, which consist of video and web-based training segments. Lastly, refresher training is critical and is offered in many jurisdictions. Examples of the diverse areas in which jailers must develop expertise include the use of CPR, fire safety, aerosol pepper spray and taser proficiency, riot-control techniques, and human relations skills.

Requirements of the Peace Officer Standards and Training (POST) are used to guide agencies and training facilities in a number of states toward ensuring that jail officers have the requisite skills needed to perform their job assignment(s). Ensuring that jailers receive well-developed training that is top quality, serious, and professional improves the jail systems in the area while also providing a clear message to jailers in training: Your job is important, and we take it seriously. The result is that the jailers take their jobs seriously as well, and employee morale tends to be much higher in local agencies throughout the region.

SPECIALIZED TYPES OF JAIL SENTENCES

Extended confinement of persons who, by legal standards, are presumed innocent until proven guilty, such as with pretrial detention of the nonadjudicated offender, creates a serious custody problem for jails that prisons are not forced to contend with. The defendant who is truly innocent but is required to spend long periods of time in jail confinement will eventually develop a sense of resentment toward the criminal justice system. Further, this simply does not smack of fundamental fairness and instead creates a situation where innocent persons feel victimized, even though that is not the intent. Thus, the use of specialized jail sentences can create "middle road" forms of custody that can help to mitigate these negative effects.

Weekend Confinement

Weekend confinement is used to lessen the negative impact of short-term incarceration and to allow those in custody to maintain their employment. Jurisdictions using **weekend confinement** have implemented methods of confinement that are restricted to the weekends or other times when the person in custody is off from work. In some respects, this can be likened to serving a sentence on an installment plan, whereby other aspects of life that keep the person functional in society are not impaired, yet, at the same time, the offender is required to pay for his or her crime through a series of small stints in jail. This type of confinement typically is assigned to misdemeanants and usually requires that they check in to the jail on Friday evening and leave on Sunday at a specified time. Through these stints, the offender adds up credit toward confinement until the total sentence is served. Weekend confinement is used with minimum-security facilities and with offenders who have committed minor or petty crimes.

Weekend confinement: Confinement that is restricted to the weekends or other times when the person in custody is off from work.

Shock Incarceration/Split Sentences

For certain offenders, the mere subjection to a loss of freedom is all that is needed to get their lives in order. As a result, one mechanism used for instilling fear—the goal being to deter future criminal behavior—is shock incarceration. **Shock incarceration** is short-term incarceration followed by a specified term of community supervision in hopes of deterring the offender from recidivating. Because the brief stint of incarceration is meant to provide a sense of punitive reality to the offender, most shock incarceration programs are designed for juvenile offenders and those who have never been incarcerated before. Shock incarceration results in the offender being sentenced to incarceration for 30, 60, or 90 days in most cases and, upon completion of this time served, being resentenced by a judge to probation supervision. The idea is that this brief period of incarceration will sufficiently "shock" the offender so he or she will desist from committing further offenses. It is presumed that this short period of incarceration will have a similar deterrent effect on this type of offender as a longer one would. This method is sometimes utilized when the concern is that a longer period of incarceration might result in the hardening of the offender.

Shock incarceration: A short period of incarceration followed by a specified term of community supervision.

With split sentencing, offenders are sentenced to a specified term of confinement that can include up to half of their original sentence (i.e., it might be for years as opposed to only 30, 60, or 90 days), after which they finish the remainder of their sentence on probation. Unlike shock incarceration, this sentencing is all done upfront with no need for later resentencing. In some cases, this type of sentence may be utilized due to a shortage of jail beds, while in other cases it may be used for criminal offenses that are too serious to allow for a shorter period of incarceration but when, due to plea agreements, more flexibility is permissible in the amount of time that is actually to be served in jail or prison.

SPECIAL ISSUES IN JAILS

As if the various categories of offenders and their movement into and out of the facility were not enough to complicate the job of jail staff and administrators, offenders also have a host of issues with which jail systems must contend. Indeed, many jail inmates have problems with illiteracy, substance abuse, mental illness and stability, general medical conditions, communicable diseases, and suicide ideation. Because many of these issues go untreated in the community, the jail becomes the dumping ground for these offenders. Thus, jails must service a diverse offender population with specialized problems and needs. We now turn our attention to some of these specialized needs that emerge in the jail setting.

Substance-Abusing Offenders in Jails

Throughout the literature, it is commonly noted that substance abuse issues are rampant within the jail population. This observation is true regardless of the region of the United States and/or the time of year at which one considers this point. Simply put, substance abuse is a major problem that impacts the day-to-day operation of jails in terms of both security and programming. According to the Center on Addiction and Substance Abuse (2010), the largest increase in the percentage of substance-involved inmates was in the jail population, with nearly 85% of all jail inmates reporting substance abuse problems (p. 10). In fact, over half of all inmates reported being under the influence of drugs or alcohol at the time of their offense, while nearly one fifth indicated that they had engaged in criminal behavior as a means of supporting their drug habit. Of those in local jails, it was found that 50.2% of those having substance abuse problems had at least one prior incarceration, while the same was true of only 27.0% of inmates without a drug problem. Further, drug abusers in jail had over twice the number of prior arrests than their counterparts who did not have a drug problem (Center on Addiction and Substance Abuse, 2010, p. 20). In addition, 25.2% of jail inmates were found to have both a substance abuse disorder and a mental health disorder (Center on Addiction and Substance Abuse, 2010, p. 26).

The drugs of choice for abusers and users vary and include, by prevalence of use, marijuana, cocaine or crack, hallucinogens, stimulants, and inhalants. As would be expected, those who indicated substance abuse problems were also more likely to have a criminal record. In addition, homelessness is very common among substance abusers who end up in jail. In fact, as noted above, it is not uncommon for many addicts who live on the streets to deliberately commit crimes during the winter months as a means of having a warm place to stay during the coldest periods of the year.

Treatment within the jail itself for drug abuse is somewhat unusual, unfortunately. Much of this has to do with the fact that offenders tend to move in and out of the jail facility much more quickly than in a prison facility, so there is little time between implementation of treatment and the offender's departure from the facility. Treatment programs are usually focused on jail inmates who have a substance abuse problem and have longer-term sentences, but, even in these cases, detoxification tends to be the primary approach to treatment, coupled with the support group interventions of Alcoholics Anonymous and/or Narcotics Anonymous.

Detoxification: The use of medical drugs to ease the process of overcoming the physical symptoms of dependence.

Detoxification is designed for persons dependent on narcotic drugs (e.g., heroin, opium) and is typically found in inpatient settings with programs that last for 7 to 21 days. The rationale for using detoxification as a treatment approach is grounded in two basic principles (Hanson, Venturelli, & Fleckenstein, 2011; Myers & Salt, 2000). The first is a conception of *addiction* as drug craving accompanied by physical dependence that motivates continued usage, resulting in a tolerance to the drug's effects and a syndrome of identifiable physical and psychological symptoms when the drug is abruptly withdrawn. The second is that the negative aspects of the abstinence syndrome discourage many addicts from attempting withdrawal, which makes them more likely to continue using drugs. The main objective of chemical detoxification is the elimination of physiological dependence through a medically supervised procedure.

While many detoxification programs address only the addict's physical dependence, some provide individual or group counseling in an attempt to address the psychological problems associated with drug abuse. Many detoxification programs use medical drugs to ease the process of overcoming the physical symptoms of dependence that make the detoxification process so painful. For drug offenders in jails and in prisons, the mechanism of detoxification varies by the offender's major drug of addiction. For opiate users, methadone or clonidine is preferred. For cocaine users, desipramine has been used to ease the withdrawal symptoms. Almost all narcotic addicts and many cocaine users have been in a chemical detoxification program at least once (Inciardi, Rivers, & McBride, 2008). However, studies show that in the absence of supportive psychotherapeutic services and community follow-up care, nearly all abusers are certain to suffer from relapse (Ashford, Sales, & Reid, 2002).

In all detoxification programs, inmate success depends upon following established protocols for drug administration and withdrawal. In essence, detoxification should be viewed as an initial step, after the intake process, of a comprehensive treatment process. Because jails are not typically long-term facilities, they will not usually offer much more than detoxification programs. Of course, this does not address the needs of those inmates who are kept in jails on a long-term basis, such as state prison inmates who are housed in local jails. In such cases, the services for these inmates should be on par with those for inmates who are kept long-term in prisons. While some of the larger jail systems in the United States might offer more comprehensive services, most do not. Thus, offenders who are kept on a long-term basis in a jail facility find themselves without suitable intervention services.

Mental Health Issues in Jails

Jails in this country are full of the mentally ill, and in many cases these offenders are homeless. Data from the Bureau of Justice Statistics indicate that roughly 64% of all jail inmates have a mental health problem (James & Glaze, 2006). This contrasts with mental health problems in prisons, where only about 11% of inmates present with mental illnesses. Further still, for nearly every manifestation of mental illness, more jail inmates than state or federal inmates are likely to exhibit symptoms, particularly hallucinations and delusions. Jail inmates tend to have mental health problems with specific diagnoses that include mania (approximately 54%), major depression (30%), and psychotic disorders (24%), according to Stohr et al. (2008). The specific identification of a mental illness for each inmate is based on the *Diagnostic and Statistical Manual of Mental Disorders (DSM-IV-TR)*. The *DSM-IV-TR* is a manual used by mental health clinicians that provides the specific criteria and symptoms to be considered when qualifying a person with a diagnosis and is the accepted standard among members of the American Psychiatric Association.

A host of problems can be associated with mental illnesses in jail systems, including homelessness, greater criminal engagement, prior abuse, and substance use. James and Glaze (2001) found that those with a mental illness designation were almost twice as likely as other inmates to have been homeless prior to being placed in jail. In addition, inmates with mental illnesses tend to have more incidents of incarceration than do those who have no record of mental illness. Roughly three times as many jail inmates with a mental health problem have a history of physical or sexual abuse than do those without such a problem (James & Glaze, 2001; Stohr et al., 2008). Likewise, nearly 75% of inmates with a mental health issue also have a substance abuse problem. Thus, those inmates with mental health problems tend to have a multiplicity of other issues that affect them as well. Naturally, these issues affect the inmate's behavior and thus place a drain on the time and effort of jail staff.

One study in particular demonstrates how for mentally ill jail inmates there is a confluence of factors that aggravate their circumstances. McNeil, Binder, and Robinson (2005) found in their study in San Francisco County that jail inmates with mental illness were prone to have multiple issues that seemed to dovetail with their mental illness. They found that mental illness, substance abuse, and prior jail incarcerations were connected life events (Stohr et al., 2008). In addition, among those inmates who were mentally ill and/or had substance abuse or dependence issues, homelessness was a much more common phenomenon. Thus, the mentally ill inmate presents with a number of challenges for jail staff—challenges that are not easily separated from each other. While it is not clear if the mental illness itself spawned these other issues or if these issues served to increase the likelihood of mental illness, it is clear that these offenders are in need of help that extends well beyond the intended scope of most jail facilities.

Jail Suicide

When discussing jail suicide, it is important to understand that during any given year the total number of incidents fluctuates but is at or around 300 inmates (Noonan & Ginder, 2014). Thus, when we discuss jail suicide, we are usually referring to approximately 270 to 315 inmates out of 10 to 14 million individuals who are processed yearly through jails around the nation. Though this is a small number, suicide is a concern for jail administrators and is important for humanitarian reasons, if nothing else. Remaining data regarding jail suicide for this section have been drawn from a 2010 report conducted by the National Institute of Corrections and authored by Lindsey Hayes. As indicated from these data, those incarcerated in jails often enter while intoxicated, and many also have some sort of mental disability. Further, this booking may be their first experience with the incarceration process and may, therefore, be one of the lowest points in their life. In such circumstances, jail inmates are at heightened risk of suicidal ideation and attempts (Hanser, 2002; Hayes, 2010).

When considering the characteristics of jail suicide decedents, the overwhelming majority (93.1%) are male (Hayes, 2010, p. 12). It is interesting to note that although White inmates account for about 44% of the total jail population throughout the country, in the report they represented the majority (67%) of inmates who committed suicide, whereas African American inmates, who account for nearly the same percentage of the total jail population as Whites (39%), constituted only 15% of jail suicide victims (Hayes, 2010, p. 12). Further, the majority of those who committed suicide in jail were single (42%), with only 21.4% being married (p. 12). Lastly, the overwhelming majority (90.1%) of suicide victims were in detention facilities at the time of their death (Hayes, 2010, p. 14).

Interestingly, only 38.1% of inmates who committed suicide were identified as having a history of mental illness during the intake process (Hayes, 2010, p. 17). Most of these inmates suffered from depression or psychosis. Further, Hayes found that 33.8 % of inmates who committed suicide reported a history of suicidal behavior during the intake process, which was substantially more than what had previously been reported in other federal statistics by James and Glaze (2006). This demonstrates that more detailed data collection methods can identify cases that may otherwise be overlooked (Hayes, 2010, p. 19).

When considering interventions for jail suicide risks, it is first important to note that the vast majority of inmates (over 90%) successfully complete their suicide by hanging, usually using either clothing items or bed sheets (Hayes, 2010). It is clear that suicide watches and other similar precautions are effective since only 7% of all jail suicides occur while such a watch is implemented (Hayes, 2010, p. 28). In those cases that are successful despite the watch, the issue usually has to do with the length of time between observations (e.g., every 15 minutes as opposed to continuous) of the individual with suicidal ideations. Lastly, many mental health clinicians often develop no-harm contracts with potentially suicidal inmates, seeking assurance that their clients will not engage in self-injurious behavior (Hanser & Mire, 2010). This is sometimes viewed as a buffer against liability in the event that the inmate commits suicide but it is questionable as to whether this holds any legal merit (Hanser & Mire, 2010).

It is clear that administrators understand the importance of staff training regarding jail suicide (Hanser & Mire, 2010). Research by Hayes (2010, p. 35) shows that over 74% of all jail administrators provide training on at least an annual basis, with most others indicating that such training is provided biannually. Of those who did not respond positively to these inquiries, most were rural jails. Thus, it may well be that larger jail facilities with greater resources are better able to provide training and resources for staff. Other advantages of larger jails also exist. For example, younger and less experienced offenders who are apprehensive of being housed with older offenders who may victimize them can often be segregated in larger jails. In smaller jail facilities, space may not exist for such protective considerations, and inmates who are in fear of victimization and already at a low point in their life would likely be at an increased risk of suicide ideation. The ability to alleviate some of the anxiety of inmates can often reduce the likelihood for suicidal thoughts and actions.

CONCLUSION

Jail facilities are perhaps the most complicated of facilities within the field of corrections and are often not appreciated for the vital role they play within the criminal justice system. The volume of persons processed in jail facilities, in and of itself, is sufficient to warrant a closer examination of these facilities. In addition, jails perform many different types of tasks, such as the holding of persons prior to their court date and providing a series of unique sentencing variations. They are also sometimes used for the incarceration of persons who are technically part of the larger prison system.

Jails and jail staff must attend to a variety of offenders who present with a number of issues. The range of problems and challenges can be quite varied, and this creates a demanding situation for jail staff and administrators. The variety of offender typologies, needs, and issues presents staff with problems that are not easily rectified, and, correspondingly, there is a need for staff to be well trained and well equipped. Unfortunately, jails frequently do not have sufficient funding for this. This is especially true with small rural jails where funds and expertise are limited. Overall, it appears that jails have been given short shrift in the world of corrections but that they will be given much more attention in the future.

Want a Better Grade?

Get the tools you need to sharpen your study skills. Access practice quizzes, eFlashcards, video, and multimedia at **edge.sagepub.com/hanserbrief**

Interactive eBook

Visit the interactive eBook to watch SAGE premium videos. Learn more at **edge.sagepub.com/hanserbrief/access**.

 Career Video 4.1: Drug Investigator

DISCUSSION QUESTIONS

Test your understanding of chapter content. Take the practice quiz at edge.sagepub.com/hanserbrief.

1. Compare issues for large metropolitan jails to those for small rural jails.
2. Provide a discussion regarding the demographics and characteristics of the jail inmate population.
3. What are some of the challenges associated with training and motivating jail staff?
4. What are some special types of sentencing options in jails?
5. What is *booking*, and why is this area of the jail so important?
6. Discuss the dynamics of suicide within the jail facility.
7. How have communicable diseases impacted jail operations during recent years?
8. How can a first-time jail experience result in labeling a person and thereby potentially increasing his or her likelihood of continuing in a life of crime?

KEY TERMS

Review key terms with eFlashcards at edge.sagepub.com/hanserbrief.

Detoxification, 106

Gaols, 84

Health Insurance Portability and Accountability Act (HIPAA), 101

Jail, 85

Jail reentry programs, 90

National Commission on Correctional Health Care (NCCHC), 102

Podular jail, 90

Public Safety Realignment (PSR), 95

Rural jail, 87

Sally port, 92

Shock incarceration, 105

Short-term jail, 91

Tower of London, 84

Weekend confinement, 105

APPLIED EXERCISE 4.1

For this exercise, students will need to determine how they would informally address the issue of sexually transmitted diseases (STDs) and other sex-related illnesses in their institution.

You are the assistant warden of a large metropolitan jail in your city. Your jail contends with a constant ingress and egress of inmates who tend to be drawn from population groups that are traditionally high risk for various STDs as well as HIV/AIDS. Your jail system naturally has an identification policy, and the members of your staff are trained in dealing with these types of illnesses, diseases, and viruses. Occasional in-service training is also provided to staff when and where time and resources permit.

Recently, the city of Sodom, where your jail facility is located, has experienced an outbreak of herpes that has reached epidemic proportions. In addition, your county, the county of Gomorrah, was identified last year as having the highest incidence of HIV/AIDS in the state. Thus, issues related to STDs and HIV/AIDS are taken very seriously in your jurisdiction.

Recently, in response to community concerns regarding public health, the mayor of Sodom met with the chief of police of Sodom and the sheriff of Gomorrah County to discuss the issue. It was decided that a very quick and aggressive crackdown on vice crime would be instituted in areas with high levels of prostitution, other illicit sex industry activity, and intravenous drug use. Police officers and sheriff's deputies were mobilized for this major enforcement activity. Amid this, the mayor, police chief, and sheriff implemented plans for increased jail space in facilities throughout the city and the county; it was understood that this major enforcement effort would result in a larger number of persons jailed in the area. Thus far, resources for containing this population have been sufficient.

However, staff in the jail facilities—and your facility is the largest in the system—tend to be very stern and unsympathetic to the plight of those inmates who are being locked up for various types of criminal behavior that correlate with the spread of STDs and HIV/AIDS. While there is no specific requirement that they be particularly empathetic, the deliberate heckling of these inmates and the calloused approach that seems to predominate does not help the situation. This attitude is exacerbated by local TV and printed news media that have showcased the public health hazards, essentially dramatizing the situation to epic proportions. In fact, the media have been critical of law enforcement and the mayor's office for being a bit slow and ineffective in their response. Thus, many jail employees feel as if their agency is considered lax on this problem.

Your concern, as the assistant warden, is centered on the increased number of accidents and altercations and the general upheaval within your facility. The warden of the facility (your boss) has explained that lawsuits are likely to emerge in the near future. The city and the county both have their legal budgets stretched way beyond what allocations would typically allow. So the warden asks you to implement some type of intervention program that will ameliorate the problems that are occurring between staff and STD-positive inmates.

The issue, it appears, has to do with the informal culture of members of the jail staff and their perceptions of the problem as well as their role in addressing that problem. It seems that by handling these offenders in a rough manner, staff are actually aggravating the long-term problem for the city and the county. While at the day-to-day level of operations the consequences of their actions do not seem to be serious, some very costly problems are likely to develop in the future. Further still, their behavior will not aid the process of trying to get these offenders to change their sexual and drug-using behaviors. It is important to the warden that something be done to change this informal culture and, if that is not possible, that some type of documentation exist to at least address this issue since the warden is quite convinced that it will be relevant in civil court in the next few months.

TO THE STUDENT: The task before you is to design some type of intervention method to address and change the behavior of staff who seem to be calloused and antagonistic toward offenders who have STDs or HIV/AIDS within your facility. This must go beyond simply making a policy statement or providing a training segment since both of these types of programs already exist in your jail system. While you may add to the policy and the training segment, explain what else you might do to address this issue, both formally and informally.

Here you are again, the assistant warden of the largest jail facility in the city of Sodom, located in the county of Gomorrah. Recently, you have found that many of the offenders who have been locked up due to a crackdown on vice crime, sex industry offenses, and intravenous drug use were on some type of drug and/or alcohol at the time of their arrest. Numerous "johns" and other typically noncriminal persons have been arrested and jailed in the process of implementing the crackdown. Needless to say, many of these persons are not used to being incarcerated, and they seem to be having a difficult time with the experience. Additionally, many of these current inmates suffer from high levels of guilt and embarrassment, not to mention problems with their employer and/or spouse. While many are able to get out quickly through the bail/bonds offices in the area, many stay in the facility for several days.

An unexpected outcome is that several suicide attempts have been made by these persons. In fact, two attempts have been successful. Four other serious attempts were detected in time to prevent their completion. One of the decedents was from a very affluent family, and it is thought that the family may sue the jail facility. The sheriff has met with members of the family to attempt to mitigate some of their concerns and complaints regarding the incident, but it is not yet clear how the situation may unfold.

Your facility does have a suicide policy that requires additional supervision of inmates at risk of suicide, the use of suicide contracts, and additional services. However, the sheriff, the mayor, and the chief of police all believe that much more can be done if staff are given training on how to provide informal crisis prevention skills for new inmates, particularly those determined to be at heightened risk via jail intake protocols. The warden of the jail has talked with you and noted that she has set aside money from the budget so that you can develop and implement some type of crisis response training for jail staff. You have also been given an additional administrative staff member to help with this process.

You must first determine who is at increased risk, and then you must identify what might be done among jail staff to minimize this risk. You are given 30 days to implement this program.

What would you do?

5

Probation and Intermediate Sanctions

Learning Objectives

1. Discuss how probation impacts the jail and prison systems of a jurisdiction.
2. Describe briefly the history of probation.
3. Identify the qualifications and characteristics of most probation officers.
4. List some of the reasons probation would be revoked.
5. Identify the various types of intermediate sanctions and their placement within the continuum of sanctions.

Probation Officer Stress

Jillian Jackson, a consultant in organization efficiency and human resources, was recently hired to address problems with poor morale within a very large probation agency.

Recently, it had been discovered that recidivism rates in the agency's jurisdiction were higher than average, much to the dismay of the administration and the chief judge of the criminal court. Nobody really knew what to do; extensive budget cuts had been made during the past 3 years and, on top of that, the jail and prison systems were full beyond capacity most of the time. It was also clear that the types of offenders on the caseloads had issues that were often more serious and complicated than offenders in the past. And even though the number of offenders was not rising, it was not going down either.

Jillian examined the small group around her. It consisted of eight probation officers who had been brought together as a focus group to discuss workplace stress. Jillian had asked these officers for feedback and thought at first that they might be hesitant to speak out. *Boy, was I wrong*, she thought to herself.

She motioned to one of the officers, Darrel Hayes, a somewhat gruff-appearing, barrel-chested man who, though professional, spoke with a commanding presence.

Once acknowledged, Hayes said, "I can have somebody's file pulled because they are a higher-risk person and I want to focus on his supervision and service needs, and then the 16 stupid things that I shouldn't have to be dealing with at all come walking in the door. At the end of the day, I realize that the one person I really should've been spending time with didn't get any time. That frustrates me because, let's face it, those are the cases that potentially are going to blow up in your face."

Jillian responded, "Yes, Mr. Hayes, I can see how this would get pretty frustrating." Another officer, Zora Gonzales, raised her hand, and Jillian indicated for her to speak.

"I often feel intense pressure because you have such a responsibility to the community," Gonzales said. "These high-risk offenders pose such a potential threat to public safety that you feel pressure to find the right treatment for this person and to make sure that they're getting something out of it."

"So if you all were able to spend more time with the more risky offenders to ensure that they received the programming that they needed, you would feel much better about your role in maintaining public safety and be able to use some of the tools that are designed to prevent some of these recidivism problems?"

"Yeah, exactly!" said Hayes. Gonzales and four other officers nodded their heads as well.

Jillian asked, "Have you guys ever made all of this known to the chief judge and other external officials?"

Jason Booker replied, "Naw, I don't think any of us are really prone to talking about this stuff beyond our own group here . . . especially not to people outside of our organization. They don't understand, and they'll also probably worry even more that we don't have a handle on things."

Nodding in agreement with Booker, Susan Grundstrom added, "You know, Ms. Jackson, most probation officers' lives are kind of chaotic due to the size of our caseloads and the nature of who we have to deal with. It's hard not to reduce everything to self-preservation, routing people without getting in-depth, just to survive. On a personal level, I think most officers stay pretty closed up—like police officers. What do you do—go home and tell your husband about the child abusers and rapists you saw in the office today? It's difficult for officers not to carry their work home, and yet difficult not to be able to talk about it there."

Jillian considered Grundstrom's words. After a moment she said, "I think that we're going to have to work on the agency's idea of communication and collaboration with outside partners, and I think this is going to require many partners coming together to ensure that programs and supervision are paired up correctly, comprehensively, and consistently with these high-risk offenders. The agency staff simply can't do *everything*."

And I think that this is going to require much more work than anyone in this administration realizes, she then thought to herself.

INTRODUCTION

Probation, as implemented by county and state jurisdictions around the nation, is the most common sanction administered in the United States. For this reason, if nothing else, attention should be given to it. However, probation is specifically important to corrections because it is, by its very nature, an option that facilitates the process of correcting offenders. In addition, probation impacts the jail and prison systems through revocation processes and by acting as a filtering device as offenders are inputted into the jail and prison systems. Since space in incarceration faculties can often be scarce, probation alleviates overcrowding problems at the front end of the criminal justice system. Thus, probation serves as an important tool for jurisdictions that operate local jails.

It is with this in mind that we will examine the use of probation and how this sanction greatly impacts the inmate flow within correctional institutions. This is particularly true within the jail setting, and it is therefore appropriate that our discussion of probation occurs just after our discussion of jail facilities and the issues that impact jail systems. Now that you, the reader, have a better understanding of jail facilities and the issues inherent to their operation, the following discussion related to probation and jail population flow should be clear.

Probation: A control valve mechanism that mitigates the flow of inmates sent directly to the jailhouse.

When we speak of **probation** we are referring to its use as a control valve mechanism that mitigates the flow of inmates sent directly to the jail. However, probation is also a sentence whereby people are given a less restrictive sanction with the understanding that they will be incarcerated if they do not comply with the terms of this type of supervision within the community. We will talk later in this chapter about the common terms and conditions of probation sentences, but for now we will focus more on the history of probation and the role that it has played, systemically, in the correctional system.

A BRIEF HISTORY OF PROBATION

Probation is a uniquely American invention (see Focus Topic 5.1). At its inception, probation was used as an alternative to incarceration in the United States. John Augustus, a cobbler and philanthropist of Boston, is often recognized as the Father of Modern Probation. During the time that Augustus provided his innovative contribution to the field of community corrections, the temperance movement against alcohol consumption was in full swing. Augustus, aware of many of the issues associated with alcoholism, made an active effort to rehabilitate prior alcoholics who were processed through the police court in Boston.

While acting as a volunteer of the court, Augustus observed a man being charged for drunkenness who would have, in all likelihood, ended up in the Boston House of Correction if it were not for Augustus's intervention. Augustus placed bail for the man, personally guaranteeing the man's return to court at the prescribed time. Augustus helped the man to find a job and provided him with the guidance and support that was necessary so that the defendant was able to become a functioning and productive member within the community. When the court ordered the return of the offender 3 weeks later, the judge noticed a very substantial improvement in the offender's behavior. The judge was so impressed by this outcome that he granted leniency in sentencing (Augustus, 1972/1852; Barnes & Teeters, 1959). From this point in 1841 until his death in 1859, Augustus continued to bail out numerous offenders, providing voluntary supervision and guidance until they were subsequently sentenced by the court. Students should recall that corrections, as used in this text, is intended to do more than simply *punish* the offender but instead seeks to *reform* the offender. This is just as true with the use of probation as it would be with any other form of correctional sanction. While probation may act as a valve that mitigates the flow of inmates into the jailhouse, this is not and should not be its primary purpose. Rather, consistent with the earlier presented definition of corrections, probation should place primary emphasis on correcting criminal human behavior. Issues related to jail logistics and other such concerns should be secondary. This was the original intent when administering probation since Augustus was primarily concerned with the malicious treatment of offenders and the desire for revenge that could easily disrupt society.

While Augustus was aware that jail and prison conditions were barbaric in many cases, his actual goal was not necessarily to spare individuals from the misery of jail but to attempt to reform offenders. He selected his candidates with due care and caution, generally offering aid to first-time offenders. He also looked to character, demeanor, past experiences, and potential future influences when making his decisions. Thus, whether you use the word *reform, rehabilitate,* or *reintegrate,* it is clear that the initial intent of probation was to provide society with people who were more productive after sentencing than they had been prior to it. This intent stood on its own merit and purpose, regardless of jail or prison conditions that might have existed, thereby establishing the original mission of community corrections as a whole. Nevertheless, in contemporary corrections, probation often plays the role of a control valve that handles inmates that the jail facility cannot hold. We will talk about this more in the section that follows.

PHOTO 5.1 John Augustus was a volunteer of the court in the Boston area during the mid-1800s. He is regarded as the Father of Modern Probation.

NYC Department of Probation

FOCUS TOPIC 5.1

Historical Developments in Probation in the United States

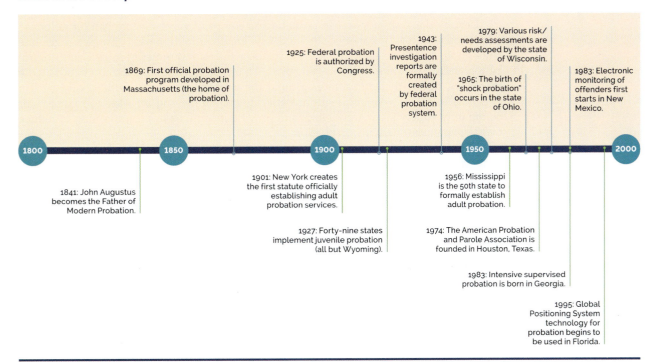

1869: First official probation program developed in Massachusetts (the home of probation).

1925: Federal probation is authorized by Congress.

1943: Presentence investigation reports are formally created by federal probation system.

1979: Various risk/needs assessments are developed by the state of Wisconsin.

1965: The birth of "shock probation" occurs in the state of Ohio.

1983: Electronic monitoring of offenders first starts in New Mexico.

1800   1850   1900   1950   2000

1841: John Augustus becomes the Father of Modern Probation.

1901: New York creates the first statute officially establishing adult probation services.

1956: Mississippi is the 50th state to formally establish adult probation.

1927: Forty-nine states implement juvenile probation (all but Wyoming).

1974: The American Probation and Parole Association is founded in Houston, Texas.

1983: Intensive supervised probation is born in Georgia.

1995: Global Positioning System technology for probation begins to be used in Florida.

CONTEMPORARY PROBATION: WHEN THE JAIL IS FULL

In many cases, probation sentences are meted out at the county level of government. This is the same level of government that tends to administer jail facilities. Thus, probation is typically administered by the same courthouse that oversees the jail facility in a given jurisdiction. As a result, these two justice functions—probation and jailing—tend to work in tandem with one another. However, this is not to imply that coordination and communication between jail administrators and probation administrators is optimal; in many cases, these two functions operate in a manner that is disjointed, despite the fact that the same courthouse may impact both probation and jail agencies.

While jails and probation agencies may (or may not) have a collaborative relationship, it is undeniable that the district attorney (the office that prosecutes criminals) will have a close working relationship with the local sheriff or sheriffs in the region as well as city police chiefs who collectively oversee law enforcement activities. These activities result in the flow of criminals before the courthouse and ultimately require that a judge sentence an offender to one of three likely options: community supervision (usually probation), a jail sentence (sometimes with additional community supervision requirements), or a prison sentence (if incarceration is to exceed a year in duration).

Because the majority of offenders tend to commit crimes that are petty, nonserious, or nonviolent, this means that they will tend to qualify for jail or probation. Those offenders who do commit serious crimes will, naturally, be sentenced to prison and then are therefore the concern of state prison system authorities. However, the bulk of the offender population will remain at the county level, and it is in this manner that the jail population develops. In many instances, the jail facility will fill quite quickly with offenders who commit crimes, particularly in large urban areas of the nation. The use of probation becomes a critical tool to monitor and alleviate the flow of inmates into the jail.

Thus, as noted previously, one function of probation is to act as a control valve mechanism that mitigates the flow of inmates sent directly to the jailhouse. Without the use of probation sanctions, jails would simply collapse in their operation because the current jail facility structure throughout the United States could not even come close to containing the total offender population. This problem is exacerbated when state prison systems are full and jail facilities are required to keep inmates who, legally speaking, should be housed within a state prison facility. This demonstrates how the jail facility can feel pressure from inmate flow from the front end and the back end of its operational system.

Despite the tendency to use probation as a control valve mechanism for jailhouse admissions, the total number of offenders who are on probation has continued to decline since 2009. As can be seen in Table 5.1, in 2009, the total number of individuals on probation around the nation was 4,198,200, with a steady decline to 3,864,100 in 2014. One reason for this is that crime rates around the country are down. Additionally, such a decline also occurs when corrections departments are looking for ways to reduce their community supervision populations and provide incentives in programming to reduce overall sentences, including probation.

There is one noteworthy exception to this decline in probation rates, however: the state of California. As students may recall from Chapter 4, California passed what is known as the Public Safety Realignment (PSR) policy, which was designed to reduce the number of offenders in the state prison system to 110,000. Since this policy was implemented, entries to probation increased nearly 15%, from an estimated 149,000 offenders in 2010 to 295,475 in 2014. Thus, one way of keeping the state's prison population from growing larger seems to be the more frequent use of probation.

Lastly, it would appear that a handful of states stand way in front in relation to the number of probationers who are supervised therein. Indeed, five states (see Table 5.1) have populations that are well above 200,000 probationers. All other states have less than 200,000 each, with only about six of them having over 100,000. The state of Michigan comes in sixth in terms of the size of the probation population, with 180,583 probationers. Thus, the top five states account for approximately 1.6 million out of the nearly 4 million probationers around the country.

TABLE 5.1

Probation Among Notable Ranking States

10 STATES WITH THE LARGEST PROBATION POPULATIONS	NUMBER SUPERVISED	NUMBER PER 100,000 ON PROBATION	PERCENT CHANGE IN 2016	10 STATES WITH THE SMALLEST PROBATION POPULATIONS	NUMBER SUPERVISED	NUMBER PER 100,000 ON PROBATION	PERCENT CHANGE IN 2015
Georgia	410,964	5,570	−10.2	New Hampshire	3,861	366	2.0
Texas	378,514	1,805	−1.1	Wyoming	4,860	1,046	−4.0
California	238,911	791	1.0	Vermont	5,164	969	−5.0
Ohio	236,375	2,624	0.2	District of Columbia	5,546	1,034	5.3
Florida	221,446	1,288	−3.3	North Dakota	6,343	1,090	−Less than 0.05%
Pennsylvania	183,868	1,783	−1.8	Alaska	6,513	1,193	1.7
Michigan	175,189	2,276	1.0	Maine	6,702	632	1.7
New Jersey	136,137	2,015	3.3	South Dakota	6,959	1,009	−5.0
Illinois	122,125	1,154	−6.7	West Virginia	7,008	448	−6.9
Indiana	111,709	2,135	−3.0	Montana	8,818	1,115	3.6

Source: Kaeble, D. (2018). *Probation and parole in the United States, 2016.* Washington, DC: U.S. Department of Justice.

CHARACTERISTICS OF PROBATIONERS

Probationers are criminal offenders who have been sentenced to a period of correctional supervision in the community in lieu of incarceration (Kaeble, 2018). During the past couple of years, the overall population of probationers has continued to decline (Kaeble, 2018). Indeed, at the close of 2016, about 3,673,100 adults were on probation (down from 3,789,800 in 2015). During the prior year, entries on probation declined from an estimated 2,065,800 to about 1,966,100; this amounted to a 5% decline that year alone. Likewise, almost two thirds (65%) of probationers completed their terms of supervision or were discharged early during 2015 (Kaeble & Bonczar, 2017). This trend resulted in an increase in probation exits from 2,043,200 in 2015 to 2,071,400 in 2016. Despite this, probationers continue to account for the largest proportion of persons under corrections supervision, with 58% of all convicted offenders being on probation. Figure 5.1 illustrates the current decline in the probation population.

Again, though probation has experienced a decline in numbers, probationers make up the largest segment of the correctional population. Thus, not only has the probationer population declined during the past few years, but the other categories of the correctional population have experienced either a stable count or a slightly increased growth rate, with proportions of jail inmates rising and prison inmates remaining stable during the past 3–4 years (Glaze & Bonczar, 2011).

Table 5.1 provides an overview of the number of probationers in the top 10 states with the largest probation populations and the 10 states with the lowest probation populations. Interestingly, only three of the states with probation populations that rank in the top 10 also show increases in the probation population during the year prior, these states being California, Ohio, and New Jersey. In 2016, the state of Virginia had the highest rate of growth (9.6%), followed by Oklahoma (7.3%). Across the nation, most states reported declines in their probationer populations (Kaeble, 2018).

Probationers: Criminal offenders who have been sentenced to a period of correctional supervision in the community in lieu of incarceration.

FIGURE 5.1

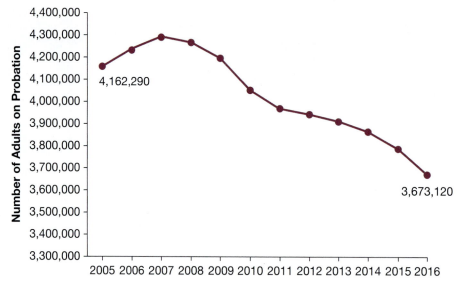

Percent Change of Adults on Probation at Year End, 2005–2016

Source: Kaeble, D. (2018). *Probation and parole in the United States, 2016.* Washington, DC: U.S. Department of Justice.

Among the probation population, almost 1 out of 4 (25%) is a female offender. Nationwide, women represented a slightly larger percentage of the probation population in 2015 than during the 10 years prior, going from 23% to 25% (Kaeble & Bonczar, 2017). This represents a slow but steady growth pattern within the female offending population. Further, more than half of all probationers are Caucasian, almost a third are African American, and one eighth are Latino American. Persons of other racial orientations comprise about 2% of all probationers in the United States.

Overall, roughly 57% of all probationers are convicted of a felony offense, with the misdemeanor population consisting of 41%. The remaining 2% have been convicted of other types of offenses, such as city ordinances, county codes, and so forth. As Kaeble and Bonczar (2017) indicate, the largest body of the probation population has been convicted of some type of property crime (28%), followed by drug-related violations (25%), and driving while intoxicated, or DWI (13%). Though most probationers are on probation for nonviolent offenses, 4% are serving for sexual assault, another 4% are serving for some form of domestic violence, and 13% are on probation for miscellaneous assault convictions (Kaeble & Bonczar, 2017). Thus, about 1 in 5 probation offenders has committed an act of violence; among these, many are not extreme acts. The remaining offenses tend to be fairly evenly distributed and include crimes such as fraud, burglary, and minor (but probably repetitive) traffic violations (Kaeble & Bonczar, 2017).

Not all probationers are required to report to their probation officer. Indeed, a certain proportion are simply kept on caseloads administratively and are checked periodically, required to pay fees during the set amount of time, and so forth. So long as no word of violation is received or detected by other agencies, these offenders finish out their probation with relative ease. In 2005, roughly 7 in 10 probationers were under active supervision, being required to regularly report to a probation authority in person, by mail, or by telephone (Kaeble & Bonczar, 2017). This means that roughly 30% of all probationers were on inactive administrative caseloads. The percentage of probationers required to report regularly has declined steadily, from 79% in 2005 to 70% in 2015 (Kaeble & Bonczar, 2017). In addition, roughly 7% of all probationers absconded during 2005. Though these individuals are still on probation caseloads, their whereabouts are completely unknown. This is a concern, as it means there are roughly 400,000 probation absconders across the nation. The rate of probation absconders has, during the past decade, slightly increased from 9% to 10% and indicates a general need for improved methods of security for probationers.

Finally, while the total state probationer population is 3,859,135 (as of 2015), the total federal probation population is a minuscule 19,062 (Kaeble & Bonczar, 2017). Thus, in comparison, the total population of federal probationers is less than ½ of a percent of the number that comprises state-level probation. To make the point even further, federal probation decreased by 3.7% in 2015, indicating that the use of federal probation is much less frequent than is the case for state probation (Kaeble & Bonczar, 2017). Moreover, many of the services of federal probation officers are associated with pretrial functions. Thus, federal probation is not a central aspect of the overall probation picture around the nation and therefore will be discussed sparingly throughout this chapter and text. Rather, this chapter will focus on the other 99.5% of the probationer population and the systems that are responsible for their supervision.

The Probation Agency

When examining the means of operation within a probation agency, one key characteristic to consider is the degree of centralization that exists within that agency. Indeed, adult probation in one state may be administered by a single central state agency, by a variety of local agencies, or by a combination of the two. When considering local levels of administration, agencies may operate at the county or even municipal level. However, these supposedly smaller jurisdictions should not be underestimated. Consider, for example, the probation departments in New York City, where felony and misdemeanor caseloads are larger than those of many entire state systems.

PRACTITIONER'S PERSPECTIVE

"Probation is this wonderful opportunity for intervention in people's lives and it offers a chance to stay connected in the community."

Interested in a career in corrections or criminal justice? Watch Jessica Johnston's video on her career as a probation officer.

The Presentence Investigation

The **presentence investigation report** is a file that includes a wide range of background information on the offender. This file will typically include demographic, vocational, educational, and personal information on the offender as well as records on his or her prior offending patterns and the probation department's recommendation as to the appropriate type of sentencing and supervision for the offender. According to the Michigan Department of Corrections (2003), "A probation file may also consist of other reports written by counselors, psychologists, and case workers. Therefore, a large amount of personal and confidential information is maintained by the probation officer which should not be disclosed arbitrarily" (p. 33). In many respects, the presentence investigation (PSI) report is the initial point of assessment, and it will often be utilized when the offender is first brought into a prison facility. In other words, the PSI report is not just used in probation sentences; it also is used when offenders are imprisoned. While writing the PSI

Presentence investigation report: A thorough file that includes a wide range of background information on the offender.

PHOTO 5.2 The judge of the court is the final authority on rulings related to probation sentencing and conditions. The judge will usually work closely with probation officers who deal with offenders from his or her court of jurisdiction.

AP Photo/Zach White

report, a probation officer may review sentencing recommendations with the offender and even perhaps with the offender's family. Note that this is separate from any other arrangements that might be made throughout the plea-bargaining process between the offender's counsel and the prosecutor's office. In some cases, probation officers may also be required to testify in court as to their findings and recommendations.

The primary purpose of the PSI report is to provide the court with the necessary information from which a sentencing decision can be derived. The PSI is conducted after a defendant is found guilty of a charge (whether by pleading or by court finding) but prior to sentencing. This information in the PSI report, along with a sentencing recommendation, will aid the judge, who must ultimately fashion a sentence as well as any corollary obligations attached to that sentence.

The PSI report also tends to serve as a basic foundation for supervision and treatment planning throughout the offender's sentence, both when on probation and later if the offender is incarcerated. Quite often this document will serve as a reference point for placing the offender in a variety of programs. This can happen when the offender is on supervision or in a jail or detention facility.

Among other things, the PSI report will contain information related to the character and behavior of the offender. This means that the probation officer's impressions of the offender can greatly impact the outcome of the PSI. The PSI is typically conducted through an interview with the offender. Because the PSI report information is largely obtained from the interview process, it is naturally important that probation officers have good interviewing skills. This cannot be overstated given the fact that probation officers are in contact with persons on a routine basis where they must collect and record information. While procedures do vary from region to region, a sentencing phase will be conducted at some point during the processing of a criminal conviction. At this point, the defense counsel can have an impact on the overall process for the offender. Defense counsel will usually challenge any inaccurate, incomplete, or misleading information that ended up in the PSI report. This function of the defense counsel is actually quite critical since the PSI report will be used to classify the offender if he or she should be incarcerated and will also be used in future decisions regarding supervision issues within the community. Thus, verification of the PSI report's validity is crucial to the welfare of the defendant and keeps from creating scenarios that make an already bad situation worse.

From the standpoint of the probation officer, the two most important sections of the PSI report are the evaluation and the recommendation. There is typically a high degree of agreement between the probation officer's recommendations and the judge's decision when sentencing, and this means that the PSI report is very important in helping to determine the offender's fate.

Granting Probation

According to Neubauer (2007), the public perceives the judge as the principal decision maker in criminal court. But the judge often is not the primary decision maker in regard to an offender's sentencing and/or the granting of probation. This is not to say that the judge does not have ultimate authority over the court, nor does it mean to imply that judges have diminished importance when presiding over their court. Rather, Neubauer demonstrates the collaborative nature of the various courtroom actors when processing offender caseloads. Throughout this process, judges will often voluntarily defer to the judgment of other members of the court, namely prosecutors, defense attorneys, victim's rights groups, and the probation agency.

During a typical day in criminal court, judges may accept bail recommendations offered by the district attorney, plea agreements that are struck by the defense and the prosecution, and even sentences recommended by a probation officer (though there is some debate as to the actual weight given to the probation officer's recommendation, at least in some courts). The main point is that though judges do of course retain their power over the courtroom, they often share influence over the adjudication process with a variety of courtroom actors (Neubauer, 2007). This is an informal process that often takes place among participants who, after working together for a time, know each other in both a professional and a more informal sense (Neubauer, 2007).

There are some challenges that can emerge when judges do not allow the input of other courtroom actors. For instance, jail overcrowding may be worsened if the judge is not receptive to the input of the sheriff and/or the police chief who will administer the local county or city jail. Or probation officer caseloads can become too burdensome to ensure public safety if the judge does not consider the recommendations of the chief probation officer.

When defense attorneys and their defendants seek to have probation considered as a sentencing option, it may behoove the defense counsel to consider the specific judge who presides over the court as well as the dynamics of a given courtroom. Because of this, in larger court jurisdictions, a technique of judge selection may be common (Neubauer, 2007). Through a process of implementing motions of continuances and motions for a change of judge, defense attorneys may maneuver to have their case heard by a judge who is expected to be the most receptive to the offender's plight. Though judges do strive to adhere to common guidelines in decisions and rulings, the fact of the matter is that they do tend to differ in terms of the sentences that are given, including the granting of probation and/or the conditions attached to a probation sentence (Neubauer, 2007). An understanding of these tendencies can aid the defense in achieving a more favorable outcome for the offender.

Conditions of Probation

As one can tell, the dynamics of sentencing in the courtroom can allow for some degree of leeway in the final decision-making process. Further illustrating the fluid nature of this process, consider that the setting of conditions during probation can be, at least in part, agreed upon prior to the judge's actual formal sentencing. Though the bargaining process may impact the final outcome of the probationer's sentence, the length of probation, and the conditions of that probation, the judge is always free to require additional conditions as he or she sees fit.

The conditions that may be required are quite lengthy, but some of the more commonly required ones are the following:

1. Refrain from associating with certain types of people (particularly those with a conviction) or frequenting certain locations known to draw criminal elements.

2. Remain sober and drug free; restrictions include using or being in possession of alcohol or drugs.

3. Obey restrictions on firearm ownership and/or possession.

4. Obey requirement to pay fines, restitution, and family support that may be due.

5. Be willing to submit to drug tests as directed by the probation officer and/or representatives of the probation agency.

6. Maintain legitimate and steady employment.

7. Refrain from obtaining employment in certain types of vocations (e.g., an embezzler would be restricted from becoming a bookkeeper, or a computer hacker would be restricted from working with automated systems).

8. Maintain a legal and legitimate residence with the requirement that the probation officer is notified of any change in residence prior to making such a change.

9. Obey the requirement that permission be requested to travel outside of the jurisdiction of the probation agency and/or to another state.

10. Refrain from engaging in further criminal activity.

Many of the conditions listed above may be statutorily authorized by state legislators as a means of validating their application to probation sentences. This is reflective of the fact that most legislators desire some degree of uniformity and consistency in the supervision requirements and process (Hanser, 2010b). Some states have only a few such requirements, while others have an extensive list that clearly requires judges and probationers to structure probation sentences according to a certain prescribed template of conditions. Further, and related to the use of discretionary conditions imposed by judges, some legislators may also clearly note that judges are to be given deference in assigning specialized conditions on certain types of offenders; this is especially true with sex offenders and/or substance abuse offenders (these offenders, the terms and conditions of their supervision, and their therapeutic programming will be discussed in later chapters of this text).

PROBATION OFFICERS

No chapter on the probation process would be complete without a thorough discussion of the job and function of probation personnel. As we have seen from the chapter vignette, the job of a probation officer is quite stressful and challenging and does not pay nearly as well as many other professions (see Figure 5.2 for information on annual salaries for probation officers throughout various areas of the United States). Further still, the qualifications for probation officers tend to be fairly high, at least in relation to the demands and pay that are associated with the position. This is truly an unfortunate paradox within the criminal justice arena since it is the probation officer who supervises the lion's share of offenders in the correctional system.

FIGURE 5.2

Annual Mean Wage of Probation Officers, May 2018

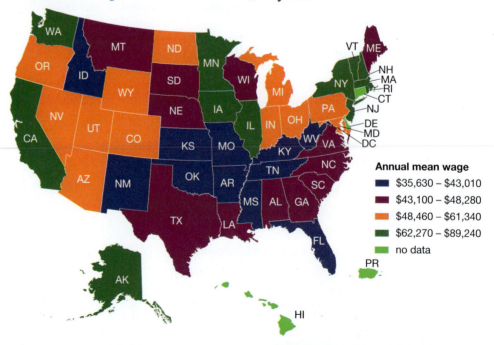

Annual mean wage
- $35,630 – $43,010
- $43,100 – $48,280
- $48,460 – $61,340
- $62,270 – $89,240
- no data

Source: Bureau of Labor Statistics. (2018). *Occupational employment statistics.* Washington, DC: Author. Retrieved from https://www.bls.gov/oes/current/oes211092.htm

Because of the stress involved with probation work, there is a great deal of turnover in the field. Naturally, this can have negative effects on the correctional system since personnel with expertise are hard to keep. This can impact the service delivery that agencies are able to provide, and, in some respects, is likely to affect outcomes among offenders on community supervision. Indeed, it may be likely that the prognosis for recidivism can be affected (at least in part) by the longevity of the probation officer and his or her demeanor on the job.

Thus, the content of this chapter is more than a simple introduction to work in the field of probation. This chapter also presents an aspect of the corrections field that is critical for students and other persons in society to understand: the important role played by probation personnel in the security of a community. Citizens should be grateful for these personnel since it is they, just as much as police, who are largely responsible for keeping society safe from known criminals. Interestingly, most probation agencies pay lower salaries—starting, midcareer, and managerial—than do police agencies in their same region.

Demographics of Probation Officers: Gender

One interesting aspect of probation work is the fact that a large portion of probation staff tends to be female, with exact proportions of female and male officers being dependent on the area of the United States. Data from the Bureau of Labor Statistics (2017a) indicate that 63.5% of all probation officers are female. This is substantially different from fields such as law enforcement, where male officers tend to predominate and women tend to constitute less than 12% of the entire policing community (Federal Bureau of Investigation, 2013). This is perhaps partly due to the nature of probation as compared to law enforcement. Indeed, even among police officers, women have been found to be highly effective in defusing conflict situations and/or providing less contact-prone means of response. The National Center for Women and Policing (2003) notes that female police officers tend to be inherently more suited to facilitate cooperation and trust in stressful contact situations and that they are less prone to use excessive force. Likewise, there tend to be fewer citizen complaints against female officers. These same characteristics would seem to be well suited to probation work given the fact that probation has a reintegrative and supportive role with offenders on the officer's caseload.

Demographics of Probation Officers: Race

Probation officers also tend to be Caucasian. Data from the Bureau of Labor Statistics (2017a) indicate that 63.8% of all probation officers are Caucasian. This can be an important issue when one considers the fact that a disproportionate amount of minority representation can be found on most client caseloads. Given the lack of minority representation among probation officers, it is likely that diversity-related training is all the more necessary and important in cultivating a rapport between community supervision personnel and those on community supervision. An abundance of literature has examined issues related to therapist–client interactions when the two are of different racial and/or ethnic groups. Generally, the prognosis in mental health research does not tend to be as good as when there is a degree of matching or when specific training and consideration are given for racial or cross-cultural issues. Since community corrections has a reformative element, it is not unreasonable to presume that such observations could also be equally true among probation officers and their probationers.

Demographics of Probation Officers: Education

Most probation officers have a college degree. This means that this group is, as a whole, a bit more educated than much of the general workforce. It may perhaps be true that this can mitigate some of the cross-cultural differences, and this also may help to lessen job dissatisfaction and stress since higher-educated persons tend to, on the whole, be motivated by more than external reward. Though this is obviously not always the case, less emphasis on money does tend to correlate with better-educated workforce members. Somewhat supporting this is the fact that several studies have found that probation work in general tends to be more enriching and challenging, requiring more of an emphasis on problem-solving skills that are likely to mesh well with high-functioning and educated persons. From this, it is clear that probation work is becoming more professionalized and has been likened to an art form since probation officers must be skilled at matching security and treatment issues with the particular offender's needs (Bureau of Labor Statistics, 2018). This, as well as the helping aspects of the profession (despite its supervisory components), is likely to appeal to educated females who seek a professional track in their lives. Students are encouraged to examine Figure 5.4 to get an idea of the number of probation officers who are employed in each state.

Tasks and Nature of Work for Probation Officers

Probation officers supervise offenders who are placed on some form of probation and tend to spend more time monitoring the activities of these offenders than anything else. Probation officers most frequently maintain this supervision through personal contact with the offender, the offender's family, and the offender's employer. In addition to making contact with the offender through a combination of field visits and/or officer interviews, probation officers make routine contact with the offender's therapist(s), often having therapeutic reports either faxed or delivered to their office. These reports, which provide the clinician's insight as to the offender's emotional progress and/or mental health, can be very important to the probation officer's assessment of the offender's progress.

Working Conditions

The daily working conditions for probation officers can be quite safe when in the office but can be fairly dangerous when conducting field visits. Some of the offenders on a probationer's caseload may be more dangerous than their arrest record or actual conviction may indicate. Further, these offenders may still (in violation of their probation) continue to maintain contact with other associates who are more prone to violence than is the probationer. In many instances, the probation officer may have to conduct fieldwork in high-crime areas.

FIGURE 5.3

Percentage of Probation Officers Who Report Traumatic Caseload Events

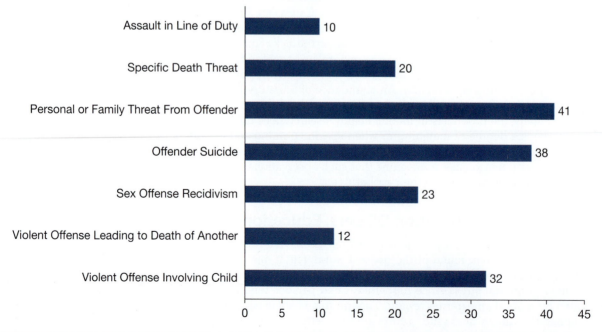

Source: Whitten, L. (2013). *Probation officers' stress and burnout associated with caseload events*. Washington, DC: National Institute of Corrections. Retrieved from https://nicic.gov/probation-officers-stress-and-burnout-associated-caseload-events-2013

This point should not be taken lightly, and it is unlikely that the average person can understand the true contextual feeling that is associated with such an experience when conducting casework. Often, members of the community may display negative nonverbal behavior toward the probation officer and may be evasive if the officer should happen to ask questions about the offender in the offender's neighborhood. In fact, in most cases, persons living next to the probationer may not disclose anything because they are also at cross-purposes with the law. In addition, family members are not always happy to have the probation officer visit the home and, while complying with the requirement, may openly resent the intrusion. Lastly, from time to time, the probation officer may make unannounced visits only to find the probationer in the company of unsavory sorts and/or engaging in acts that are violations of probation conditions (e.g., drinking, carrying a firearm, discussing various criminal opportunities). All of these issues can lead to some dangerous situations. This is even truer when one considers that most probation officers do not carry a firearm. Thus, it is safe to say at this point that there is a personal security concern when meeting probationers on their own turf.

Relevant research has shown that stressful events (e.g., a violent offense by a supervisee) and victimizations (e.g., officers being personally threatened or assaulted) were associated with higher reports of "compassion fatigue" and safety concerns, and many of them were linked with stress and burnout (Whitten, 2013). This research shows that officers who experience higher numbers of traumatic events and victimizations tend to have higher score levels of burnout, mistrust, family problems, anger, distorted world-view, and social/emotional isolation. In a study by Lewis, Lewis, and Garby (2012), 309 adult probation officers were surveyed or interviewed across three states. Those who reported aggravating incidents related to offenders in their caseloads scored significantly higher on measures of traumatic stress and burnout than officers who did not experience these incidents. The percentages of probation officers reporting these events are represented in Figure 5.3.

Probation Officers in the Role of Law Enforcers and Brokers of Services

Probation officers tend to approach their jobs from different vantage points, much of which has to do with their own perceptions of their particular role in the community corrections process. Daniel Glaser (1964) conducted seminal research on the orientation by which community supervision officers approach their job. Though Glaser focused on parole officers, his contentions apply equally well to probation officers. Thus, Glaser's work will be utilized in this chapter to provide a general framework for probation officers and the informal roles they play when supervising their caseloads. Basically speaking, Glaser contended that officers tend to operate at differing points along two spectrums: offender control (law enforcers) and offender assistance (brokers of services). These two spectrums work in seeming contradiction with one another, as they each tend to put officers at cross-purposes when trying to balance their job as reformer and public safety officer. This means that four basic categories emerge that describe the officer's general tendency when supervising offenders.

Paternal officer: Uses a great degree of both control and assistance techniques in supervising offenders.

PHOTO 5.3 Fieldwork is an important aspect of probation supervision. This requires that officers talk with the offender and others who know the offender in settings outside of the office.

© REUTERS/Lucy Nicholson

Paternal officers use a great degree of both control and assistance techniques (Glaser, 1964). They protect both the offender and the community by providing the offender with assistance as well as praise and blame. This type of officer can seem inconsistent at times. These officers are ambivalent to the concerns of the offender or the community; this is just a job that they do. Indeed, these officers may be perceived as being noncommittal due to taking the community's side in one case and the offender's in another. These officers tend not to have a high degree of formal training or secondary education, but they tend to be very experienced and thus are able to weather the difficulties associated with burnout within the field of probation.

Punitive officer: Sees himself or herself as needing to use threats and punishment in order to get compliance from the offender.

Punitive officers (pure law enforcers) see themselves as needing to use threats and punishment in order to gain compliance from the offender. These officers will place the highest emphasis on control and protection of the public against offenders, and they will be suspicious of offenders on their caseload. This suspiciousness is not necessarily misplaced or unethical, however, as this is part and parcel of the supervision of offenders, but these officers may in fact never be content with the offender's behavior until they find some reason to award some form of punitive sanction. In other words, the view is that those on the caseload are doing wrong, but they are just not getting caught. Naturally, relations between this officer and those on his or her caseload are usually fairly impaired and sterile.

Welfare worker: Views the offender more as a client rather than a supervisee on his or her caseload.

The **welfare worker** (pure broker of services) will view the offender more as a client rather than as a supervisee on the caseload. These individuals believe that, ultimately, the best way they can enhance the security and safety of the community is by reforming the offender so that further crime will not occur. These officers will attempt to achieve objectivity that is similar to that of a therapist and will thus avoid judging the client. These officers will be most inclined to consider the needs of their offender–clients and their potential capacity for change. These officers view their job more as a therapeutic service than as a punitive service, though this does not mean that they will not supervise the behavior of their caseload. Rather, the purpose of their supervision is more likened to the follow-up screening that a therapist might provide to a client to ensure that he or she is continuing on the directed trajectory that is consistent with prior treatment goals.

Passive agent: Views his or her job dispassionately as just a job and tends to do as little as possible.

The **passive agent** tends to view his or her job as just that, a job. These officers tend to do as little as possible, and they do not have passion for their job. Unlike the punitive officer and the welfare officer, they simply do not care about the outcome of their work so long as they avoid any difficulties. These individuals are often in the job simply due to the benefits that it may provide as well as the freedom from continual supervision that this type of career affords.

In reality, it is not likely that officers would best be served using one consistent type of approach rather than using each orientation when appropriate. Thus, the community supervision process can be greatly impacted by the approach taken by the community supervision officer. Further, agencies can transmit a certain tendency toward any of these orientations through policies, procedures, informal organizational culture, or even daily memos. The tone set by the agency is likely to have an effect on the officer's morale and his or her approach to the supervision process.

Some agencies may be very clear about their expectations of community supervision officers. In this case, if the agency has a strict law-and-order flavor, the officer may be best served by utilizing the approach of a punitive officer to ensure that he or she is a good fit with agency expectations. In another agency, the emphasis might be on a combined restorative/community justice model coupled with community policing efforts designed to reintegrate the offender. In an agency such as this, the officer may find that a welfare worker approach is the best fit for that agency and that community. Thus, the culture of the community service organization will have a strong impact on the officer's orientation, and, if the officer's personal or professional views are in conflict with the organizational structure, the likelihood of effective community supervision is impaired. This is important because it is another indicator of the stress encountered among most community supervision workers.

Qualifications for Probation Officers

Some basic background qualifications for probation officers are listed in the *Occupational Outlook Handbook* (Bureau of Labor Statistics, 2018). These qualifications vary by state, but generally a bachelor's degree in criminal justice, social work, or a related field is required for initial consideration. This was not always the case in times past (some states allowed for less education when combined with experience), but it is increasingly becoming the norm in most states. Some employers may even require previous experience in corrections, casework, or a treatment-related field, or a master's degree in criminal justice, social work, psychology, or a related discipline. For a broad look at the overall opportunities in the field of probation, Figure 5.4 provides an idea of how many probation officers are hired in each state.

FIGURE 5.4

Probation Officers Employed in Each State, May 2018

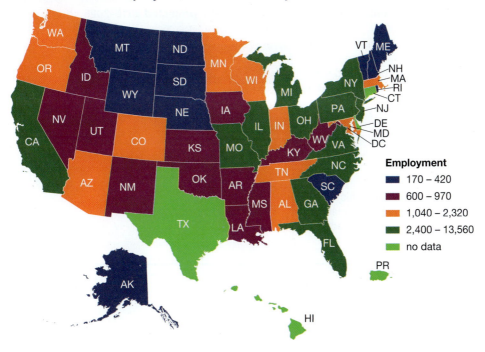

Employment

- 170 – 420
- 600 – 970
- 1,040 – 2,320
- 2,400 – 13,560
- no data

Source: Bureau of Labor Statistics (2018). *Occupational employment statistics.* Washington, DC: Author. Retrieved from https://www.bls.gov/oes/current/oes211092.htm

Entry-level probation officers should be in good physical and emotional condition. Most agencies require applicants to be at least 21 years old and, for federal employment, not older than 37 (Bureau of Labor Statistics, 2018). In many jurisdictions, persons who have been convicted of a felony may not be eligible for employment in this occupation (Bureau of Labor Statistics, 2018). Familiarity with the use of computers is typically expected given the increasing use of computer technology in probation and parole work (Bureau of Labor Statistics, 2018). Probation officers should have strong writing skills because they are required to prepare many reports. In addition, a graduate degree in a related field such as criminal justice, social work, counseling, or psychology can aid an employee in advancing into supervisory positions within the agency (Bureau of Labor Statistics, 2018).

According to the *Occupational Outlook Handbook*, applicants are usually administered a written, oral, psychological, and physical examination. Given the concern with job stress that is inherent in this field of work, it is no surprise that changes in screening mechanisms during the hiring phase have been observed (Bureau of Labor Statistics, 2018). Indeed, hiring and selection procedures may include psychological interviews and personality assessments to identify those most able to handle the stress and psychological challenges of probation and parole work (Bureau of Labor Statistics, 2018). This demonstrates that agencies are aware of the unique challenges with this type of work and wish to identify those persons hearty enough to withstand the pressures that are inherent therein. This is a wise and prudent move on the part of agencies from a liability standpoint, a public safety standpoint, and an employee–agency relations standpoint. Effective recruitment and selection at the forefront can prevent a host of problems potentially encountered by supervisors and agency leaders in the future. Given the declines in probationers that was discussed earlier in this chapter coupled with tightening state budgets, hiring for probation officers will likely not increase greatly when compared to other professions. Students are encouraged to examine Figure 5.5 to get an idea of the national growth rate in the demand for probation officers during the next few years.

FIGURE 5.5

Probation Officers and Correctional Treatment Specialists

Percentage change in employment, projected 2016–2026

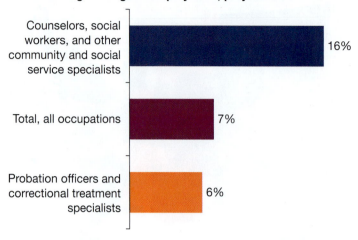

Counselors, social workers, and other community and social service specialists — 16%

Total, all occupations — 7%

Probation officers and correctional treatment specialists — 6%

Source: Bureau of Labor Statistics (2018). *Occupational outlook handbook.* Washington, DC: Author. Retrieved from https://www.bls .gov/ooh/community-and-social-service/probation-officers-and-correctional-treatment-specialists.htm#tab-6

Caseload Management

The job of a probation officer is stressful and places numerous and diverse demands upon the professional working in such a role. The workload can be difficult to quantify since much of the time that is allocated to various functions may not always be easy to truly understand or operationalize. Nevertheless, the need to quantify expectations has resulted in an analysis of community supervision caseloads. The main considerations involved with such a formal analysis are the number of offenders and the type of offenders on one's caseload. It should be clear that if community supervision officers are stretched too thin among the various offenders being supervised, the safety of the public is compromised. Table 5.2 provides an overview of the recommended caseload for probation officers, depending on the type of offenders being supervised.

The American Probation and Parole Association (APPA) has attempted to identify the ideal caseload for community supervision officers. The first official attempt to address this issue occurred in the early 1990s, when a paper issued by the APPA recommended that probation and parole agencies examine staffing needs and caseload size within their own organizations (American Probation and Parole Association, 1991; Burrell, 2006). Though this seemed to be a reasonable recommendation, it has been much harder to implement than might initially have been imagined. The quest to determine the ideal caseload size has been a tricky one that has been complicated by multiple factors that are difficult to resolve and/or include in any specific equation.

When considering prior attempts to reduce caseloads, a consensus model slowly emerged throughout the nation (Burrell, 2006). This was the result of input from experienced and thoughtful practitioners in the field of community supervision (Burrell, 2006). Though not necessarily ideal for all agencies, these generally

TABLE 5.2

Recommended Caseload Sizes When Considering Type of Offender Case

ADULT CASELOAD STANDARDS	
CASE TYPE	**CASES-TO-STAFF RATIO**
Intensive	20:1
Moderate to high risk	50:1
Low risk	200:1
Administrative	No limit? 1,000?
JUVENILE CASELOAD STANDARDS	
CASE TYPE	**CASES-TO-STAFF RATIO**
Intensive	15:1
Moderate to high risk	30:1
Low risk	100:1
Administrative	Not recommended

Source: Kaeble, D., Maruschak, L.M., Bonczar, T. P. (2015). *Probation and parole in the United States, 2014.* Washington, DC: Bureau of Justice Statistics.

agreed-upon recommendations provided a baseline from which other agencies can operate, comparing and modifying their own operations against the backdrop of the consensus that has emerged. Classifying offenders on these relevant criteria is critical since it ensures that offenders are correctly matched with the level of supervision necessary to optimize their potential for completing their community supervision requirements (Burrell, 2006; Hanser, 2007). It is important to note that as of 2018, there have been no further official updates from the APPA on probation officer caseloads. In fact, despite the concern for probation officer welfare, no significant nationwide standards have been developed beyond the APPA report by Burrell published in 2006.

The "evidence suggests that staff resources and services should be targeted at intensive and moderate- to high risk cases, for this is where the greatest effect will be had. Minimal contacts and services should be provided to low risk cases" (Burrell, 2006, p. 7). This reallocation of staff would shift supervision to higher-risk offenders and away from those who are low risk (Burrell, 2006). It is in this manner that community supervision caseloads can be structured to optimize overall public safety and also support the reintegrative aspects that serve as the basis of any correctional system that truly seeks to correct criminal behavior.

PROBATION REVOCATION

This discussion is intended to present the use of revocation as a sanction and a component of the probation process in circumstances where offenders are not able to complete their initially given probation sentence. Previous research demonstrates that roughly 33% of all probationers fail to complete the initial requirements of their probation (Herberman & Bonczar, 2015). However, some areas of the nation are more prone to probation revocation than others. Certain counties and/or communities may be more criminogenic in nature and will therefore tend to have more offending as well as more serious offenders processed through the local justice system. In such areas, it should not be surprising that probation departments will generate higher rates of revocation proceedings.

Generally, revocation proceedings are handled in three stages. First, the **preliminary hearing** examines the facts of the arrest to determine if probable cause exists for a violation. Second, the **hearing stage** allows the probation agency to present evidence of the violation while the offender is given the opportunity to refute the evidence provided. Though the agency (or the local government) is not obligated to provide an attorney, the offender does have the right to obtain legal representation, if he or she should desire. Third, the **sentencing stage** is when a judge requires either that the offender be incarcerated or, as in many cases where the violation is minor, that the offender continue his or her probation sentence but under more restrictive terms.

Lastly, it is not uncommon for offenders to have some sort of hearing or proceeding throughout their term of probation. The longer the period of probation, the more likely this is to happen. Many offenders do eventually finish their probation terms. For those offenders who do, termination of the sentence then occurs (see Table 5.3 for additional data on completion rates). In 2016, 50% of the 1,928,687 probationers who exited supervision were discharged because they either completed their term of supervision or received an early discharge. These offenders are free in society without any further obligation to report to the justice system. It is at this point that their experience with community corrections ends, presuming that they lead a conviction-free life throughout the remainder of their days.

Court Decisions on Revocation

Essentially, there are two primary cases that established due process rights for probationers. The first was *Morrissey v. Brewer* (1972), which dealt with revocation proceedings for parolees, not probationers. However, this case was followed by another Supreme Court case, *Gagnon v. Scarpelli* (1973), which extended the rights afforded to parolees under *Morrissey* to offenders on probation as well.

The *Morrissey* court ruled that parolees facing revocation must be given due process through a prompt informal inquiry before an impartial hearing officer. The Court

Preliminary hearing: Initial examination of the facts of the arrest to determine if probable cause does exist for a violation.

Hearing stage: Stage of a revocation proceeding that allows the probation agency to present evidence of the violation, which the offender is given the opportunity to refute.

Sentencing stage: When a judge determines if the offender will be incarcerated or continue his or her probation sentence under more restrictive terms.

TABLE 5.3

Offenders Who Completed Probation and Those Who Did Not

RATE OF PROBATION EXITS, BY TYPE OF EXIT, 2010–2015						
TYPE OF EXIT	2010	2011	2012	2013	2014	2015
Total exit rate[a]	55	54	52	54	55	53
Completion	36	36	36	36	35	33
Incarceration[b]	9	9	8	8	8	8
Absconder	1	1	1	1	1	1
Other unsatisfactory[c]	6	5	5	6	7	7
Other[d]	2	2	2	2	2	2

Source: Bureau of Justice Statistics (2016). *Annual Probation Surveys, 2010–2015.* Washington, DC: U.S. Department of Justice.

Note: Rates are per 100 probationers. Detail may not sum to total due to rounding. Rates based on most recent data and may differ from previously published statistics.

[a]The ratio of the number of probationers exiting supervision during the year to the average daily probation population (i.e., average of the January 1 and December 31 populations within the reporting year). Includes 1 per 100 probationers or fewer who were discharged to custody, detainer, or warrant; 1 per 100 who were transferred to another probation agency; and less than 0.5 per 100 who died.

[b]Includes probationers who were incarcerated for a new offense and those who had their current probation sentence revoked (e.g., violating a condition of supervision).

[c]Includes probationers discharged from supervision who failed to meet all conditions of supervision, including some with only financial conditions remaining, some who had their probation sentence revoked but were not incarcerated because their sentence was immediately reinstated, and other types of unsatisfactory exits. Includes some early terminations and expirations of sentence.

[d]Includes, but not limited to, probationers discharged from supervision through a legislative mandate because they were deported or transferred to the jurisdiction of Immigration and Customs Enforcement; transferred to another state through an interstate compact agreement; had their sentence dismissed or overturned by the court through an appeal; had their sentence closed administratively, deferred, or terminated by the court; were awaiting a hearing; or were released on bond.

required that this be through a two-step hearing process. The reason for this two-step process is to first screen for the reasonableness of holding the parolee since there is often a substantial delay between the point of arrest and the revocation hearing. This delay can be costly for both the justice system and the offender if it is based on circumstances that do not actually warrant full revocation. Specifically, the Court stated that some minimal

> inquiry should be conducted at or reasonably near the place of the alleged parole violation or arrest and as promptly as convenient after arrest while information is fresh and sources are available. . . . Such an inquiry should be seen as in the nature of a "preliminary hearing" to determine whether there is probable cause or reasonable ground to believe that the arrested parolee has committed acts that would constitute a violation of parole conditions. (p. 485)

The Court also noted that this would need to be conducted by a neutral and detached party (a hearing officer), though the hearing officer did not necessarily need to be affiliated with the judiciary, and this first step did not have to be formal in nature. The hearing officer is tasked with determining whether there is sufficient probable cause to justify the continued detention of the offender.

The initial hearing is then followed by the revocation hearing. During the revocation hearing, the parolee is entitled to contest the charges and demonstrate that he or she did not violate any of the conditions of his or her parole. If it should turn out that the parolee did, in fact, violate his or her parole requirements but that this violation was necessary due to mitigating circumstances, it may turn out that the violation does not warrant full revocation. The *Morrissey* Court specified additional procedures during the revocation process, which include the following:

1. Written notice of the claimed violation of parole
2. Disclosure to the parolee of evidence against him or her

3. An opportunity to be heard in person and to present witnesses and documentary evidence

4. The right to confront and cross-examine adverse witnesses

5. A "neutral and detached" hearing body, such as a traditional parole board, members of which need not be judicial officers or lawyers

6. A written statement by the fact finders as to the evidence relied on and reasons for revoking parole

PHOTO 5.4 The probation officers are taking this probationer into custody due to his continued use of drugs and alcohol while on community supervision.

AP Photo/Rich Pedroncelli

The *Morrissey* case is obviously an example of judicial activism, much like *Miranda v. Arizona* (1966), that has greatly impacted the field of community corrections. The Court's clear and specific guidelines set forth in *Morrissey* have created specific standards and procedures that community supervision agencies must follow. Rather than ensuring that revocation proceedings include a just hearing and means of processing, the Court laid out several pointed requirements that continue to be relevant and binding to this day.

The next pivotal case dealing with revocation proceedings and community supervision is *Gagnon v. Scarpelli* (1973). In the simplest of terms, the Court ruled that all of the requirements for parole revocation proceedings noted in *Morrissey* also applied to revocation proceedings dealing with probationers. However, this case is also important because it addressed one other key issue regarding revocation proceedings. The Court noted that offenders on community supervision do not have an absolute constitutional right to appointed counsel during revocation proceedings. Such proceedings are not considered to be true adversarial proceedings and therefore do not require official legal representation.

Common Reasons for Revocation

There are a number of reasons that offenders may have their probation revoked. Perhaps the most frequent reason that revocation hearings are initiated is due to a probationer's failure to maintain contact with his or her probation officer (Glaze & Bonczar, 2011). Of those probationers who experience a disciplinary hearing, the most frequent reason tends to be absconding or failing to contact their probation officer (Glaze & Bonczar, 2011). Other reasons may include an arrest or conviction for a new offense, failure to pay fines/restitution, or failure to attend or complete an alcohol or drug treatment program (Glaze & Bonczar, 2011). Among probationers who have revocation hearings initiated against them, almost half are generally permitted to continue their probation sentence. For those who are allowed to continue, they will almost always have additional conditions imposed upon them, and their type of supervision will typically be more restrictive (Glaze & Bonczar, 2011).

Further complicating the picture is that some conditions of probation result in what are often called technical violations. **Technical violations** are actions that do not comply with the conditions and requirements of a probationer's sentence, as articulated by the court that acted as the sentencing authority. Technical violations are not necessarily criminal, in and of themselves, and would likely be legal behaviors if the offender were not on probation. For instance, a condition of a drug offender's probation may be that he or she stay out of bars, nightclubs, and other places of business where the selling and consumption of alcohol is a primary attraction. Another example might be if a sex offender is ordered to remain a certain distance from schools. For most citizens, going to nightclubs and/or setting foot on school grounds is not a violation of any sort, and neither of these acts is considered criminal. However, for the probationer, this can lead to the revocation of probation.

Any number of other behaviors can be technical violations. Additional examples might include the failure to attend mandated therapy, failure to report periods of

Technical violations: Actions that do not comply with the conditions and requirements of a probationer's sentence.

unemployment, or failure to complete scheduled amounts of community service. Though these violations are substantially different from those that carry a new and separate criminal conviction, they still can lead to a revocation (Hanser, 2010b) and are important in demonstrating whether the offender is making genuine progress in the corrections process. Excessive technical violations would seem to indicate that reform is not a priority for an offender.

TYPES OF INTERMEDIATE SANCTIONS

When it comes to finding alternatives to punishment and rehabilitation for offenders, there is no shortage of options, particularly in terms of community-based services. On a simplistic level, intermediate sanctions could be defined as alternatives to traditional incarceration that consist of sentencing options falling anywhere between a standard prison sentence and a standard probation sentence. The National Institute of Corrections (1993) has also defined *intermediate sanctions* as "a range of sanctioning options that permit the crafting of sentences to respond to the particular circumstances of the offender and the offense; and the outcomes desired in the case" (p. 18). For this text, we will use a blend of both definitions. Thus, **intermediate sanctions** are a range of sentencing options that fall between incarceration and probation, being designed to allow for the crafting of sentences that respond to the offender, the offense, or both, with the intended outcome of the case being a primary consideration. Table 5.4 provides an examination of the various intermediate sanctions and sentencing measures.

The definition of intermediate sanctions just presented provides a perspective that is highly consistent with the emphasis on reintegration found in this text. It is clear that these types of sanctions allow for a great deal of flexibility and can be adjusted to accommodate treatment and supervision considerations. These sanctions allow for consideration of the various needs, challenges, and issues associated with a particular offender and the type of offending that he or she is prone to committing. This permits the calibration of sentences so that specific details are a better fit with the type of offense committed as well as the individual variables associated with the offender. Such flexibility is what

Intermediate sanctions: A range of sentencing options that fall between incarceration and probation.

TABLE 5.4

Summary Listing of Coercive Intermediate Sanction Measures and Sentencing Options

WARNING MEASURES (NOTICE OF CONSEQUENCES OF SUBSEQUENT WRONGDOING)	ADMONISHMENT/CAUTIONING (ADMINISTRATIVE; JUDICIAL) SUSPENDED EXECUTION OR IMPOSITION OF SENTENCE
Injunctive measures (banning legal conduct)	Travel (e.g., from jurisdiction to specific criminogenic spots)
	Association (e.g., with other offenders)
	Driving
	Possession of weapons
	Use of alcohol
	Professional activity (e.g., disbarment)
	Restitution
Economic measures	Costs
	Fees
	Forfeitures
	Support payments
	Fines (standard; day fines)
	Community service (individual placement; work crew)

WARNING MEASURES (NOTICE OF CONSEQUENCES OF SUBSEQUENT WRONGDOING)	ADMONISHMENT/CAUTIONING (ADMINISTRATIVE; JUDICIAL) SUSPENDED EXECUTION OR IMPOSITION OF SENTENCE
Work-related measures	Paid employment requirements Academic (e.g., basic literacy, GED)
Education-related measures	Vocational training Life skills training Psychological/psychiatric
Physical and mental health	Chemical (e.g., methadone; psychoactive drugs)
Treatment measures	Surgical (e.g., acupuncture drug treatment)
Physical confinement measures	Partial or intermittent confinement: Home curfew Day treatment center Halfway house Restitution center Weekend detention facility/jail Full/continuous confinement: Outpatient treatment facility (e.g., drug/mental health) Full home/house arrest Mental hospital Other residential treatment facility (e.g., drug/alcohol) Boot camp Detention facility Jail Prison
Monitoring/compliance measures (may be attached to all other sanctions)	Required of the offender: Electronic monitoring (telephone check-in; active electronic monitoring device) Mail reporting Face-to-face reporting Urine analysis (random; routine) Required of the monitoring agent: Sentence compliance checks (e.g., on payment of monetary sanctions; attendance/performance at treatment, work, or educational sites) Criminal records checks Third-party checks (family, employer, surety, service/treatment provider; via mail, telephone, in person) Direct surveillance/observation (random/routine visits and possibly search; at home, work, institution, or elsewhere) Electronic monitoring (regular phone checks and/or passive monitoring device—currently used with home curfew or house arrest, but could track movement more widely as technology develops)

Source: National Institute of Corrections. (1993). *The intermediate sanctions handbook: Experiences and tools for policymakers.* Washington, DC: Author.

provides the field of community corrections with its greatest source of leverage among the offender population, in terms of treatment and supervision.

This chapter will discuss the most commonly implemented and researched intermediate sanctions of intensive supervision, day reporting centers, day fines, home detention, electronic monitoring, shock incarceration, boot camps, community service, and various methods of compliance assurance. Due to the overcrowding problems in prison systems around the nation, it is clear that there is simply a pragmatic need for more space. As a result of the public's outcry for increased traditional sentencing of offenders and the legislative action that has responded to this demand, the use of intermediate sanctions has grown. Recent reports indicate that many parolees and probationers are supervised under some form of intermediate sanctioning, and this trend is likely to increase in the future (Mitchell, 2011; Office of Program Policy Analysis and Governmental Accountability, 2010). While we must remember that the applicability of such programming is dependent upon the potential for harm in local communities and the ability of intermediate sanctions to reduce recidivism, there is little doubt that in today's era of state-level budget cuts, many prison systems are using intermediate sanctions as alternatives to imprisonment (Mitchell, 2011). Students should refer to Table 5.5 to see the cost difference between intermediate sanctions and prison terms.

Fines

Most offenders convicted of a criminal offense are assessed a fine as a punishment for committing the offense. A fine can be defined as a monetary penalty imposed by a judge or magistrate as a punishment for having committed an offense. In most cases, the fine is a certain dollar amount established either by the judge or according to a set schedule, depending upon the offense committed. The logic behind the fine is that it will deter the offender from committing another offense in the future for fear of being fined again. In most jurisdictions, the fines are assessed and paid in monthly installments to the receiving agency. In contemporary community supervision agencies, offenders are now able to pay their fines via credit or debit cards. The feelings are mixed regarding allowing an

TABLE 5.5

Intermediate Sanctions Are Less Costly Than Incarceration

INTERMEDIATE SANCTION	DAILY COST PER OFFENDER	ANNUAL COST PER OFFENDER	TOTAL ANNUAL COST FOR 100 OFFENDERS	POTENTIAL ANNUAL SAVINGS PER 100 OFFENDERS
Prison—minimum security[a]	$79.46	$29,002.90	$2,900,200	$0
Intermediate probation or parole supervision	$4.85	$1,770.25	$177,000	$2,723,200
Electronic monitoring with GPS[b]	$11.08	$4,044.20	$404,400	$2,495,800
Day reporting centers[c]	$15.29	$5,580.85	$558,000	$2,341,400
Residential drug treatment[d]	$64.51	$23,546.15	$2,354,600	$545,600

Source: North Carolina Department of Public Safety. (2017). *Cost of corrections.* Raleigh: Retrieved from https://www.ncdps.gov/Adult-Corrections/Cost-of-Corrections

[a]Note that minimum security is the least expensive security level of incarceration. Medium security is $89.78 per day and Close Custody is $106.92 per day.

[b]This sanction includes and subsumes probation or parole supervision, as well.

[c]Day reporting centers provide services that include assessments, screenings, counseling, alcohol and drug treatment programs/services, educational programs/services, vocational programs/services, and employment programs/services.

[d]A 90-day residential substance abuse treatment facility.

offender to pay in this manner. Officers sometimes feel that by delaying the effect of the monetary fine, it does not allow the offender to accept personal responsibility.

As the offense seriousness increases from misdemeanor to felony, presumably the fines increase as well. This assessment of fines is totally dependent upon judicial discretion. Traditionally, there was one set fine for certain offenses regardless of the financial standing of the offenders. As time has passed, many judiciaries have begun to understand that one set fine is more punishing to the offender who happens to earn the least amount of money and a "cakewalk" to those offenders who happen to be financially blessed. Given this, there has been a push for graduated fines that are dependent upon the income of the offender at time of sentencing. In this day of drastic budget cuts and pushes for alternatives to incarceration, fines are frequently used. They allow the offender an opportunity to pay for his or her treatment and punishment as opposed to the traditional method of placing the financial burden on the state.

Community Service

Perhaps the most widely known yet least likely to be used form of intermediate sanctioning is community service. Community service is the work that one is required to perform in order to repay his or her debt to society after being found at fault for committing a criminal or deviant offense.

Community service serves a dual purpose: to rehabilitate and punish offenders. In terms of rehabilitation, community service affords the offender the opportunity to participate in something constructive, allowing him or her to build "sweat equity" in something that is beneficial to the community. Community service is punitive, too, in that the offender is forced to give up his or her own time to work off a criminal debt without being paid. Due to the low cost of overhead in funding community service programs and the need for labor in communities, community service options will continue to be frequently utilized. One of the most pressing problems in evaluating community service is that such opportunities vary so widely, and often offenders participate in multiple community service sites and types throughout their time under supervision. In most cases, community service is completed anywhere the offender can get the hours. For example, an offender may begin community service at the local courthouse and complete his or her hours at the homeless mission. Despite the lack of research in this area, community service is clearly an integral part of intermediate sanctioning and provides a positive avenue through which offenders and the community can learn the rehabilitative and punitive ideals.

Intensive Supervision Probation

Perhaps the most commonly known form of intermediate sanction is **intensive supervision probation (ISP)**. ISP is the extensive supervision of offenders who are deemed the greatest risk to society or are in need of the greatest amount of governmental services (e.g., drug treatment). In most cases, ISP is the option afforded to individuals who would otherwise be incarcerated for felony offenses. Early forms of ISP operated under the conservative philosophy of increasing public safety via strict offender scrutiny. Today's ISP programs are focused on a host of components.

When examining the various facets of modern-day intensive supervision, one finds that they are quite diverse. Some ISPs focus on specific offense and offender types (e.g., sex offending and younger offenders). Others differ in their level of supervision, which may range from 5 days per week to once every 2 weeks. The types of supervising officers vary from untrained community supervision officers to specialized officers who have been well schooled in the supervision of at-risk offenders. In most cases, officers are afforded a lighter than normal caseload of approximately 10 to 20 offenders. Placement into intensive supervision is dependent upon

Intensive supervision probation (ISP): The extensive supervision of offenders who are deemed the greatest risk to society or are in need of the greatest amount of governmental services.

PHOTO 5.5 Some offenders must complete community service as part of their sentence. Offenders may be given community supervision or a split sentence (which includes both jail time and community supervision). Both of these sanctions usually require the completion of some type of labor, as seen here with this offender, who is preparing to do yard work.

Robert Hanser

the sentencing judge, the supervision officer, the parole board, or some combination of these. In today's agencies, the decision to place an offender on intensive supervision is made based on the level of assessed offender risk of rearrest.

There is research that shows that ISP programs can be particularly effective. Indeed, one study was designed to evaluate the use of ISP with DWI offenders in three different states (Wiliszowski, Fell, McKnight, Tippetts, & Ciccel, 2010). This study provided evidence that strongly supported the use of ISPs with the following common features:

- Screening and assessment of offenders for the extent of their alcohol/substance abuse problem

- Relatively long-term, close monitoring and supervision of the offenders, especially for alcohol and other drug use or abuse

- Encouragement by officials to successfully complete the program requirements

- The threat of jail for noncompliance

ISPs are an alternative to jail, which is very costly. Offenders who remain out of jail can be employed and can contribute to society and the well-being of their families. In some ISPs, offenders who remain out of jail pay some or all of the costs of their participation in the ISP. Wilisowsky et al. (2010) note that the monitoring of alcohol and other drug use by offenders to maintain abstinence is very effective in reducing offender recidivism. They note that there are several methods utilized to monitor alcohol use by offenders:

- Frequent contact by probation officers, the judge, or other officials (observation)

- Surprise visits in the home and blood alcohol content (BAC) testing (and sometimes drug testing via a urine sample)

- Daily call-in with random testing (sometimes the offender must report for a test, sometimes not)

- Electronic monitoring and home confinement with remote BAC testing (e.g., Sobrietor)

- Use of the alcohol ignition interlock record of the offender

- Regularly scheduled testing (e.g., twice/day like the 24/7 program; p. 385)

Thus the cost/benefit of ISP programming seems to be effective in reducing rearrests with alcoholics. It is important to understand that each rearrest for a DWI offense costs thousands of dollars in law enforcement resources, court costs, and jail sanctions. Most importantly, it is the public that stands to benefit the most, in terms of savings due to preventing potential injuries and deaths due to the reduction in DWI recidivism. Thus, ISP provides benefits both behind bars and in the outside society.

Electronic Monitoring

Electronic monitoring:
The use of any mechanism worn by the offender for the means of tracking his or her whereabouts through electronic detection.

Perhaps the most widely used but least understood intermediate sanction is electronic monitoring. **Electronic monitoring** includes the use of any mechanism that is worn by the offender for the means of tracking his or her whereabouts through electronic detection. Electronic monitoring includes both active and passive monitoring systems. With both types of electronic monitoring devices, offenders are required to wear an ankle bracelet with a tracking device. The active system types are used in conjunction with the local telephone line. At random times throughout the day, the offender's home phone will ring, and the offender has a certain amount of time to answer. Once the offender answers the phone, a signal is transmitted via the tracking device, which validates that the offender is at home. With the passive system, the ankle bracelet transmits a continuous signal to a nearby transmitter, which transmits the signal to a monitoring computer. With each type, the supervising officer is sent a readout each morning of the offender's compliance.

FIGURE 5.6

Electronic Monitoring and Supervision Costs Much Less Than Prison

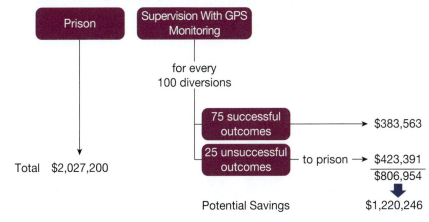

Prison

Supervision With GPS Monitoring

for every
100 diversions

75 successful outcomes → $383,563

25 unsuccessful outcomes — to prison → $423,391
                                        $806,954

Total $2,027,200

Potential Savings $1,220,246

Source: Office of Program Policy Analysis and Governmental Accountability. (2010). *Intermediate sanctions for non-violent offenders could produce savings.* Tallahassee, FL: Author.

Note: Unsuccessful exits are program outcomes that denote noncompliance with program requirements and result in termination from the program. For the purpose of estimating cost savings in this report, all offenders that unsuccessfully exit are assumed to be sent to prison.

If an offender attempts to alter the connection, most devices have alarms that will sound and send an immediate message to the monitor. It is also not uncommon for an offender to be at home, not having tampered with the ankle bracelet, and yet it appears that he or she is noncompliant. Of the two electronic monitoring types, the active system has the lower rate of false alarms. But, even though the passive system has the higher rate of false alarms, it is the fastest in determining noncompliance because it assesses the offender's whereabouts much more quickly.

Those in favor of electronic monitoring base their argument on the ability of the system to increase public safety due to the knowledge of the whereabouts of each offender. Simply knowing they are being personally tracked may deter offenders from committing crime. Proponents also argue that electronic monitoring provides the least punitive alternative to incarceration, as it allows for offenders to be supervised in the community. Undoubtedly, electronic monitoring is a fiscally feasible intermediate sanction in terms of saving money while diverting offenders from incarceration (see Figure 5.6). However, questions remain as to the ability of electronic monitoring to assist offenders in the desistance from criminal behavior, whether supervision officer discretion helps or hurts the outcome, and the effects of community and family involvement on electronic monitoring.

Global Positioning Systems

In this new millennium, community supervision is beginning to utilize military capabilities to keep track of offenders. A Global Positioning System (GPS) receiver uses 24 military satellites to determine the exact location of a coordinate. As we have seen earlier in this chapter, GPS tracking devices allow supervision officers to detect when an offender violates one of his or her restrictions on movement due to a condition of his or her supervision.

PHOTO 5.6 An offender on supervision is being fitted with an ankle bracelet that will be used for tracking purposes.

© ZUMA Press, Inc / Alamy Stock Photo

PHOTO 5.7 The equipment in this photo is used for GPS tracking. In some cases, equipment has specialized functions, such as the emission of alarms or noises that can be sounded by a community supervision officer from a distance. When an offender cannot be found physically or when he or she enters a restricted or off-limits area, an alarm can be emitted by the probation officer merely pushing a button, even from a distance of several miles. The noise can be deafening and serves as a deterrent for most offenders and as a warning to community members.

Robert Hanser

Home detention:
The mandated action that forces an offender to stay within the confines of his or her home for a specified time.

Despite the advances that GPS technology has provided for community corrections, there are also some disadvantages. GPS tracking devices often lose their signal during bad weather or in areas densely populated with trees. Further, this equipment is expensive and can be cost prohibitive for smaller, less affluent agencies. Because of this, only a small minority of offenders are tracked by GPS technology when compared with the entire offender population under community supervision. Nevertheless, when used in a strategic manner, this feature does provide an extra level of maintenance for those offenders where it is deemed appropriate. Table 5.6 provides a comparison of some advantages and disadvantages of GPS electronic monitoring.

Home Detention

Also known as house arrest, **home detention** is the mandated action that forces an offender to stay within the confines of his or her home or on the property until a time specified by the sentencing judge. It is the Father of Modern Science, Galileo, who provides an example of the first offender placed on home detention; after he proposed that the earth rotated around the sun, he was confined to his villa by the Roman Inquisition until his death. However, it wasn't until the late 20th century's War on Drugs that home detention gained notoriety. As a result of the massive numbers of drug offenders being sentenced to jail or prison, officials were seeking an alternative to supervision that would allow an offender to be supervised prior to trial or just before being placed into a residential treatment facility.

Many offenders sentenced to home confinement are required to complete community service and pay a host of fines, fees, and victim restitutions, while others are forced to wear electronic monitors or other detection devices to ensure that they are remaining in their residence during the specified time. In many cases, home detention is used for

---

TABLE 5.6

Advantages and Disadvantages of GPS Electronic Monitoring

	ADVANTAGES	DISADVANTAGES
Active GPS systems	• Seek to alleviate prison overcrowding • Immediate response capability • Data reporting in near-real time	• High daily cost • Reliance on wireless data service coverage • Labor intensive • Require immediate agency response • Greater agency liability • Tracing device size and weight
Passive GPS systems	• Small, lightweight device • Can be independent of wireless data services • Lower daily cost • Less labor intensive	• "After-the-fact" tracking data • No immediate notification of zone violations

Source: International Association of Chiefs of Police. (2008). *Tracking sex offenders with modern technology: Implications and practical uses with law enforcement.* Alexandria, VA: Author.

offenders during the pretrial phase or just prior to an offender being let out of prison on a work or educational release program. If an offender leaves his or her residence without permission or against the policies set forth, the offender is seen as having technically violated the conditions of his or her supervision (Government Accounting Office, 1990). Home detention can also provide sufficient deterrence for some offenders who may contemplate returning to crime. Rather than seeing this type of sanction as being easy on the offender, it is best to view it as commensurate with the type of crime that was committed.

Day Reporting Centers

Day reporting centers are treatment facilities to which offenders are required to report, usually on a daily basis. These facilities tend to offer a variety of services, including drug counseling, vocational assistance, life skills development, and so forth. The offenders assigned to day reporting centers are generally one of two types: those placed on early release from a period of incarceration or those on some form of heightened probation supervision. For those released early from a jail or prison term, the day reporting center represents a gradual transition into the community during which they are supervised throughout the process.

The advantage of day reporting centers is that they do not require the use of bed space and therefore save counties and states a substantial portion of the cost in maintaining offenders in their custody (see Figure 5.7). For the purpose of estimating cost savings in this report, all offenders that unsuccessfully exit are assumed to be sent to prison.

Day reporting centers are similar to residential treatment facilities except that offenders are not required to stay overnight. In some jurisdictions, the regimen of the day reporting center is designed so that offenders attend 8- to 10-hour intervention and treatment classes. This is an important element of the day reporting center since it provides added human supervision. The implementation of creative and versatile forms of human supervision serves to optimize both treatment and security characteristics of offender supervision, and day reporting centers facilitate this concept. Indeed, one staff person conducting some form of instruction class (e.g., a life skills class, a psychoeducational class on effective communication, or perhaps a parenting class) can essentially watch over several offenders at the same time. Further, the offender's time is spent in prosocial activities with little opportunity to engage in any form of

Day reporting centers: Treatment facilities to which offenders are required to report, usually on a daily basis.

FIGURE 5.7

Day Reporting Centers Are Less Costly Than Prison

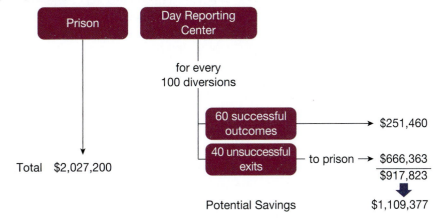

Source: Office of Program Policy Analysis and Governmental Accountability. (2010). *Intermediate sanctions for non-violent offenders could produce savings.* Tallahassee, FL: Author.

Note: Successful completions in substance abuse treatment programs are used as a proxy for successful completions in the day reporting program. Unsuccessful exits are program outcomes that denote noncompliance with program requirements and result in termination.

undetected criminal activity. Thus, day reporting centers enhance security processes while filling up the leisure times in an offender's day or evening with activities that are constructive and beneficial to the offender and to society; little time is left for distractions or unregulated activity.

CONCLUSION

This chapter illustrates the importance of probation as a sanction within the correctional system. Whether one is considering community- or institution-based corrections, the use of probation affects the overall correctional system quite significantly. This is especially true when one considers the impact that probation has on the jail facility. Without probation, jail facilities would be even more overcrowded than they already tend to be. The true purpose of probation, however, is to facilitate the reformation of offenders. This has been the case since its earliest inception, when John Augustus first established this sanction in the United States. Thus, students should consider probation to have reformative value. This means that the purpose of probation is consistent with this text's definition of corrections, whereby the ultimate goal is to correct criminal tendencies among the offender population.

This chapter also highlights the importance of the presentence investigation and the report it generates for judges who sentence offenders and for later issues that may arise during an offender's sentence. The PSI report is even used in institutional settings when making early release, treatment-planning, and custody-level decisions. Due to the impact the PSI report can have on the correctional process as a whole, probation officers tend to spend a significant amount of time constructing these reports and/or referring to the information contained therein.

Probation officer qualifications and standards of training were also discussed. It is clear that the job of probation officer is not an easy one. This job requires a degree in most cases but pays little in comparison to other professions. Further, the work is stressful given the high caseloads that tend to exist for many probation officers. Qualifications and characteristics of probation officers demonstrate that this area of correctional employment may be quite different from institutional corrections and that this dimension of the correctional system has been underappreciated.

Further, revocation procedures were covered in detail. Leading cases related to revocation for probation and the corresponding procedures for revocation were discussed. The process of revocation is structured to ensure that capriciousness does not occur when determining whether an offender should or should not remain on probationary status. The issue of technical violations, as opposed to those related to new criminal convictions, was also touched on, illustrating that the types of infractions can vary considerably depending on the offender and the particular circumstances involved.

Lastly, this chapter has provided an overview of several types of intermediate sanctions that are used around the country. Specific examples have been offered to demonstrate the variety of sanctions that exist and how they are utilized. The flexibility of intermediate sanctions gives community supervision agencies a range of potential responses to offender criminal behavior. These responses fall along a continuum according to the amount of liberty that is denied the offender. These penalties vary by level of punitiveness to allow community supervision agencies to calibrate the offender's punishment with the severity of the specific offense committed and/or the offender's tendency toward recidivism. In addition, intermediate sanctions help to connect the supervision process with the treatment process. Various intermediate sanctions, such as community service and the payment of fines, create a system of restoration, while the use of flexible supervision schemes, such as electronic monitoring and GPS tracking, allows the offender to engage in employment activities. Other programs, such as day reporting centers, ease the transition of offenders from incarceration to community membership and provide a series of constructive activities to ensure that the offender remains on task with respect to his or her reintegration process. Each of these sanctions can be used in conjunction with others to further augment the supervision process, all the while being less expensive and more productive than a prison term.

Want a Better Grade?

Get the tools you need to sharpen your study skills. Access practice quizzes, eFlashcards, video, and multimedia at **edge.sagepub.com/hanserbrief**

Interactive eBook

Visit the interactive eBook to watch SAGE premium videos. Learn more at **edge.sagepub.com/hanserbrief/access.**

 Career Video 5.1: Probation Officer

 Prison Tour Video 5.1: Electronic Monitoring

 Prison Tour Video 5.2: Day Reporting Centers

DISCUSSION QUESTIONS

Test your understanding of chapter content. Take the practice quiz at edge.sagepub.com/hanserbrief.

1. Who is the Father of Modern Probation, and how did he initially implement the practice of probation?
2. How does probation impact the jail and prison systems of a jurisdiction?
3. Discuss the means by which probation agencies might be organized.
4. Why is the presentence investigation report so important?
5. Identify and discuss the qualifications and characteristics of most probation officers.
6. Discuss the importance of *Morrissey v. Brewer* (1972) and *Gagnon v. Scarpelli* (1973).
7. How does critical criminology dovetail with probation supervision?

KEY TERMS

Review key terms with eFlashcards at edge.sagepub.com/hanserbrief.

Day reporting centers, 139

Electronic monitoring, 136

Hearing stage, 129

Home detention, 138

Intensive supervision probation (ISP), 135

Intermediate sanctions, 132

Passive agent, 126

Paternal officer, 125

Preliminary hearing, 129

Presentence investigation report, 119

Probation, 114

Probationers, 117

Punitive officer, 126

Sentencing stage, 129

Technical violations, 131

Welfare worker, 126

APPLIED EXERCISE 5.1

Read each of the case scenarios below, and select the type of intermediate sanction that you think is best suited for each offender. The list of possible sanctions is provided below. Once you have made your selection, write a 50- to 150-word essay for each scenario that explains why you chose a particular intermediate sanction or combination of sanctions.

Students should remember that intermediate sanctions operate on a continuum, and they will need to ensure that the sanction they choose is proportional to the offense. Similarly, students should not overpredict the likelihood of an individual committing a future crime. Total word count for this assignment is approximately 500 to 1,500 words.

Grading Rubric (This assignment is worth a maximum total of 100 points.)

1. Student provides a reasonable match between scenario and intermediate sanction(s). (Each match is worth 3 points.)

2. Student provides adequate justification for each match between scenario and intermediate sanction(s). (Each essay justification is worth 7 points.)

Applied Exercise Case Scenarios

SCENARIO #1: A male juvenile who constantly sneaks out of the house despite his parents' attempts to prevent him. He leaves home at night to meet friends, use drugs, and commit acts of vandalism.

SCENARIO #2: A female offender who "keyed" the car of a neighbor who kept parking on the curb nearest to her own side of the street.

SCENARIO #3: A male delivery driver who is on community supervision for failing to appear in court for proceedings related to thefts in a neighborhood.

SCENARIO #4: A male gang offender who physically assaulted a man who smarted off to him.

SCENARIO #5: A female drug addict who has been busted for prostitution.

SCENARIO #6: A young male who has been committing petty acts of vandalism.

SCENARIO #7: A woman convicted of writing hot checks.

SCENARIO #8: A male teenager who made threatening prank calls to various people in the community.

SCENARIO #9: A middle-age male who continues to drive while drunk. This is his second DWI offense. He has never been to prison.

SCENARIO #10: Several young people who were playing pranks on elderly people in the community. They egged multiple homes and caused some light property damage to porch lights and other such components of victims' homes.

Applied Exercise Intermediate Sanction Choices

A. Intensive supervised probation/parole

B. GPS tracking and probation

C. Standard probation and restitution

D. Home detention and electronic monitoring

E. Standard probation and community service

F. Shock incarceration, probation, and community service

G. Restitution

H. Community service

I. Day reporting center and the use of ISP

J. Standard probation

K. Home detention, electronic monitoring, and ISP

L. Drug court with ISP

NOTE: Students may use any of the above options more than once to apply to the scenarios provided. Likewise, there is no requirement that students use every one of the choices presented above.

You are a probation officer in a small rural agency. The chief probation officer explains to you that services have not been optimal and that many offenders are having difficulty finding employment. Further, many of them cannot travel the distances that are required to make their appointments. The local jail is full to capacity, and the state is not taking low-risk offenders. Your supervisor explains that this issue is not one of offenders purposefully failing to meet the conditions of probation but is instead one where they are simply unable to meet these conditions.

Your supervisor has established a steering committee made up of you, a member of the social services office, a person from city hall, a deputy from the sheriff's office, and a representative from the regional hospital. She wants you to come up with some sort of action plan that will allow you to aid persons on community supervision in finding employment, making therapeutic and medical meetings, and meeting the conditions of their restitution requirements. She tells you she knows that it is a difficult task, particularly since your jurisdiction spans a large tri-county area with a total population of 25,000 people. Though resources are scant, she encourages you to take this commission seriously and notes that she will provide any support that is possible. In the meantime, she has lightened your caseload to make your new task more manageable.

What would you do?

©iStockphoto.com/Gatsi

6 Facility Design and Classification in Jails and Prisons

Learning Objectives

1. Identify the different types of prison facility designs, including supermax facilities.

2. Identify key legal precedents related to facility design, supermax facilities, and classification systems.

3. Discuss the various features that improve the perimeter security of an institution.

4. Evaluate the strategies used to improve internal security.

5. Analyze technological developments in prison security.

6. State the main goals of classification systems.

7. Identify different custody levels and the methods used for housing assignments and reclassification.

8. Describe the different categories of special housing assignments.

9. Discuss how classification processes are central to prison service operations.

The Challenges of Design and Security

Warden Thompson walked the long hallway that ran down the middle of the prison, nodding his head as inmates passed by on their way from breakfast, some of them giving him a friendly "Mornin', warden." Exiting the prison, he made his way to the sidewalk outside and looked up at the new camera that had been installed to observe inmate traffic. The camera was positioned so that it would be able to observe movement in an area of the external part of the building where an alcove existed.

This "nook," as this recess had been informally named by officers and inmates alike, was an area impossible to observe by security from the towers and also difficult to see by officers on the ground from a distance. In order to observe the area, one had to stand right in front of it. This was a poor design feature that was not originally part of the floor plan for the prison. Though this largely brick-and-mortar structure was over 70 years old, it was not designed with many of these types of blind spots. However, a decade ago, an addition to this wing was built, and an oversight led to an area that was not closed in, leaving a gap that was about 20 feet by 10 feet. This space was not observable by security, and no cameras had been installed that covered it.

Initially, this was not such a big problem, as it was not an area from which an inmate could attempt any type of escape. It also was not an area where inmates were housed, so they did not have many opportunities to loiter there. However, the inmates began using it as a location to meet each other and potentially exchange contraband. This section of the facility was near a storage area for cleaning supplies; thus some trusty inmates passed near the nook. Trusties tend to have more mobility in prisons and, as a result, can more easily engage in trafficking and trading throughout the facility.

Later that afternoon, Thomson hosted a meeting with several regional executives who were visiting the prison. At the meeting, Thomson showed the administrators a PowerPoint presentation on how the amount of trafficked contraband in the facility had gone down in the past 6 months.

"I believe that we have identified a key weakness in our security, and, as a result, our response has led to a 53% reduction in the number of incidents where contraband has been brought into our prison," Thomson stated proudly. "The cameras we installed were hidden from inmate view so we have been able to nab several inmates involved in bringing contraband inside the prison. From there, we were able to get names of outside persons who were smuggling drugs and weapons into the prison and handing them off to inmates. Outside law enforcement has brought charges against those persons."

One of the regional executives asked, "Overall, when we consider the cost of making modern additions rather than rebuilding the entire prison, and when we add these digital security features, are we being cost-effective?"

Thomson smiled. "I knew one of you would ask that." All of the attendees chuckled. "In truth, the figures are close, but, overall, the total bill is actually a bit less than if we were to open a new facility to replace this one. Our projections show this to be true for operational costs that extend out another 10 years. After that, I am not sure. But currently, our security is tighter than ever, and we are running below the costs that would be entailed if we scrapped this facility. For now, that is."

The group discussed all of the factors involved and agreed that the best course of action would be to give the warden a bit more in his budget line to improve features at the facility.

Thomson also suggested some ideas that might offset these costs but could make no guarantees. Ultimately, it was decided that it was cheaper to keep the current facility and that security could continue to be effectively maintained with the additional technological features.

Later that day, Warden Thomson drove home in his truck and thought to himself, "I have another 7 years to retirement. . . . I hope that the next warden of this facility is able to navigate the pushes and pulls of running this facility."

INTRODUCTION

This chapter provides an overview of various physical prison facilities and their design, as well as classification systems that reflect the internal running of the prison. Students should understand that both the physical design of a prison and the internal classification process used are critical to a facility's operation. Nevertheless, it is the physical features of the prison that serve as its most prominent aspect. In addition, the general organizational climate and operations of the prison are greatly impacted by the physical features of the facility. Historical aspects of prison design will be covered. Even more important and also more unique is the discussion on physical facilities related to prison services (kitchens, workshops, religious sections, recreational areas, etc.). This aspect of the facility can be very important in meeting programming and security requirements.

PHOTO 6.1 The Bastille (French for "little bastion") was originally built as a fortress in the late 1300s but was later used as a prison. The Bastille was a four-story stone structure that had eight closely spaced towers connected by a 15-foot-thick stone wall.

Wikimedia Commons

PRISON FACILITY DESIGNS THROUGHOUT HISTORY

Prison construction and development have evolved substantially in the United States. This evolution has corresponded with the needs of correctional managers throughout the ages and the availability of resources for those who provided the construction. In other words, the needs of the persons operating a prison have typically determined how the prison was designed. It is clear from Chapters 1 and 2 that prisons have come in a variety of shapes and sizes—from abandoned quarries to old decommissioned ships to huge monolithic structures, culminating in the high-tech designs of today's correctional environment. Prison construction has gone through phases in order to meet social, penal, and managerial objectives and has been affected by what a jurisdiction is willing or able to finance.

Overall, there have been very few changes in the actual inmate housing unit; the basic human needs for a place to sleep, remove waste, and maintain daily hygiene tend to remain the same. Jail cells, cell blocks, dormitory spaces, and other such units tend to comprise the same basic features. Nevertheless, the form and appeal of these arrangements have changed in terms of cleanliness and/or type of building material used. In this regard, innovations in both health and security issues have advanced the living standards for inmates in the United States.

Without question, the history of prison construction has been impacted by the use of certain well-known facilities. These facilities tend to influence construction in other areas of the world, depending on the correctional era and the corresponding needs of prison administrators. We begin our discussion of these facilities by examining the French prison known as the Bastille. This prison was most famous for its eventual demise during the French Revolution in 1789.

The Bastille

The **Bastille** was a fortification built in the city of Paris and was a symbol of tyranny and injustice for both commoners and political prisoners in France. Citizens were held within its confines for an indefinite period without formal accusation or trial. During its early years, the Bastille was a prison for the upper class, mostly nobles who had been charged with treason. Political and religious prisoners were also held within its walls. By the late 1700s, however, this facility was used to house inmates of every class and profession. Ultimately, the Bastille was attacked and captured by a mob that revolted against the king of France, and it was destroyed in 1789.

PHOTO 6.2
The panopticon prison model, created by Jeremy Bentham in 1785, influenced the design of prisons like Eastern State Penitentiary. This design consists of cells that face each other across a wide circular space with an enclosed observation post at the center of the structure.

Mike Graham from Portland, USA

Originally called the Chastel Saint-Antoine, the Bastille (French for "little bastion") was originally built as a fortress in the late 1300s. It was a four-story stone structure with eight closely spaced towers connected by a 15-foot-thick stone wall. The stone masonry of the fortress was contained within a continuous wall that surrounded Paris. The towers surrounded two enclosed courtyards, and the walls had several windows on each floor (George, 2008). The Bastille was, in many respects, a miniature castle. This facility was the blueprint for the Western State Penitentiary of Pennsylvania, near Pittsburgh, which was built in the 1830s (George, 2008).

Pennsylvania Prisons

As noted in our prior section, Western State Penitentiary, built in Pittsburgh, Pennsylvania, was based on the Bastille blueprint design. This prison was run and operated on the principle of solitary confinement, as was the norm within the Pennsylvania system of corrections. Though Eastern State Penitentiary, built in Philadelphia, was discussed in detail in Chapter 1, this section will clarify the physical features of that facility, particularly since the Pennsylvania system was so instrumental to early American correctional development. It is important to note again that Eastern State Penitentiary was built as a design improvement to its predecessor, Western State Penitentiary.

The perimeter of Eastern State Penitentiary was rectangular, and it featured cells that were arranged along the outer walls of a cell block. The cell blocks, eight in total, were all connected to a central rotunda, and each cell block jutted out in a radial fashion similar to spokes on a wheel. Thus, cells radiated from a central hub from which the open area of each cell block could be observed.

As noted in Chapter 1, this prison had many modern conveniences that did not exist in most prisons at the time. Advents such as modern plumbing systems and outside cell configurations with lavatory fixtures and showers were, until this time, unheard of within a prison environment. The design and physical features of Eastern State Penitentiary were therefore quite novel and progressive when compared to other institutions of the era.

Bastille: A fortification in Paris, France, that was a symbol of tyranny and injustice for commoners and political prisoners.

Auburn/Sing Sing

Prisons in the New York state system, which originated with the facilities at Auburn and Ossining, were based on a design quite different from the Pennsylvania system. Originally the Auburn/Sing Sing model utilized two back-to-back rows of multitiered cells arranged in a straight, linear plan. The typical cell was around 25 square feet, measuring about 3.5 feet wide by 7 feet long. This is obviously a very small amount of room, especially considering that in today's prisons inmates are afforded 7 to 8 feet by 10 feet of space (approximately 80 square feet). So these conditions were fairly cramped, and the design itself was unimaginative and inefficient. However, it was a simple design that was easy to construct and implement.

PHOTO 6.3 This drawing shows the outside of a cell block at Sing Sing Correctional Facility. As seen here, the Auburn/ Sing Sing model utilized two back-to-back rows of multitiered cells arranged in a straight, linear plan.

Wikimedia Commons

Panopticon: Designed to allow security personnel to clearly observe all inmates without the inmates themselves being able to tell whether they are being watched.

As time went on and as improvements in living standards brought in plumbing, electrical, and ventilation systems, the rows of cells were no longer joined in a back-to-back fashion; instead, a small corridor was built between them.

Unlike cell blocks in the Pennsylvania model, those in the Auburn/Sing Sing model did not face one another. The distance from the front of a cell to the outer building wall was usually from 7 to 10 feet, and this provided room for a long walkway along the front of the cells on the second tier and above. In most all cases, the outer walls of the building had windows that allowed sunlight and air into the facility. Inmates could not reach these windows if locked in their cells due to the distance between cell and outer wall.

Correctional officers working the cell block would either walk the first floor or along the walkways in front of the cells of the upper floors. This type of design, sometimes described as a "telephone pole" design, made it necessary for officers to conduct their security checks by passing in front of each individual cell; otherwise, they could not view the inmate(s) inside. Despite the inefficiency of this type of supervision, the telephone pole design proliferated throughout the United States during the 1900s. In fact, these features are so common that when people envision correctional facilities, it is this design that comes to mind (George, 2008).

Panopticon

This prison model was created by Jeremy Bentham in 1785. The **panopticon** was designed to allow security personnel to clearly observe all inmates without the inmates themselves being able to tell whether they were being watched. Indeed, the internal tower was constructed so that inmates could not tell if security personnel inside were watching them.

Oddly enough, this design was never adopted in England, Bentham's home nation, but was used in the United States. A handful of facilities using this architectural design were built in Virginia, Pennsylvania, and Illinois (George, 2008). Even today, the panopticon at Stateville Correctional Center in Illinois is in operation. This facility consists of cells that face each other across a wide circular space with an enclosed observation post at the center of the structure. The individual cells are arranged on the thick masonry perimeter walls, with narrow windows. The observation post consists of two stories and is accessible from the main floor. While movement on the ground level is easy enough, officers on the upper tiers of the cells must follow the curving, circumferential balcony for some distance to reach the stairs. If security staff must reach the observation post or an area near a cell block, the travel time can be quite significant.

Direct Supervision

Direct supervision design: Cells are organized on the outside of the square space, with shower facilities and recreation cells interspersed among the typical inmate living quarters.

The direct supervision type of prison design was implemented by the Federal Bureau of Prisons (BOP) due to the need to abate the conditions that contributed to a long, deadly disturbance at New York's Attica Correctional Facility in 1971. This design greatly differs from prior models. The **direct supervision design** features a large, open, central indoor recreational or day room space that can be effectively supervised by a single security person. Cells are organized on the outside of the square space, with shower facilities and recreation cells interspersed among the inmate living quarters. These types of cell blocks may consist of multiple levels of cells. Security staff on each level can walk around that level of the unit and see through the interior space to nearly any other area of the space.

Minimum-Security Prison Design (Modern)

Minimum-security prisons are, in many respects, not even prisons as envisioned by most of the public. On some occasions they are referred to as open institutions due to the low-key security that is provided at these facilities. **Minimum-security facilities** and/or open institutions are typically designed to serve the needs of farming areas or public transportation works rather than being optimized for the offender's reform. Inmates may often work on community projects, such as roadside litter cleanup or wilderness conservation. Many minimum-security facilities are small camps located in or near military bases, larger prisons (outside the security perimeter), or other government institutions to provide a convenient supply of convict labor to the institution. Minimum-security facilities include plantation-style prison farms as well as small forestry, forest fire–fighting, and road repair camps, depending on the type of labor that is needed.

Minimum-security inmates live in less secure dormitories that are regularly patrolled by correctional officers. These facilities typically have communal showers, toilets, and sinks. A minimum-security facility generally has a single fence that is watched, but not patrolled, by armed guards. In very remote and rural areas, there may be no fence at all. The level of staffing assigned tends to be much lighter than at other facilities. The facility is constructed and operated in such a manner that most inmates can access showers, television, and other amenities on their own without oversight of security staff.

Minimum-security facilities: These prisons usually consist of dormitory-style housing for offenders rather than cellblocks. Also, these prisons are designed to facilitate public works, such as farming or roadways, rather than being optimized for the offender's reform. A minimum-security facility generally has a single fence that is watched, but not patrolled, by armed guards. This is different from the typical prisons envisioned by the public.

Medium-Security Prison Design (Modern)

Medium-security facilities may consist of dormitories that have bunk beds with lockers for inmates to store their possessions. These facilities tend to have communal showers, toilets, and sinks. Dormitories are locked at night with one or more security officers holding watch and conducting routine patrols. Inside the dorm, supervision over the internal movements of inmates is minimal. The perimeter of these facilities, when they are not connected to another larger facility, is generally a double fence that is regularly patrolled by armed security personnel. In modern times, the external chain-link fences and other security features (e.g., cameras) may be the only clue that the grounds are, in fact, prisons for inmates.

Much of the newer prison construction during the past 4 to 5 decades has consisted of medium-security designs. Indeed, it is thought that close to one third of all state inmates are housed within medium-security facilities. These types of prisons tend to have tighter security than do minimum-security facilities, but they are not as restricted as are maximum-security facilities. Programs for inmates, opportunities for recreation, and the ability to move throughout the grounds are much better in these institutions than in maximum-security prisons. Additionally, these types of facilities tend to implement more sophisticated technology than larger prisons.

Medium-security facilities: Consist of dormitories that have bunk beds with lockers for inmates to store their possessions and communal showers and toilets. Dormitories are locked at night with one or more security officers holding watch.

Maximum-Security Prison Design (Modern)

Originally, when prisons were first designed, their primary and nearly only concern was security. Because of this, most all prisons were designed as maximum-security facilities. In most cases, they were surrounded by a high wall (up to 50 feet tall) that was usually made of brick-and-mortar material. Atop these stone walls could be found various forms of razor wire, and towers were built at each corner and at intervals along the walls so that security personnel could look down upon inmates inside as well as outside of the facility. These armed guards were instrumental in preventing escapes and open-yard riots.

Nowadays, **maximum-security facilities** are not likely to have stone walls but will instead use corrugated chain-link fence that is often topped with razor wire. These fences are lit by floodlights at night and may be electrified. Chain-link fences are as effective for security as stone walls, and they eliminate blind spots in security where inmates can hide. The open visibility makes it nearly impossible for inmates to approach the fence line without being detected. The perimeter is generally double fenced with chain-link and includes watchtowers that house armed security personnel. Add to this the use of cameras and sensors and the result is a fairly tightly maintained facility.

Maximum-security facilities: These high-security facilities use corrugated chain-link fence. These fences will be lit by floodlights at night and may even be electrified and eliminate "blind spots" in security where inmates can hide.

Security Breaches Can Happen, Even at USP Florence ADMAX

Ishmael Petty, a 46-year-old inmate serving time at USP Florence ADMAX, was convicted for an assault that he committed on prison staff in 2013.

According to court documents and evidence presented at trial, on September 11, 2013, Petty, who was serving a life sentence at ADX for killing his 71-year-old cellmate at the United States Penitentiary Pollock in Louisiana, attacked two BOP librarians and a case manager as they were delivering books to his cell. At ADX, there is an outer door, a secure area, and then an inner door before entering the actual cell. While the BOP employees believed that Petty was in his cell, he was in fact hiding in the area between the outer and inner doors. He was wearing self-made body armor, consisting of cardboard box–like material, and had a weapon, specifically a shank. When the attack began, Petty threw hot sauce in the eyes of one BOP librarian and then attacked the other librarian. The third BOP employee, a case manager, quickly came to their aid and called for help. One of the BOP employees used their baton to try to subdue Petty. Petty ultimately took control of two batons and used them in his attack. Once the call for help was made, Petty went back into his cell.

Prior to Petty's life sentence for killing his cellmate, he was sentenced to federal prison for 420 months for an elaborate armed bank robbery in Mississippi where he was wearing a police officer's uniform.

"Defendant stands convicted of a cowardly and brutal assault on defenseless staff at ADX. The defendant, who was serving time at ADX for murdering his cellmate at another prison, ambushed and brutally assaulted two staff librarians, using his much greater size to injure the older of the two severely and permanently," said U.S. Attorney John Walsh. "Only the courageous intervention of a third staff member prevented the defendant from killing that librarian."

Petty, who is currently serving a life sentence, faces not more than 20 years in prison and up to a $250,000 fine per count for each of the three counts of conviction. ●

Source: U.S. Attorney's Office. (2015, July 14). *ADX inmate convicted of three counts of assault, resisting and impeding a federal employee.* Washington, DC: Federal Bureau of Investigation.

USP Florence ADMAX:
A federal prison with a design that is nearly indestructible on the inside. Essentially, the offenders there have no contact with humans.

referred to as the administrative maximum (ADX) facility, or **USP Florence ADMAX**. This prison originally housed over 400 inmates who were the worst of the worst in the BOP. These inmates were selected because they were either extremely violent, high escape risks, or identified gang leaders. Construction for this facility cost over $60 million, and each cell had a price tag of around $150,000, making it one of the most (if not *the* most) expensive prisons ever built and maintained. Annual maintenance for inmates is roughly $40,000 per year, with the overall cost of maintenance being around $20 million a year.

USP Florence ADMAX is located on State Highway 67, 90 miles south of Denver, 45 miles south of Colorado Springs, and 40 miles west of Pueblo. The structure's design is nearly indestructible on the inside. The cell design consists of a bed, desk, stool, and bookcase, all constructed from reinforced concrete and immovable. Each cell measures 7 feet by 12 feet and has a shower stall and reinforced plumbing within. Cell windows are designed so that all views of the outside are restricted, and inmates can only see the sky above them. In addition, cells are positioned in such a manner so that inmates cannot make eye contact with one another, so they have very little contact with other people. Food is delivered by tray through a door slot, and all visits are noncontact in nature. Recreation is for 1 hour a day and is completed alone. In many ways, it seems that a Pennsylvania model of operation has been implemented, but in this case, the intent is to simply incarcerate and hold the inmate in place; reform is no longer a priority with these hardened inmates who have a poor prognosis due to dangerous and repetitive behavior.

This prison was much more expensive to build than others due to the enhanced security features, higher-quality locks and doors, and perimeter fencing designs. Facilities such as Florence ADMAX are expensive to maintain for two reasons. First, the advanced security features are costly to maintain, as noted above, and the additional staff security adds to the financial burden. Second, a large proportion of supermax inmates will never be released; their sentences are for very long terms, and this means that, over time, they will become more costly due to the need for geriatric medical care. Thus, the financial picture goes from bad to worse for these facilities.

As can be seen in Focus Topic 6.1, even supermax facilities are not completely secure. But given the assaultive nature of the inmates that are held within these facilities, it is easy to see that it is an important priority to ensure that these violent individuals are kept secure. The notion of rehabilitation is not a priority in supermax prisons; the seriousness of the inmates' offenses calls for more of an incapacitation approach.

Constitutional Issues With Confinement in Supermax Custody

It has been found that when kept in prolonged periods of confinement, inmates tend to exhibit mental health issues due to sensory deprivation. These issues can include states of paranoia, irrational fears, resentment, inability to control anger, depression, heightened anxiety, and forms of full mental breakdown. Suicides tend to be higher among inmates who are confined for prolonged periods of time. Many of these issues, as students may recall from Chapter 2, are similar to those of inmates confined in solitude for prolonged periods in Eastern State Penitentiary under the Pennsylvania model of prison operation. This is an important observation because it demonstrates that history does, indeed, repeat itself, and it validates the fact that extreme isolation does cause mental health problems. In modern correctional literature, the negative mental health effects of extended isolation may be referred to as **special housing unit syndrome**.

Supermax-specific case law to date is limited. The first case to capture national attention, *Madrid v. Gomez* **(1995)**, was a wide-ranging attack on operations at the Pelican Bay SHU in California. This was a supermax-type control unit facility where inmates identified as gang members; offenders with a history of violence, crime, or serious rule violations within prison; and other inmates considered major management threats were incarcerated. The Pelican Bay SHU was one of the first such facilities in modern American history explicitly planned and built as a state-level supermax facility.

In the *Madrid* case, the Ninth Circuit Court found widespread violations of the Eighth Amendment prohibition against cruel and unusual punishment. The court ordered the institution to professionalize the means by which security staff operated the prison and to improve the insufficient medical care at the facility. In addition, it was ordered that all inmates with mental illnesses be transferred to an institution more suited for their needs. In order to ensure that prison officials adhered to the orders of this injunction, the court appointed a federal monitor to the prison. Federal monitors are individuals who are assigned to inspect for deficiencies and ensure compliance of prison operations to legal requirements. It was not until 2011 that the injunction was removed, at which time it was determined that Pelican Bay SHU had sufficiently met the requirements that had been established in *Madrid*, nearly 20 years prior.

ADA Compliance

It is important that correctional agencies ensure that staff are trained on the specific issues related to the **Americans with Disabilities Act (ADA)** and inmates. According to the U.S. Department of Justice, Civil Rights Division, Disability Rights Section (2010), the ADA requires correctional agencies to make reasonable modifications in their policies, practices, and procedures necessary to ensure accessibility for individuals with disabilities, unless making such modifications would fundamentally alter the program or service involved. There are many ways in which an agency might need to

Special housing unit syndrome: The negative mental health effects of extended isolation.

***Madrid v. Gomez* (1995):** Case where the constitutionality of supermax facilities was questioned.

Americans with Disabilities Act (ADA): Requires correctional agencies to make reasonable modifications to ensure accessibility for individuals with disabilities.

modify its normal practices to accommodate a person with a disability. Some examples include the following:

1. Ensure that at least 2% of the total number of general holding cells and general housing cells (or at least one cell if 50 cells or less) is equipped with audible emergency alarm systems and permanently installed telephones within the cell.

2. Ensure that, when safety permits, qualified inmates with disabilities have access to all programs that they would otherwise have available. This includes educational, vocational, work release, employment, and religious programs, among others.

3. Provide for hearing aids, wheelchairs, and prostheses.

4. Authorize the purchase of special commissary items for inmates with disabilities, based on the recommendation of a medical provider.

5. Develop and implement staff training on inmates with disabilities issues.

6. Create a disability committee to ensure appropriate coordination of interdepartmental services and to address particularly complex field issues.

7. Ensure that ramps, alternate door entry mechanisms, and other architectural features are compliant with ADA standards.

8. Provide color-coded signage and large-sized signage to better assist the visually disabled and/or geriatric population.

9. Ensure that gender-specific accommodations are provided for women with disabilities.

Lastly, state and local government entities should conduct a self-evaluation to review their current services, policies, and practices for compliance with the ADA. When such an evaluation finds that the agency has deficiencies, the agency should develop a transition plan that identifies structural changes that need to be made. As part of that process, the ADA has encouraged entities to involve individuals with disabilities from their local communities. This process will promote access solutions that are reasonable and effective.

PRISON DESIGNS

Every correctional facility has its own challenges and needs that make it a bit different from other facilities in the respective correctional system. A variety of factors must be considered in design, such as if the site will be rural or urban, the security level needed, and the operational concerns for security and programming, which include the specific needs of inmates, the surrounding community support, the durability of materials needed, and so forth.

Prisons can vary widely in size in terms of both inmate capacity and the physical area encompassed by the prison buildings. The *National Directory of Corrections Construction*, written by DeWitt (1988) and published by the National Institute of Justice, classifies prisons into the following general types:

1. Campus style: a number of individual buildings that are not connected

2. Ladder, telephone pole: linear cell blocks arranged in parallel configuration off a central connecting corridor

3. Wheel, spoke, or radial: linear cell blocks that emanate from one central control area like spokes from the hub of a wheel

4. Clusters: a number of individual buildings that are interconnected

5. Courtyard: linear cell blocks interconnected around a central enclosed courtyard

Figures 6.1 and 6.2 provide a basic illustration of each type of design and the manner in which it might be organized. There are numerous reasons that one design might be used rather than another, including the security level of the institution, the location of the prison, and/or the intended function of the facility.

Various Prison Complex Designs

Campus

Ladder, telephone pole

Wheel, spoke, or radial

Clusters

Courtyard

Source: Office of Program Policy Analysis and Governmental Accountability. (2010). *Intermediate sanctions for non-violent offenders could produce savings.* Tallahassee, FL: Author.

Perimeter Security

A prison's **perimeter security system** is, ideally, a collection of components or elements that, when assembled in a carefully formulated configuration, achieve the objective of confinement with a high degree of confidence. Anyone attempting to escape the institution by crossing the perimeter must be interdicted by responding officers before the perimeter is successfully penetrated. The perimeter system must slow the escapee down so as to ensure that this confrontation can occur. While the perimeter is capable of keeping intruders out of the facility, its primary purpose is, of course, to keep inmates contained within the facility.

Figure 6.3 (not drawn to scale) shows a cross section of a basic two-fence security perimeter. It illustrates the components of a perimeter system and their relationship to each other. The perimeter itself is delineated by a chain-link outer fence and a stone, brick, or concrete wall. The outer fence has coiled razor wire that is placed within the top Y-wedge, preventing inmates from being able to climb the fence from inside the facility. This outer fence is sometimes called an oyster fence due to the shape of the cylindrical razor wire coils that are connected to the fence. Next, an isolation zone, often about 15 to 20 feet in length, exists between the outer oyster fence and a 20-foot-high stone or brick wall with a thickness at the top of 16 to 24 inches. Inside the inner brick wall is a camera that is mounted atop a tall pole, allowing a bird's eye view of the inner fence and the isolation zone between the outer oyster fence and the inner fence. The lighting system is placed farther inside and is usually designed so that illumination extends beyond the outer fence and includes enough coverage of the inside facility to allow detection of persons prior to their approaching this perimeter.

The system shown in Figure 6.3 is assumed to be solid brick, stone masonry, or reinforced concrete, which should discourage attempts to cut through the wall. Lighting

Perimeter security system: A collection of components or elements that, when assembled in a carefully formulated plan, achieve the objective of confinement with a high degree of confidence.

FIGURE 6.2

Typical Cell Layout and Organization

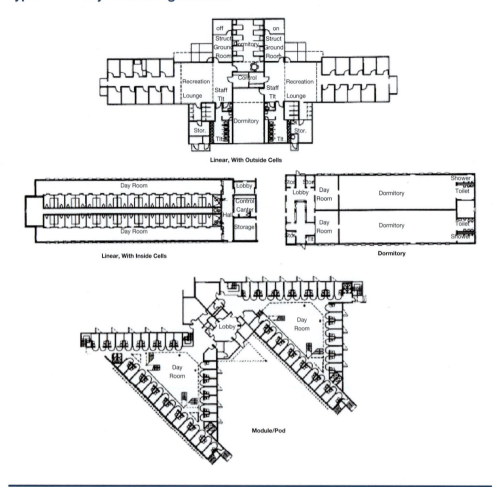

Source: Office of Program Policy Analysis and Governmental Accountability. (2010). *Intermediate sanctions for non-violent offenders could produce savings.* Tallahassee. FL: Author.

features are also shown mounted on the top of the wall since this is typical of most stone-walled facilities. Sensor equipment is mounted on the wall and placed to detect an object or a person approaching the top of the wall or something leaning against or touching the wall. This same sensor equipment is also designed to detect persons who might scale the wall from the outside since there would be no convenient way to get over the wall and inside the yard without encountering the sensor array. Thus, this security system also serves to thwart the attempt of outside accomplices.

Compassionate Prison Design

Recently, there has emerged a concept that will be defined as the compassionate prison design. This type of design is intended to remedy the foibles of prior designs that emphasized isolation, sterility, and detachment from the natural world. This design, instead, has the physical and mental health of the inmate in mind, rather than a single-purpose focus on security. One excellent example of the use of compassionate prison design is the newly built Las Colinas Women's Detention and Reentry Facility, in San Diego County, California. This facility has residential style buildings that are built around exterior courtyards. As we will see in later chapters of this text, there is abundant research that shows how isolation can breed violence and mental health problems. When physical and psychological barriers are minimized to encourage some form of

Fenced Perimeter of a Prison With a Stone or Concrete Wall

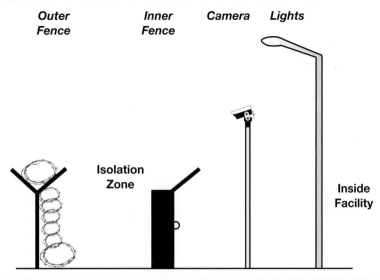

Source: Crist, D., & Spencer, D. (1991). *Perimeter security for Minnesota correctional facilities.* St. Paul: Minnesota Department of Corrections.

socialization, the less frequent that medical or psychological complaints are generated. In the case of the Las Colinas facility, the appearance is that of a college campus. The floors are shades of warm brown, and green accent walls that include materials such as stone, cork, and wood are used rather than simple concrete. There are usually large windows and ample natural light. These features humanize rather than dehumanize the individual.

While this form of design cannot be used with every facility and at every security level, it serves as an example of what designers are tasked to accomplish: a combination of security and aesthetics. Given the boom in demand for health care services for offenders and given that administrators are having to operate with a reentry mindset (not a keep 'em locked up mindset), the use of these more peaceful designs seems to be timely. While still a new concept that is most often utilized with specialized populations, this trend is worth noting as corrections continues to lean toward less institutionalization and more toward community-oriented approaches. In following this trend, the prisons themselves seem to more and more mirror the architecture and landscape of nonincarcerating milieus.

Alarm Systems

When establishing alarm systems for perimeter security, camera video systems should be placed inside the inner fence and positioned to allow for viewing of the source object that triggers a given alarm. Positioning should ensure that the entire security zone is viewable rather than being narrow in scope and focus. Attention to lighting is important as well, since this will impact the effectiveness of camera systems at night. A primary issue when implementing a video surveillance system is being able to discriminate between person-sized alarm sources that could be escaping inmates and small animals or other nuisance alarms (Crist & Spencer, 1991).

PHOTO 6.4 The above room is a lounge area of Las Colinas Women's Detention and Reentry Facility that exemplifies the compassionate prison design concept.

Robert Hanser

Isolation Zone

The primary goal of fencing is to provide enough of a delay so that security responding to the escape attempt will have time to intercept the inmate before he or she can penetrate or clear the fence (Crist & Spencer, 1991). The area between the inner and outer fences is often referred to as the isolation zone. The **isolation zone** is designed to prevent undetected access to the outer fencing of the prison facility. It also forms an area of confinement where the inmate escapee is trapped once the alarm is triggered, which allows security staff time to respond and apprehend the inmate before the escape is successful.

Lighting

When providing lighting for these types of security designs, some problems are likely to be encountered. First, if the lighting is located atop the wall, the system will usually be about 15 feet from the sensor system and 5 or 6 feet inboard of the wall. This type of configuration produces hot spots in the lit areas that will exceed typical light-to-darkness ratios for optimal viewing. Because of this, some facilities may add light sources along the top of the wall but reduce their intensity. In addition, the lighting equipment may be directed away from the surface of the wall, just a bit, to reduce intensity and potential glare. However, the use of additional lighting fixtures is expensive, and this then adds further to the cost of the prison in terms of layout and maintenance.

Razor Wire

Three coils of razor wire create the barrier delay and are placed above the sensor, along the inner area of the wall and along the top of it. Because determined inmates may be willing to concede detection with the thought that they can clear the wall prior to the response of security personnel, the razor wire is used to further prevent most inmates from taking the chance. To demonstrate how complicated security can be and how crafty some inmates may become out of desperation, consider that the lighting poles could be used to bypass this barrier. Thus, the poles should be made to give way and break if they are loaded with heavy resistance. When taken together, these collective features are customary to most basic perimeter security systems of older maximum- and medium-security-level facilities.

INTERNAL SECURITY

A number of features can be used within the facility to enhance internal custody and control of the inmate population. For instance, the use of touch-screen surveillance systems in control rooms can provide centralized security staff with the ability to oversee security throughout various sections of the institution. Showcased in Figures 6.4–6.7, this type of equipment can provide a sense of organization to the overall security maintenance throughout the facility and help to coordinate the use of security staff resources during times of both routine operation and crisis response.

Redding (2004) notes that although physical design and inmate classification may be key elements of the security process, it is the security staff who are integral to maintaining custody and control of the inmate population. Regardless of how well one builds a prison and designs its attendant security features, the physical features of an institution's perimeter alone are useless without staff properly trained to be alert to their responsibilities while operating in their assigned capacities (American Correctional Association, 1998). Prison officials are responsible for the security measures that the physical design cannot control. These duties include "access control, searching of prisoners and their belongings, and movement control both inside and outside prisons and during the transportation of prisoners" (Redding, 2004).

Given that the human element is so important to internal security, it is essential that an appropriate form of organizational culture that seeks to psychologically motivate employees toward security-minded objectives be developed within an agency (Robbins, 2005). This refers to a system of shared meaning held by members of that organization.

FIGURE 6.4

Touch-Screen Control Panel

LOG ON
Logged on: Ruffin

1. ENTER YOUR LOG ON NAME

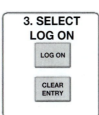

3. SELECT LOG ON

LOG ON

CLEAR ENTRY

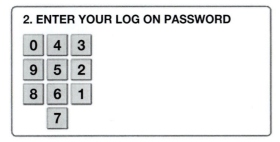

2. ENTER YOUR LOG ON PASSWORD

Source: Reprinted by permission of Cornerstone Institutional Sales & Service, www.cisupply.com.

FIGURE 6.5

Full View of Facility

MAIN SCREEN
Logged On: Ruffin

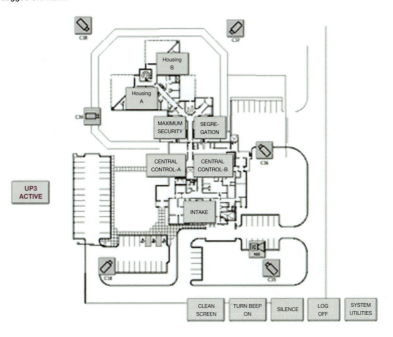

Source: Reprinted by permission of Cornerstone Institutional Sales & Service, www.cisupply.com.

FIGURE 6.6

Multilevel View of Pod

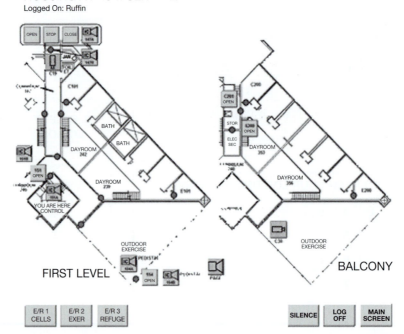

Source: Reprinted by permission of Cornerstone Institutional Sales & Service, www.cisupply.com.

FIGURE 6.7

View of Segregation Pod

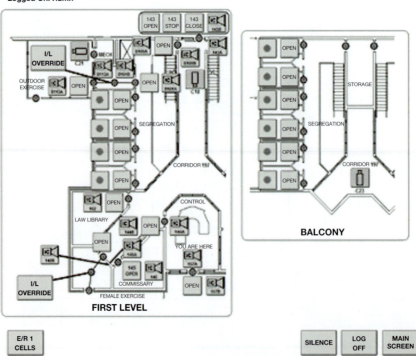

Source: Reprinted by permission of Cornerstone Institutional Sales & Service, www.cisupply.com.

One shared meaning or belief is the need for security and an attention to detail (Robbins, 2005). This attention to detail entails the degree to which employees are expected to exhibit precision, analysis, and attention to the specific routines that occur within the agency (Robbins, 2005). Such attentiveness would provide an environmental mindset that is conducive to security-related issues. Without such a mindset, truly disastrous results can occur.

Consider, for instance, that inmates in a San Antonio, Texas, prison managed to steal 14 revolvers, a 12-gauge shotgun, and a rifle before they drove away in a prison van without being stopped. The security breach was due to personnel not following agency policy and security procedure. This escape led to a Christmas Eve robbery and the death of an Irving police officer (Associated Press, 2001). This example demonstrates that problems with security, particularly human compliance with security principles, can have a crucial and deadly impact on both the internal security of the prison facility and the outside community.

PRACTITIONER'S PERSPECTIVE

"I started looking into policing and I thought, 'Oh, well, that might be interesting.' It certainly paid well at the time. For a single parent it paid extremely well."

Interested in a career in corrections or criminal justice? Watch Tracy Keesee's video on her career as a police captain.

Avoiding Blind Spots in Correctional Facilities

Blind spots in correctional facilities can occur when the design has certain areas that are obscured from easy view of security staff and/or surveillance equipment. While the primary means of achieving internal security is the prison staff, they cannot detect all things that occur. Surveillance equipment inside the facility is often used to add a security element to areas that are less frequently patrolled and/or difficult for staff to observe in person. The use of closed-circuit television (CCTV) systems can provide good visual surveillance within a facility, but these types of systems can be costly. And, as noted in the previous subsection, this equipment is only as effective as the security personnel who use it.

Thus, the best means of avoiding blind spots is to leave as few areas out of restricted view as possible during the design stage. In addition, staff should be aware of any areas where obstructed views occur and physically monitor these areas more closely. This prevents a number of problems that go beyond just the possibility of inmate escape. The blind spot areas of a prison can allow inmates to engage in inappropriate or illicit behaviors, such as trafficking and trading of contraband, inappropriate behavior within the institution, or assaultive actions against other inmates or staff. Thus, these areas are danger zones within the prison facility and should be given maximum attention by security personnel to offset the potential breach to internal or external security.

Blind spots: In correctional facilities, these occur when areas of the prison are not easily viewed by security staff and/or surveillance equipment.

Kitchen Services and Facilities

The design and development of kitchen facilities in prison environments require that equipment be able to service hundreds or thousands of individuals on a daily basis yet, at the same time, facilitate security needs of the institution. This is a difficult balance

PHOTO 6.5 Prison kitchens and dining halls tend to be busy throughout the entire day and much of the evening. The continuous feeding process within the institution requires nearly round-the-clock work on the part of staff and inmates assigned to culinary duties. Pictured here is a jail facility kitchen at Ouachita Correctional Center.

Robert Hanser

in design since any mass feeding operation is, in and of itself, a challenging process. When adding security issues to the kitchen operation process, the physical layout of the kitchen becomes very important.

In addition, the storage of cutlery equipment within the kitchen is critical. Strict monitoring of the use of culinary tools such as knives, forks, and meat saws is very important to institutional safety and security. Facilities must be designed so that the inventory of kitchen tools can be completed quickly and easily. This feature distinguishes a prison kitchen from most other types of commercial or mass-feeding kitchens.

The most common method of serving food to inmates is through a large, open, cafeteria-style room. These rooms tend to have either long tables or a series of small tables with chairs that are secured to the ground. Inmates gather at the feeding line. This line may be open so inmates can observe the persons who are serving their food or a screen may extend down several feet from the ceiling so that they do not know who prepared their tray. This feature eliminates the possibility that inmates who do not like one another will have problems regarding the food that is served on the line.

Feeding and Security

Prison kitchens can be the source for all kinds of contraband within the prison. Naturally, contraband can include any sort of kitchen tool or product: knives and cutlery, poisonous substances, or food items that can be used to make homemade alcohol, such as fruits and yeast. To maintain control over kitchen tools, an inventory is taken daily, and inmates are required to formally sign these tools out. Many kitchens use what is called a shadow board, which is usually a board with the silhouette of the item drawn on the board so that missing kitchen tools can be quickly identified.

The use of secure storage facilities for food products that can be used to manufacture alcohol is also required. Kitchen staff must be alert so that inmates are not able to sneak products out of the kitchen area. In addition to physically inspecting the inmates, staff should ensure that products have not been hidden in the kitchen, placed in the trash (so that they can be taken out of the kitchen area undetected), or hidden in items that go in and out of the dining facility. It is important to note that junior and/or inexperienced staff may not realize when certain foods can be problematic and therefore may not prevent much of the contraband that leaves from the kitchen. Because of this, they must be trained on these issues.

Food Service Facilities and Equipment

Food service facilities and equipment vary from one jail or prison to another (Johnson, 2008). Some facilities may have equipment that is not up to date and is therefore difficult for the workers to use. Most prison dining areas are set up in a cafeteria-style arrangement, as noted above, and inmates are usually fed three meals a day. It is important to note that the feeding times for inmates may not be during the hours that most people would expect. There may be some flexibility in the times that inmates are fed depending on programming requirements, work schedules, and other factors that may impact scheduling.

In addition, some areas of the prison may require that inmates be given a food tray within their cell. For instance, inmates in administrative segregation and/or solitary confinement will usually be given a plastic tray that is kept in a heated food cart or otherwise stored for portable delivery. In cases where inmates in lockdown facilities are being fed, officers in the department will usually provide the inmates with their food tray, utensils, and required liquid beverage.

Food Supplies and Storage

Johnson (2008) notes that food should be of the best quality possible within budgetary constraints. This is true for multiple reasons; one of these is that providing decent food minimizes inmate discontent. Also, high-quality nutritional food keeps inmates in good health and therefore reduces the amount of health problems that might arise. This is important because lower medical costs translate to savings for prison administrators.

Food service equipment within a prison kitchen is typically similar to what one would find in any large-scale cafeteria. However, the products may have security features not found in free-world kitchens. For instance, the company Jail Equipment World markets a maximum-security convection oven. The description of this product explains that this oven prevents "the tampering of control and product settings, protect[s] ovens and essential components from vandalism and abuse, eliminate[s] hiding places for contraband and prohibit[s] the removal of parts and components for weapon fabrication" (Jail Equipment World, 2010, p. 1). These ovens carry a price tag of $5,200 to $12,000 each (Jail Equipment World, 2010, p. 1). This example demonstrates that kitchen appliances must meet functional requirements as well as security requirements and that such equipment can carry a hefty price tag.

PRISON TOUR VIDEO: A lot of preparation and planning goes into delivery of food services in prison. Go to the IEB to watch a discussion about prison food service.

Laundry Facilities

The need to continually wash clothes and bedding is also a routine issue in prison facilities. Prisons and jails must have reliable washers and dryers and other machinery, such as ironing equipment, that is both heavy duty and reliable. Laundry facilities must be suitable for continual use and must also allow for adequate security over the inmate population. The constant wear-and-tear of equipment requires that it be oversized, commercially rated, and industrial in strength. Such equipment must be made to take abuse, as it is operated by inmates who, for the most part, have little incentive to adequately take care of it.

As with kitchen facilities, laundry facilities should be designed so that security staff are able to maintain effective custody and control over inmates working in them. All products must be accounted for, including items such as irons and products such as bleach and cleaning soap; these items may be traded throughout the inmate population and can be used as weapons. Laundry facilities must be designed with the need for inventory control of tools and equipment firmly in mind.

PHOTO 6.6 Laundry facilities in a prison or jail must be suitable for continual use and allow for adequate security to be exercised over the inmate population.

Robert Hanser

Recreational Facilities

Though the issue of inmate recreation may be controversial, it is inevitable that prison administrators will have to attend to it. Recreation for inmates comes in many shapes and forms and can be an effective behavior management tool inside the prison. The threat of exclusion from recreation programs usually gains compliance from most of the inmate population, and this alone demonstrates some operational value for such programs. When the leisure time of inmates is used in the construction of arts and crafts, multipurpose activities, and participation in sports activities or physical exercise, they are less likely to be lured into other activities that are not prosocial. Recreational activities for inmates might include any of the following:

1. Intramural sports (e.g., basketball, handball, flag football, soccer)
2. Games (cards, dominoes, bingo)
3. Individual exercise (weight lifting and jogging)
4. Arts and crafts (leather working, woodworking, painting)
5. Clubs (Key Club, Jaycees, Toastmasters)

Each of the above activities will have different requirements for facility space, and some of them can be held outdoors. As with any additional prison service, effective security maintenance is key to the implementation of these programs. Naturally, some types of recreation include tools or instruments that can, if not kept under tight inventory, be trafficked among the inmate population. For instance, leather working tools, wood-cutting saws, and other types of instruments should be subjected to very close security scrutiny.

Religious Facilities

The impact of religion and faith-based programming in prison is undeniably important. In some prisons, an actual religious gathering facility may exist, but this is rare. More typically, a general meeting room will be used, with seats and speaking equipment provided. Given that different religious orientations may be found in prisons and that states do not wish to discriminate between different religions, it is probably prudent for administrators to simply provide space for religious services on an as-needed basis.

Tool Shop Facilities

Tool shops tend to exist in many larger prisons where some type of industry is performed. Such facilities allow inmates to be productive while serving their sentence. Inmates can learn and perform repair work of appliances, lawn equipment, and even motor vehicles. Such shops produce tangible benefits for the prison and the inmate population. Nevertheless, security risks exist, as such equipment can be used to make weapons and/or effect escape attempts. Thus, the design of tool shops must accommodate security concerns and allow for the easy inventory and control of equipment and tools.

TECHNOLOGY SYSTEMS IN PRISONS

The use of technology in prisons is becoming widespread and has impacted a number of operational aspects. Surveillance equipment is essential and has undergone improvements in recent years, but other developments have been equally innovative and just as important. For instance, the physical security of the institution itself has been strengthened through improvements in building and fencing technology as well as technology related to ingress and egress from correctional institutions.

For example, in California, Colorado, Missouri, and Alabama, innovations involving lethal forms of electric fencing have been effective in deterring escape attempts (Carlson & Garrett, 2008). While these types of fences should not be seen as a full replacement of security staff, they do eliminate the need for so many staff to concentrate on perimeter security. Carlson and Garrett (2008) note that these developments

> truly do offer the opportunity to cut back on staffing in towers and external mobile patrols. California's Department of Corrections has found the installation of these fences has facilitated the deactivation of nearly all towers and has enabled administrators to redeploy staff to other important posts. (p. 417)

Thus, such technology can allow administrators to maximize the use of their employees by diverting them to areas of the prison that can benefit from more personalized and human-oriented job functions.

Prisoner Identification

New processes regarding inmate identification are being developed that greatly enhance institutional security and cost-effectiveness. For instance, mug shots are now taken with digital cameras and stored in computers; this allows for photographic images of inmates to be readily available for prison staff at a fraction of the cost that once was required and enhances inmate recognition by staff. Likewise, fingerprints are now maintained in online data systems, and retina imaging and iris-scanning equipment have undergone innovations and improvements.

In addition, some facilities have begun to use wristbands with bar codes that are used to track inmates throughout the prison. In this way, prison staff can monitor the movement of individual inmates in real time by using touch-key control systems. Similar tracking technology is being used in Los Angeles, where the tracking of inmates in the nation's largest jail system is now conducted with radio-linked wristbands (Etter, 2005). These devices allow security staff to pinpoint inmates' location within a few feet. Los Angeles County uses these devices at Pitchess Detention Center in Castaic, about 40 miles northwest of downtown Los Angeles, and has plans to expand the program to another 6,000 inmates at the county's central jail and then to other facilities (Etter, 2005).

In addition to tracking inmates around cell blocks, this technology has the potential to allow work-release crews to roam within an electronic fence construction that could be easily moved wherever needed (Etter, 2005). This is an important innovation because it can make work processes more secure for prison farms, minimum-security facilities, and prison camps where inmates work on wilderness and highway construction projects. Further still, with the use of surveillance systems inside the facility, a specific inmate can be tracked by both central security and that in the immediate vicinity of the inmate.

PRISON TOUR VIDEO: Recent innovations in technology have greatly changed how corrections officers can monitor inmates. Go to the IEB to watch an assistant warden describe prisoner identification and tracking.

Modern Classification Systems

Due to litigation and overcrowding issues that have cropped up over the past decade, prisons have experienced increased pressure to improve their systems of classifying inmates according to custody, work, and programming needs. Classification systems are viewed as the principal management tool for allocating scarce prison resources efficiently and minimizing the potential for violence or escape. These systems are also expected to provide greater accountability and forecast future prison bed-space needs. In other words, a properly functioning classification system is seen as the "brain" of prison management. It governs many important decisions, including such fiscal matters as staffing levels, bed space, and programming.

In times past, subjective forms of classification, based on the clinical opinions of professionals, were largely utilized. However, this has been eliminated in most modern systems of classification. Instead, objective forms of classification are used in which decisions can be measurable and the same criteria are used with all inmates in the same manner. The late 1990s in particular witnessed significant improvements in classification practices: The level of overclassification has been reduced, custody decisions are made more consistently, criteria for custody decisions have been validated, inmate program needs are assessed systematically, and institutional safety for both staff and inmates has been enhanced.

Despite these improvements, additional issues remain unresolved regarding prison classification systems. In particular, decisions at the institutional or internal level that guide housing, program, and work assignments need to be as structured and organized as those made at the system or external level. As correctional facilities become more crowded, internal classification decisions play a more significant role. The widespread use of double celling in high-security units and the expanding use of dormitories for low- and medium-custody inmates have triggered the need for a systematic process for assigning inmates to beds or cells. As inmate populations continue to increase, decisions governing housing and programs, especially for inmates with extremely

long sentences, will become increasingly difficult. Classification used within a prison facility is often referred to as an internal classification system. An internal classification system is used to devise appropriate housing plans and program interventions within a particular facility for inmates who share a common custody level (minimum, medium, closed, or maximum). On the other hand, external classification processes have more to do with the placement of inmates at different facilities and/or on community supervision.

PRISON TOUR VIDEO:
Prisoner classification helps determine what jobs and responsibilities an inmate can obtain. Go to the IEB to watch a work release director and an assistant warden describe classification systems.

Morris v. Travisono (1970): Resulted in a detailed set of procedures for classifying inmates that was overseen by the federal court system.

Palmigiano v. Garrahy (1977): Ruling that attested to the importance of effective and appropriate classification systems.

RATIONALE FOR CLASSIFICATION

In *Holt v. Sarver* (1969), the Supreme Court found that the totality of prison conditions should not become so deplorable as to violate constitutional expectations for inmates. Shortly after this came ***Morris v. Travisono* (1970)**, the first federal court ruling to establish that a state must design and implement a classification system. In this case, inmates of the Adult Correctional Institutions (ACI) system brought allegations against Rhode Island prison authorities in a class action lawsuit for alleged violations of the Eighth and Fourteenth Amendments of the Constitution. This case resulted in a detailed set of procedures for classifying prisoners that was overseen by the federal court system.

The Court approved and incorporated the procedures while continuing to implement a consent decree where jurisdiction was maintained for 18 months while parties in litigation established the final legal scheme that would enforce compliance with these procedures. Later, in 1972, the district court entered a judgment as a final consent decree that established the so-called Morris rules as the minimal procedural safeguards to which inmates would be entitled. This federal case led to the development of further precedent. The Morris rules, among other things, established that classification is essential to a well-run prison and contributes to treatment objectives as well as those related to custody, discipline, work assignments, and other facets of inmate existence. However, it is perhaps the case of ***Palmigiano v. Garrahy* (1977)** that best exemplifies the importance of classification processes to the operation of a prison system. The decision in this case stated that

> classification is essential to the operation of an orderly and safe prison. It is a prerequisite for rational allocation of whatever program opportunities exist within the institution. It enables the institution to gauge the proper custody level of an inmate, to identify the inmate's educational, vocational, and psychological needs, and to separate nonviolent inmates from the more predatory. . . . Classification is indispensable for any coherent future planning. (p. 22)

It is clear that classification serves as the central organizing component when processing offenders within a correctional system. In order to avoid the potential for unconstitutional operations and to avoid liability, states must have effective classification criteria.

The use of classification is imperative not just in prisons but in jails as well, and courts have recently become more involved in reviewing jail operations. The primary responsibility of the jail is to safely and securely detain all individuals remanded to its custody, and classification is an essential management tool for performing this function. Solomon and Baird (1982) note that

> corrections must recognize that classification is first and foremost a management tool. It should, in fact, be perceived as the veritable cornerstone of correctional administration. As a means of setting priorities, its purposes are to promote rational, consistent, and equitable methods of assessing the relative needs and risk of each individual and then to assign agency resources accordingly. (p. 98)

In addition to providing a consistent and documented rationale for assignment decisions, there are economic advantages of classifying inmates efficiently. Accurate classification allows for the redistribution of personnel according to the custody requirements of inmates and permits the facility to implement better crisis management response (see Figure 6.8). The reduction of false overpredictions (predictions of dangerousness that end up not being true) will save the facility money due to the fact that inmates will not be inappropriately placed in costly, high-security housing but will instead be located within a less secure and less expensive environment, all without jeopardizing public safety.

THE GOALS OF CLASSIFICATION SYSTEMS

Essentially, there are four broad goals of classification. These goals are fairly straightforward and commonsense but are nonetheless critical to effective prison management. Classification is intended to do the following:

1. Protect the public from future victimization by offenders who are already in the correctional system

2. Protect inmates from one another by separating likely predators from less serious inmates

3. Control inmates' behavior, which includes their ability to trade in contraband, engage in prison gang activities, or harm themselves (as with suicidal inmates)

4. Provide planning so that agencies can better allocate resources and determine programming accountability

These goals are what day-to-day prison operations hinge upon. In one way or another, all of the operational aspects of the prison are designed to augment these goals. Thus, classification sets the tone by which the institution operates through the matching of inmates with security and treatment services offered within the organization. We now devote our attention to examining these four goals in more detail.

Protect the Public

Public safety is an important element of corrections. In fact, it is the first and most paramount element. When we sentence offenders, we seek not only to hold them accountable for their injury to the public but also to protect the public from further injury. The key issue in classification is to determine how likely an offender is to cause further injury and to match that likelihood with a commensurate level of security.

Protect Inmates

It is important to note that classification cannot and should not be expected to totally eliminate prison violence. As with protections for the public, the qualities of the risk-prediction instrument used to classify offenders are important when protecting inmates from potential victimization. Effective instruments can identify factors that might lead to prison violence, such as the placement of mentally disturbed inmates within the general population (where they are often victimized), and mitigate them. Solutions might include the separation of predatory inmates from likely victims or separating rival gang members. Prisons that use effective screening and

FIGURE 6.8

Overview of External and Internal Classification Systems

Admission to prison

Initial classification
- Custody assessment
- Program needs assessment
- Facility designation

Transfer to facility

Internal classification
- Housing assignment
- Program assignment
- Work assignment

Transfer to designated housing area

Reclassification
- External classification
  - Custody
- Internal classification
  - Programs – Facility
  - Housing   – Community
  - Work   programs

Source: Hardyman, P. L., Austin, J., Alexander, J., Johnson, K. D. & Tulloch, O. C. (2002). *Internal prison classification systems: Case studies in their development and implementation.* Washington, DC: National Institute of Corrections.

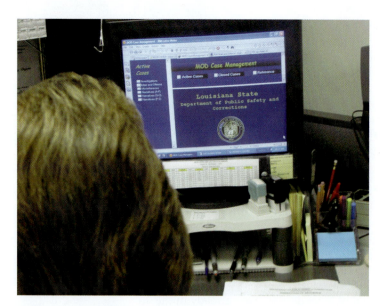

PHOTO 6.7 The use of automated classification systems is now standard practice in both prisons and community-based correctional agencies. These systems keep track of programming, sentence time served, and other important aspects of the offender who is serving time.

Robert Hanser

classification and base their security decisions on the outcomes of these processes will likely be effective in providing reasonable protection of inmates in their custody.

Control Inmate Behavior

The control of inmate behavior often occurs by providing rewards or punishments for actions that are committed. Inmates who are repetitive rule violators can be reclassified to a level of custody that provides them with fewer privileges and prosocial activities. Inmates who maintain good behavior can be placed in less restrictive correctional facilities (e.g., minimum security) or may be given a special status, such as an inmate mentor on a given cell block, in which capacity they will assist both staff and fellow inmates. In addition, screening for offenders who have special risks (e.g., suicidal inmates or those with medical or mental health conditions) can help to identify those inmates who will require special security or treatment assignments. Effective security and custody placement can reduce fear, violence, escapes, and potential litigation.

Provide Planning and Accountability

Classification systems allow correctional agencies to allocate their resources where they are most needed while not overspending on inmates who do not need as much supervision or intervention. Thus, these systems can save tax dollars and thereby help prisons to operate more efficiently. As mentioned before, the elimination of overclassification is very important when determining budgetary allocations. If inmates are repeatedly overclassified, they will be placed under restrictions that are more costly than is necessary, or they will be provided services that are not appropriate and are thereby wasted.

Classification systems are extremely important for developing prison budgets, determining the types of units to build in the future, determining staffing levels, and providing services. Classification systems can also provide information on the characteristics of a system's prison population as a means of making security, custody, and staffing projections or even in determining if a new prison may need to be built within a system. This can help administrators avoid the error of overestimating the need for expensive maximum-security cells while underestimating the need for less expensive minimum-security beds. Thus, classification processes are critical to the management of any correctional system, whether it be a prison, a jail, or even a community-based program or facility, since the underlying goal of effective resource allocation applies to all spectrums of the correctional industry.

Security and Custody Issues

Level of security: The type of physical barriers that are utilized to prevent inmates' escape and are related to public safety concerns.

Custody level: Related to the degree of staff supervision that is needed for a given inmate.

The **level of security** provided for a particular inmate refers to the type of physical barriers utilized to prevent that inmate's escape and is related to public safety concerns. As long as the inmates are kept within the prison grounds, it is less likely that they can harm persons in the community. On the other hand, the **custody level** is related to the degree of staff supervision needed for a given inmate. This is more related to safety and security within the prison facility itself, though this can also impact the likelihood that the inmate will be able to engage in activities that affect the public.

SPECIAL HOUSING ASSIGNMENTS

Special housing assignments are provided for a number of reasons, but in this section, we focus on those related to security. In this chapter, we look at the rationale and reasons for placing an inmate in special housing, which includes protective custody. The other primary form of unique housing that requires special security considerations is administrative segregation. This type of housing assignment is not punitive but is a designation intended to prevent potential disruption of institutional security. It seeks to ensure the safety of staff and inmates who may be in danger if a given inmate is allowed to freely intermingle within the general population of the prison.

We begin with protective custody and follow with a discussion of administrative segregation. It is perhaps worth mentioning that the author of this text worked for several years in an administrative segregation department with violent and gang-related offenders, with occasional tours of duty in the protective custody section of the Texas prison facility where he was employed. When appropriate, information regarding these types of special housing may be derived from professional experience gained during that employment.

PHOTO 6.8 Individual housing units such as the ones pictured here consist of individual cells that are adjacent to a common dayroom, in a barred section of the jail facility.

Robert Hanser

Protective Custody

According to the National Institute of Corrections (1986), **protective custody** includes "special provisions to provide for the safety and well-being for inmates who, based on findings of fact, would be in danger in the general population" (p. 41). This general definition, provided by the premiere clearinghouse on correctional research, will serve as the definition of protective custody for this text. The designation of protective custody as an official classification did not emerge until the 1960s, but its usage began to grow in the mid-1980s. During this period, the number of inmates in protective custody nationwide exceeded 25,000; this number declined to around 7,500 by 2001. This trend is consistent with two other important occurrences within the field of corrections during these time periods. First, the 1980s saw a rise in the use of protective custody as a result of a number of federal court rulings and injunctions during the late 1970s and 1980s that required many prison systems to revamp their programs. Second, prisons during the 1980s began to see more gang-related membership; this was followed by an influx of drug offenders in the 1990s, when another small increase (in 1990) was observed among those in protective custody. These trends impacted both protective custody and administrative segregation populations.

Protective custody:
A security-level status given to inmates who are deemed to be at risk of serious violence if not afforded protection.

Administrative Segregation

Administrative segregation is a type of classification that is nonpunitive in nature but requires the separation of inmates from the general population due to the threat they may pose to themselves, staff, other inmates, or institutional order (see Figure 6.9 for rates of inmates placed in this custody level). Classification officials may cite a number of reasons for imposing this custody level upon an offender. One of the most common reasons is membership in a prison gang. Thus, many administrative segregation programs consist of a large number of gang members.

Inmates in administrative segregation are usually housed in single cells where they receive the basic necessities and services, including food, clothing, showers, medical care, and even visitation privileges. Usually these inmates are provided 1 hour of recreation time daily, on a confined recreation yard that has some type of fencing to prevent inmates from having physical contact with other inmates. Though they may be able to see one another through the fencing (such as with chain-link fencing), they

Administrative segregation:
A nonpunitive classification that requires the separation of inmates from the general population for safety.

FIGURE 6.9

Inmates in Administrative Segregation Nationwide

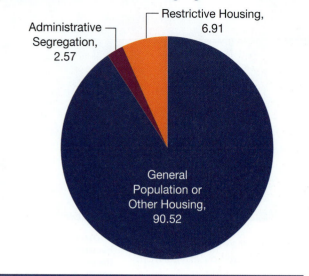

Administrative Segregation, 2.57

Restrictive Housing, 6.91

General Population or Other Housing, 90.52

Source: Association of State Correctional Administrators. (2004). *Fact sheet: Corrections safety: What the research says.*

Note: Most recent data available.

are not allowed to engage in recreational activities on the same yard. Though many of the amenities available for general population inmates are also provided to inmates in administrative segregation, some privileged activities and/or items may be restricted due to an inmate's gang status and the likelihood of dangerousness or because certain items may be used in the manufacture of homemade prison weapons. Indeed, the author of this text has a collection of prison weapons made by inmates while in administrative segregation; the types of deadly creations that can be made are limited only by one's imagination.

Lastly, as with inmates in protective custody, inmates in administrative segregation receive routine reviews of their classification status. They may be eligible for reclassification, depending on the circumstances and whether they are affiliated with a gang. Typically, regardless of behavior, if an inmate is a known gang member, he or she will remain in administrative segregation due to the danger posed to the institution. But even in these cases, programs may exist to help inmates eventually get out of administrative segregation, if they so desire.

Specialized Offender Categories

Many inmates have a variety of characteristics, issues, or specialized needs that set them apart from the remaining prison population (Hanser, 2007). From a classification standpoint, the various needs that inmates may have can affect the decision to place them within a given facility, and, within that facility, these issues can affect individual housing decisions. It is important to note that each of these categories of the offender population creates the need for effective assessment and classification processes.

Further, one issue is very prevalent within the inmate population and directly relevant to the majority of the offender population: drug use. Because this issue is related to specific offender categories discussed later in this text and because this issue affects a large number of offenders in jail and prison, a section on drug use and the need for screening and classification has been included in this chapter. Thus, we now turn our attention to screening, assessment, and classification issues when faced with placement decisions for substance abuse treatment programs.

Substance Abuse Issues, Assessment, and Classification

Many offenders in state and federal prisons are convicted of violating drug laws. In fact, 22.4% of all federal inmates and 32.6% of all state inmates reported being under the influence of drugs or alcohol at the very time that they committed the offense for which they were incarcerated (Bureau of Justice Statistics, 1998). In addition, more than 83% of all state inmates and more than 73% of all federal inmates reported past drug use, with most of that past drug use occurring during the year prior to their offense (Centers for Disease Control and Prevention, 2003). The problem is compounded by the fact that substance abuse is closely related to recidivism; inmates with prior convictions are significantly more likely than first-time offenders to be regular drug users.

Given the association between injection drug use and HIV/AIDS, detoxification also provides counseling to reduce AIDS-related risk behaviors (McNeece, Springer, & Arnold, 2002). This largely is associated with drug abusers who use intravenous drugs and those who engage in risky sex behaviors, particularly when prostitution and drug use are combined. In addition, many cases of mental illness in jails and prisons co-occur with substance abuse disorders.

Screening and Placement Criteria

Every form of treatment program involves some sort of screening. According to Myers and Salt (2000), screening serves two major purposes: It attests to the presence of a condition that may go unrecognized if not detected, and it provides data to decide whether a client is appropriate for a specific treatment program, or vice versa.

In the first use of screening, social, health, and criminal justice workers determine if there are sufficient grounds for referral of an inmate to a particular drug/alcohol treatment program. This screening is very important because the earlier the intervention takes place, the better the prognosis for the client. Obviously, the odds of reforming a drug experimenter are much better than when treating a compulsive user. The second use of screening is to determine client appropriateness for a given treatment modality. It should be pointed out that the discretion in placement may not only consider the client's individual characteristics; the ability of a given agency to provide these services must also be considered. For example, fiscal constraints can be a factor despite the fact that the treatment program may be ideal for the client. In any case, it is this use of screening that provides the placement criteria for drug offenders in the criminal justice system.

Placement criteria are very important when processing drug offenders. The initial placement is important for both public safety and treatment-oriented concerns. When deciding upon placement criteria, a match must be made between the severity of the addiction and the level of care needed, which can include (from most serious to least serious) medical inpatient care, nonmedical inpatient care, intensive outpatient care, or outpatient care (Hanson, Venturelli, & Fleckenstein, 2011). Further, matching the client's profile to a treatment modality is more likely to achieve lasting success, which can translate to the enhanced evaluation of program effectiveness. For example, a client with attention deficit/hyperactivity disorder might be unsuited for the regimentation of a therapeutic community. Conversely, a person with low self-esteem, insecurities, and a fragile sense of self-worth would not be appropriate for a highly confrontational style of intervention.

Goldberg (2003) notes that several questions should be asked before placing clients in treatment and that these questions should also be considered before a program begins to go into operation—or, if a program is already being administered, officials should be willing to change the program if necessary. According to Goldberg, these questions are as follows:

1. Which treatment produces the best outcome for a specific group or person?

2. Do members of certain ethnic or socioeconomic groups respond similarly to certain types of treatment?

3. Is the effectiveness of a specific program linked to the age of participants?

4. Do females and males differ in their response to treatment? (p. 297)

The matching of treatment to gender, culture, ethnicity, language, and even sexual orientation has been shown to improve the odds of achieving positive outcomes in a variety of treatment settings for a variety of treatment issues (Goldberg, 2003). The utilization of culturally competent programs for various racial/ethnic groups should be particularly addressed given the fact that minorities are highly prevalent within the correctional population. Goldberg (2003) also points out the importance of addressing issues relevant to female drug offenders, among these the need for prenatal care and treatment and the need for contact with their children. As will be seen in Chapter 8, a high proportion of female offenders on community supervision (72%) are the primary caretakers of children under 18 years of age (Bloom, Owen, & Covington, 2003). Many female offenders may initially be motivated by the desire to reduce drug-related harm to their expected babies or to improve their relationship with and ability to attend to their children. This should not be overlooked, as it can be used to encourage these offenders to complete their treatment, thereby improving their chances for long-term recovery.

The main point is that treatment programming must be effectively matched to the specific inmate. Certain groups of inmates may also have needs that are common to that group (e.g., as based on ethnicity, age, or gender), and they may have other, more individualistic factors (e.g., divorce, abuse trauma) that may significantly affect their likelihood of success in treatment. In either event, treatment programs must rely on effective screening, assessment, and classification processes for inmate selection if they are expected to have successful outcomes. However, there is often a widespread tendency to simply place all inmates into drug treatment because substance abuse issues are so prevalent among the inmate population.

CONCLUSION

This chapter demonstrates that the physical features of a prison require a great deal of forethought prior to the ground being broken at the construction site. Issues related to the location of the prison facility, the types of custody levels and security, the function of the facility, and even logistical support for the facility are all important considerations. The design of the prison should be such that security is not compromised, but there is no true consensus on the best way to design a prison facility to address logistic issues while optimizing security features. In modern times, the use of large penitentiaries with gothic architecture has become antiquated. The formidability of modern security designs is obvious, particularly when one considers the introduction of the supermax facility to the realm of prison structures.

The various services within a facility and the different functions of the prison are again emphasized. Attention to the design of kitchen and laundry spaces is important, both for security purposes and for institutional living standards. Other services, such as religious services, recreational services, and tool shops where inmates can develop experience in specific trades, are also important for many reasons. These additional services eliminate idle inmate hands and minds and also give administrators some leverage in the control of inmate behavior; failure to comply with institutional rules can lead to restriction from these programs. Thus, the allocation of space and materials for these prison features can be important to future operational considerations. Though the cost to accommodate such features can be high, the likely reductions in prison offenses and misbehavior may offset such financial concerns.

Lastly, technological developments in the field of corrections have led to numerous improvements in security. Improvements in cell block and electric fence construction have made facilities even more secure, in terms of both internal and external security. Progress has also been made with regard to inmate identification and tracking. These advents in technology make the prison facility a part of a broader system that incorporates many other aspects of criminal justice and the surrounding community, enhancing security both inside and outside the prison.

The classification process is essential in determining both security needs and the specific needs of the inmate toward reformation. Without an adequate scheme for these types of decisions, corrections would likely be little more than guesswork. Offender placement in various security levels within a prison can be important for the safety of staff, other inmates, and the offenders themselves. Thus, the welfare of all hinges on the classification process. Appropriate classification is also important with regard to treatment approaches.

The ability of classification and treatment staff to correctly place clients into the appropriate substance abuse intervention program is critical to assisting the offender. As we move into future chapters on prison culture and the processing of specialized offenders, it will become clear that effective assessment and classification are critical to the correctional process. Society counts on the field of corrections to ensure public safety and to maintain security over offenders within its jurisdiction. This is an impossible mission without effective classification processes. Building a program on faulty assessment and classification processes is, arguably, a form of negligence to our obligation to protect society.

Want a Better Grade?

Get the tools you need to sharpen your study skills. Access practice quizzes, eFlashcards, video, and multimedia at **edge.sagepub.com/hanserbrief**

Interactive eBook

Visit the interactive eBook to watch SAGE premium videos. Learn more at **edge.sagepub.com/hanserbrief/access.**

 Career Video 6.1: Police Captain

 Prison Tour Video 6.1: ADA Compliance

 Prison Tour Video 6.2: Prison Food Service

 Prison Tour Video 6.3: Prisoner Identification

 Prison Tour Video 6.4: Security Classifications

 Prison Tour Video 6.5: External and Internal Security, Inmate Escape Attempts, and Chase Efforts

DISCUSSION QUESTIONS

Test your understanding of chapter content. Take the practice quiz at edge.sagepub.com/hanserbrief.

1. What is a panopticon, and how is this design used today?
2. What are the various levels of security that classify prison facilities? What are the various levels of security used to classify cell blocks within facilities?
3. What is the supermax or Marion Model of prison facility construction?
4. What were some of the constitutional issues that emerged as a result of *Madrid v. Gomez*?
5. What are the various components of perimeter security of a correctional facility?
6. What are the primary components of internal security within a correctional facility?
7. What are the four broad goals of classification?
8. What are the different custody levels of inmate housing, and what methods are used to determine the housing assignments of inmates?
9. Define the term *administrative segregation*. How is it usually used (and with what population is it usually used) within a prison?
10. What are four reasons for inmates being placed in protective custody in a prison facility?

KEY TERMS

Review key terms with eFlashcards at edge.sagepub.com/hanserbrief.

Administrative segregation, 169

Alcatraz, 151

Americans with Disabilities Act (ADA), 153

Bastille, 147

Blind spots, 161

Custody level, 168

Direct supervision design, 148

Isolation zone, 158

Level of security, 168

Maximum-security facilities, 149

Medium-security facilities, 149

Minimum-security facilities, 149

KEY CASES

APPLIED EXERCISE 6.1

Determine whether the following inmates should be given minimum-, medium-, or maximum-level security or if they should be given either administrative segregation or protective custody. Be sure to explain your answer.

Inmate Al is a member of a known street gang that is well represented in your correctional system. He denies being a member of a gang but has been identified by law enforcement as a bona fide member when operating in the community.

Inmate Nancy is a female offender who has two kids and is incarcerated for check fraud. She has served the majority of her sentence and seems to be doing well in educational and vocational programs offered in the prison. She often reflects on getting out, getting a good job, and taking care of her children.

Inmate Butch is a known sexual predator who has sexually assaulted a number of inmates and even one female correctional officer throughout his stint in prison. Butch is not liked by many inmates but is considered very masculine within the prison subculture. He seems to have no guilt or remorse for any of his acts.

Inmate Tom was a priest who was convicted of sexually abusing dozens of young boys within his church. He is slight of build and is a fairly docile-appearing man. He just arrived at your facility a couple of days ago.

Inmate Conway minds his own business and does not talk much with prison staff. He keeps out of trouble, for the most part, but has an independent quality that keeps him a bit at odds with the programming aspects of prison. He has not been identified as dangerous (he was convicted of burglary), but he also does not seem to make great strides toward getting released early.

WHAT WOULD YOU DO?

You are the public relations officer for the state department of corrections. Recently, the state system has identified an area for a new prison site, but once word of this decision was presented to the local community, a number of letters were sent to the governor's office expressing concern and resistance to this idea. You have been assigned the task of addressing these concerns and have been asked to mitigate public reactions to the idea of building a prison in the area. It is very clear that the state intends to build the facility regardless but would like to resolve the public relations issue that has emerged.

The community is midsize; it has a population of approximately 75,000, with some outlying towns also existing in the area. The nearest major metropolitan area is well over 3 hours' driving distance away. The community's economy is somewhat stable but lacks significant industrial development and is mostly agricultural in nature.

The prison is slated to be a minimum- to medium-security facility that will house 1,800 inmates. Many of these inmates will likely be in various industrial and educational programs. Security officers, educational specialists, and other employees within the prison will be needed, which might mean jobs for community members. This additional employment might also draw persons from outside of the area, which will increase the need for various goods and services (e.g., dry cleaning, groceries, restaurant services, home purchases) from business owners in the local area.

The governor's office has tasked you with showcasing the positive impact that this facility can have on the community while also alleviating concerns about the potential pitfalls to having a prison in the area.

What would you do?

Kevork Djansezian/Getty Images News/Getty Images

7

Prison Subculture and Prison Gang Influence

Learning Objectives

1. Compare importation theory with exportation theory.
2. Discuss the modern inmate subculture standards.
3. Evaluate the impact of prison culture on corrections staff.
4. Describe the process of prisonization.
5. Identify various aspects of guard subculture, including the unique subculture of female correctional officers.
6. Discuss the impact that prison gangs have had on prisons, including the traditional prison subculture.
7. Explain what prison systems do to control gang problems that occur in their facilities.

The Gang Is My Family; the Gang Is My Purpose

I will stand by my brother

My brother will come before all others

My life is forfeit should I fail my brother

I will honor my brother in peace as in war.

Aryan Brotherhood Oath

James Harris, known as "Cornfed" to his family and friends, was a member of "The Brand" (a term for the Aryan Brotherhood, or AB). He was hanging on the recreation yard with "Blinky," another member of The Brand. Blinky was family to Cornfed; they had known each other for years, both inside and outside of prison.

"So if you keep your nose clean, you get on parole; you're going on a mission, right?" Cornfed asked Blinky.

Blinky nodded. "Yep, that's right. In fact, I am taking the mission, and then I'll be free to conduct *rahowa* all that I want. I know I'll end up back here, but at least I'll have paid my debt and taken a couple of them with me."

Cornfed looked at the ground and thought about what Blinky had said. Blinky had been "down" (doing time) for many years, and he had benefitted from the AB substantially throughout those years. His "mission" would be to operate a meth lab for the AB, as he was known to be a very good "cook"—someone skilled in manufacturing methamphetamine.

Cornfed also understood why Blinky would be going on *rahowa*, the term in the AB for "racial holy war," which would ultimately entail some form of hate crime against African Americans. He knew that Blinky would be highly motivated to engage in *rahowa* because many of the AB members would be expecting it due to his past experiences.

Initially, Blinky had stayed out of the gang life, but then he had been sexually assaulted by two black men while working in the laundry area of the prison. The men had told him from that point on he would be their *punk*, a term for someone who has been "turned out" and given the identity of a woman in prison. Blinky fought during the assault, but they had caught him off guard and he was unable to defend himself. The second time they attempted to rape him, he was ready and was able to hold them off until "Big John" unexpectedly entered the shop. Big John did not work in the laundry but had been sent to gather some towels to clean up blood from another assault that had occurred just 10 minutes prior.

Big John saw what was going on and said, "My only regret today has been that I wasn't involved in what La Eme just finished. . . . I hate being the one who has to do the cleanup for somebody else's handiwork." He reached behind his back with one hand and pulled out an 8-inch blade. "I want you to meet my friend here, Aryan Express. Like American Express, I never leave home without it."

Big John carved both of the men up, holding one in the air by the throat while he repeatedly gouged the other in the abdomen. He then reached around the man he was holding and buried the knife into the small of his back. With a satisfied grunt, Big John threw the man down on the ground. Then he glared at Blinky. "You better not say nuthin,' you understand?"

Blinky nodded vigorously as Big John stepped over to the first man and slashed the man's throat. Then he moved to the other man and kicked him repeatedly in the head until his face was a bloody, unrecognizable mass of pulpy flesh.

When he had finished, Big John turned back to Blinky. "You now owe the AB, unless of course you like being a punk, in which case you can be the AB's punk. They're gonna be locking me up in just a bit, but my brothers will come for you. This completes my *rahowa* for now." Kneeling, he wiped the blade, Aryan Express, across the shirt of the first man he killed and hid it under a spare mattress in the shop.

"When they come to talk with you about joining the family, let them know where I put the metal, okay?"

Blinky agreed. Leaning against one of the laundry tables, Big John pulled out a cigarette and handed another to Blinky. They smoked together in silence until the guards came.

Now Cornfed looked at Blinky and said, "My brother, when you're out there, just know that I'm proud of you and I'll honor your name here on the inside."

Blinky smiled and gazed through the chain-link fence at the world beyond. Soon he would be on the outside with his family, the Aryan Brotherhood of Texas.

INTRODUCTION

This chapter examines a unique aspect of the world of corrections. Students will learn that within the institutional environment, a commonality of experiences arises between those who are involved—both inmates and staff. Many people may not be aware that in fact the mindset and experiences of inmates often affect the mindset of security personnel who work with those inmates. In essence, there is an exchange of beliefs and perspectives that produces a fusion between the two groups. This creates a singular subculture that is the product of both inmate norms being brought in from the outside and norms being taken from the prison to the outside community.

It is important for students to understand that prison staff are not immune to the effects of the profound social learning that occurs in prison, and, over time, as they become more enmeshed in the prison social setting, they begin to internalize many of the beliefs and norms held by the prison subculture. While this may seem to be counterproductive and/or even backward from what one might wish within the prison environment, this is an inevitable process as prison staff find themselves interacting with the street mentality on a day-to-day basis. In actuality, this has a maturing effect on correctional workers, as they begin to see a world that is not necessarily black and white but instead has many shades of gray. When staff get to know inmates on a personal level, issues become more complicated than being simple "good guy–bad guy" situations. The nuances and differences between different offenders tend to complicate what initially might seem like obvious, easy decisions.

Because correctional staff interact with offenders on a daily basis, a sense of understanding develops both among correctional staff and between staff and the inmate population. Inmates come to expect certain reactions from correctional staff, and staff come to expect certain reactions from inmates. Informal rules of conduct exist where loyalty to one's own group must be maintained, but individual differences in personality among security staff and inmates will affect the level of respect that an officer will get from the inmate population, or, for inmates, the amount of respect that they gain from others serving time. Correctional staff learn which inmates have influence, power, or control over others, and this may affect the dynamics of interaction. Further still, some inmates may simply wish to do their time, whereas others produce constant problems; to expect security staff to maintain the same reaction to both types of inmates is unrealistic.

The dynamics involved in inmate–inmate, inmate–staff, and staff–staff interactions create circumstances that do not easily fall within the guidelines of prison regulations. In order to maintain control of an inmate population that greatly outnumbers the correctional staff, many security officers will learn the personalities of inmates and will become familiar with the level of respect that they receive from other offenders. Likewise, and even more often, inmates watch and observe officers who work the cell block, the dormitory, or other areas where inmates congregate. They will develop impressions of the officer, and this will determine how they react to him or her. The officer is essentially labeled by the inmate population, over time, as one who deserves respect or one who deserves contempt. In some cases, officers may be identified as being too passive or "weak" in their ability to enforce the rules. In such cases, they are likely to be conned, duped, or exploited by streetwise convicts.

The various officer–inmate interactions impact the daily experiences of those involved. Understanding how a myriad of nuances affect these interactions is critical to understanding how and why prisons operate as they do. In prisons that have little technology, few cameras, and shortages of staff, the gray areas that can emerge in the inmate–staff interactional process can lead to a number of ethical and legal conundrums. It is with this in mind that we now turn our attention to factors that create and complicate the social landscape of American prisons.

THEORIES OF PRISON SUBCULTURE

Importation Theory

The key tenet of **importation theory** is that the subculture within prisons is brought in from outside the walls by offenders who have developed their beliefs and norms while on the streets. In other words, the prison subculture is reflective of the offender subculture on the streets. Thus, behaviors respected behind the walls of a prison are similar to behaviors respected among the criminal population outside of the prison. There is some research that does support this notion (Wright, 1994). Most correctional officers and inmates know that the background of an offender (or of a correctional officer) has a strong impact on how that person behaves, both inside and outside of a prison; this simply makes good intuitive sense.

There are two important opposing points to consider regarding importation theory. First, the socialization process outside of prison has usually occurred for a much longer period of time for many offenders and is, therefore, likely to be a bit more entrenched. Second and conversely, the prison environment is intense and traumatic and is capable of leaving a very deep and lasting influence upon a person in a relatively short period of time. While this second point may be true, most inmates in prison facilities have a history of offending and will tend to have numerous prior offenses, some of which may be unknown to the correctional system. This means that inmates will likely have led a lifestyle of dysfunction that is counter to what the broader culture may support. Thus, these individuals come to the prison with years of street experience and bring their criminogenic view of the world to the prison.

It has been concluded by some scholars (Bernard, McCleary, & Wright, 1999; Wright, 1994) that though correctional institutions may seem closed off from society, their boundaries are psychologically permeable. In other words, when someone is locked up, they still are able to receive cultural messages and influences from outside the walls of the prison. Television, radio, and mail all mitigate the immersion experience in prison. Visitation schedules, work opportunities outside of the prison, and other types of programming also alleviate the impact of the prison environment. Indeed, according to Bernard, McCleary, and Wright (1999), "Prison walls, fences, and towers still prevent the inside world from getting outside, [but] they can no longer prevent the outside world—with its diverse attractions, diversions, and problems—from getting inside" (p. 164). This statement serves as a layperson explanation of importation theory that is both accurate and practical.

Importation theory: Subculture within prisons is brought in from outside by offenders who have developed their beliefs and norms while on the streets.

Indigenous Prison Culture and Exportation Theory

In contrast to the tenets of importation theory is the notion that prison subculture is largely the product of socialization that occurs inside prison. It was the work of Gresham Sykes (1958) that first introduced this idea in a clear and thorough manner. His theory has been referred to as either the deprivation theory of prisonization or the indigenous model of prison culture. Sykes referred to the pains of imprisonment as the rationale for why and how prison culture develops in the manner that it does. The **pains of imprisonment** is a term that refers to the various inconveniences and deprivations that occur as a result of incarceration. According to Sykes, the pains of imprisonment tended to center on five general areas of deprivation, and it was due to these deprivations that the prison subculture developed, largely as a means of adapting to the circumstances within the prison. Sykes listed the following five issues as being particularly challenging to men and women who do time:

1. The loss of liberty
2. The loss of goods and services readily available in society
3. The loss of heterosexual relationships, both sexual and nonsexual
4. The loss of autonomy
5. The loss of personal security

Inmates within the prison environment essentially create value systems and engage in behaviors that are designed to ease the pains of deprivation associated with these five areas. Research has examined the effects of prison upon inmates who are forced to cope with the incarceration existence.

For instance, Johnson and Dobrzanska (2005) studied inmates who were serving lengthy sentences. They found that incarceration was a painful but constructive experience for those inmates who coped maturely with prison life. This was particularly true for inmates who were serving life sentences (defined to include offenders serving prison terms of 25 years or more without the benefit of parole). As a general rule, lifers came to see prison as their home and made the most of the limited resources available there; they established daily routines that allowed them to find meaning and purpose in their prison lives—lives that might otherwise seem empty and pointless. The work of Johnson and Dobrzanska points toward the notion that humans can be highly adaptable regardless of the environment.

Many of the mannerisms and behaviors observed among street offenders have their origins within the prison environment. Indeed, certain forms of rhyming, rap, tattoos, and dress have prison origins. For example, the practice known as "sagging," when adolescent boys allow their pants to sag, exposing their underwear, originates from jail and prison policies denying inmates the use of belts (because they could be used as a weapon or means to commit suicide). This practice is thought to have been exported to the streets during the 1990s as a statement of African American solidarity as well as a way to offend white society.

Another example might be the notion of "**blood in—blood out**," which is the idea that in order for inmates to be accepted within a prison gang, they must draw blood (usually through killing) in an altercation with an identified enemy of the gang. Once in the gang, they may only leave if they draw blood of the enemy sufficient to meet the demands of the gang leadership or by forfeiting their own blood (their life). This same phrase is heard among street gangs, including juvenile street gangs, reflecting the fact that these offenders mimic the traditions of veteran offenders who have served time in prison. Consider also certain attire that has been popular off and on during the past decade, such as when Rhino boots became popular footwear, not due to their stylishness or functionality but because they were standard issue for working inmates in many state prison systems.

THE INMATE SUBCULTURE OF MODERN TIMES

In all likelihood, the inmate subculture is a product of both importation and indigenous factors. Given the complicated facets of human behavior and the fact that inmates tend to

Pains of imprisonment: The various inconveniences and deprivations that occur as a result of incarceration.

Blood in—blood out: The idea that for inmates to be accepted within a prison gang they must draw blood from an enemy of the gang.

cycle in and out of the prison system, this just seems logical. In fact, attempting to separate one from the other is more of an academic argument than a practical one. The work of Hochstellar and DeLisa (2005) represents an academic attempt to negotiate between these two factors. These researchers used a sophisticated statistical technique known as structural equation modeling to analyze the effects of importation and indigenous deprivation theories. They found evidence supporting both perspectives but found that the main factor that determined which perspective was most accountable for inmates' adaptation to prison subculture was their level of participation in the inmate economy (Hochstellar & DeLisa, 2005).

PHOTO 7.1 This inmate proudly displays his gang-affiliated tattoos.

Andrew Lichtenstein/Corbis Premium Historical/Getty Images

This is an important finding because it corroborates practical elements just as much as it navigates between academic arguments regarding subculture development. The prison economy is one of the key measures of influence that an inmate (and perhaps even some officers) may have within the institution. The more resources an inmate has, the wealthier he or she will be in the eyes of the inmate population. Oftentimes, those inmates who are capable of obtaining such wealth are either stronger, more cunning, or simply smarter (usually through training and literacy, such as with jailhouse lawyers) than most other members of the inmate population. Thus, these inmates are likely to be more adept at negotiating the prison economy, and they are likely to have more influence within the prison subculture. They are also more likely to be successfully adapted to the prison culture (the influence of indigenous prison cultural factors) while being able to procure or solicit external resources (this being a source of exportation outside the prison walls). In short, those who master the economy often have effective and/or powerful contacts both inside and outside of the prison. This is consistent with the findings of Hochstellar and DeLisa (2005).

The Convict Code and Snitching

The prison subculture has numerous characteristics that are often portrayed in film, in academic sources, and among practitioners (see Focus Topic 7.1). Chief among these characteristics is the somewhat fluid code of conduct among inmates. This is sometimes referred to as the **convict code**, and it is a set of standards in behavior attributed to the true convict—the title of *convict* being one of respect given to inmates who have proven themselves worthy of that title. Among academic sources, this inmate code emphasizes oppositional values to conventional society in general and to prison authorities in particular. The most serious infraction against this code of conduct is for an inmate to cooperate with the officials as a snitch. A **snitch** is the label given to an inmate who reveals the activity of another inmate to authorities, usually in exchange for some type of benefit within the prison or legal system. For example, an inmate might be willing to tell prison officials about illicit drug smuggling being conducted by other inmates in the prison in exchange for more favorable parole conditions, transfer to a different prison, or some other type of benefit.

Among all inmates, it is the snitch who is considered the lowest of the low. In traditional "old school" subcultures (i.e., those of the 1940s through the 1970s), snitches were rare and were afforded no respect. Their existence was precarious within the prison system, particularly because protection afforded to snitches was not optimal. During riot situations and other times when chaos reigned, there are recorded incidents where inmates specifically targeted areas where snitches were housed and protected from the general population. In these cases, snitches were singled out and subjected to severely gruesome torture and were usually killed. Perhaps the most notorious of these incidents occurred at the New Mexico Penitentiary in Santa Fe. This prison riot occurred in 1980 and resulted in areas of the prison being controlled by inmates. These inmates

Convict code: An inmate-driven set of beliefs that inmates aspire to live by.

Snitch: Term for an inmate who reveals the activity of another inmate to authorities.

(Continued)

At this point, the hyena is obviously crossed out among others on the dorm and is held in even lower regard by other inmates as a coward who sidesteps his charge and allows others to "pay his lick," so to speak. In some ways, this process of maintaining some semblance of decency and tangible respect for institutional authority is a means of redemption for the inmate and the dorm. In this regard, it is viewed that these men are capable of reform on at least a base level, contrasting with the typical saying that "there is no honor among thieves." In essence, even the inmate subculture has standards that the hyena has failed to maintain. Thus, the lack of respect goes to the hyena, and, so long as the officer conducts his job in a firm but fair manner, the officer is afforded his due respect.

This information was drawn from an interview with Jonathan Hilbun. It is worth mentioning that Mr. Hilbun has been incarcerated for 18 and a half years in a variety of institutions, including Angola and Richwood Correctional Center. ●

Source: Jonathan Hilbun. (2011, December 2). Personal interview. Used with permission.

that homosexual, transgender, or even just more effeminate males are at great risk for being assaulted behind bars. With this concern in mind, the BJS issued additional tabulations for its PREA work titled *Supplemental Tables: Prevalence of Sexual Victimization Among Transgender Adult Inmates*. This work found the following:

1. An estimated 35% of transgender inmates held in prisons and 34% held in local jails reported experiencing one or more incidents of sexual victimization by another inmate or facility staff in the past 12 months or since admission, if less than 12 months.

2. About a quarter of transgender inmates in prisons (24%) and jails (23%) reported an incident involving another inmate. Nearly three quarters (74%) said the incidents involved oral or anal penetration, or other nonconsensual sexual acts.

3. When asked about the experiences surrounding their victimization by other inmates, 72% said they experienced force or threat of force and 29% said they were physically injured.

4. Transgender inmates reported high levels of staff sexual misconduct in prisons (17%) and jails (23%). Most transgender inmates who had been victimized reported that the staff sexual misconduct was unwilling on their part (75%) and that they experienced force or threat of force (51%) or were pressured by staff (66%) to engage in the sexual activity.

5. Among those victimized by staff, more than 40% of transgender inmates in prison and jails said they had been physically injured by the staff perpetrator.

PHOTO 7.3 When weather permits, these female inmates are allowed to walk and jog in the recreation yard of a women's prison.

Keith Myers//*The Kansas City Star*/Newscom

Two issues with this data are worth noting. First, the majority of the transgender population within the custodial environment is male-to-female; the female-to-male transgender population is miniscule within the correctional population around the nation. Second, the notion of being effeminate, gay, or transgender has severe and profound negative consequences within the male prison subculture, but this is not so much the case within female facilities. Thus, this is an important area of concern within male prisons, and the results from recent PREA data confirm this.

The Strong, Silent Type and the Use of Slang

Also common within the prison subculture is the belief that inmates should maintain themselves as men who show no emotion and are free from fear, depression, and anxiety. Basically, the "strong, silent type" is the ideal. Today's young offenders may attempt to maintain this exterior image, but the effects of modern society (the prevalence of mental health services, reliance on medications, technological advances, and a fast-changing society) often preclude this stereotypical version of the offender because a reliance on those services and medications is viewed as being "weak" in many prisons. Inmates are also expected to refrain from arguments with other inmates. The general idea is that inmates must do their own time without becoming involved in the personal business of others. Getting involved in other people's business is equated to being a gossip, and this is considered more of a feminine behavior. Thus, to be manly in prison, inmates must mind their own business.

Inmates who stick to these two rules of behavior are generally seen as in control, fairly wise to the prison world, and not easily manipulated. In some cases, these inmates may be referred to as a *true convict* rather than an inmate. In the modern prison culture, the title of **convict** refers to an inmate who is respected for being self-reliant and independent of other inmates or the system. Convicts are considered mature and strong, not weak and dependent on others for their survival. They are considered superior to the typical inmate, and, while not necessarily leaders of other inmates (indeed, most do not care to lead others but simply wish to do their own time), they are often respected by younger inmates becoming acculturated into the prison environment.

Convict: An inmate who is respected for being self-reliant and independent of other inmates or the system.

> ### PRACTITIONER'S PERSPECTIVE
>
> *"For my generation, law enforcement was an honored field."*
>
> Visit the IEB to watch Patricia Wooden's video on her career as a division chief.
>
>

Within prison systems, there evolves a peculiar language of slang that often seems out of place in broader society. This slang has some consistency throughout the United States but does vary in specific terminology from state to state. The language often used by inmates, including slang, is affected by their racial and gang lineage. For instance, members of the Crips or the Bloods have certain terms that usually are only used by their group, often as a means of identifying or denigrating the other group. Likewise, Latino gang members in the Mexican Mafia and/or the Texas Syndicate tend to have their own vernacular, much of which is either Spanish phraseology or some type of unique slang. Terms like *punk*, *shank* (a knife), *bug juice* (referring to psychotropic medications), and *green light* (referring to clearance to assault another inmate) are commonly used in most prison systems.

Maintaining Respect

In nearly every prison around the United States, one concept is paramount among inmates: respect. This is perhaps the most important key to understanding the inmate subculture.

Respect: An inmate's sense of standing within the prison culture.

Prison status revolves around the amount of respect given to an inmate and/or signs of disrespect exhibited toward an inmate. **Respect** is a term that represents an inmate's sense of masculine standing within the prison culture; if inmates are disrespected, they are honor bound to avenge that disrespect or be considered weak by other inmates. If an inmate fails to preserve his sense of respect, this will lead to the questioning of his manhood and his ability to handle prison and will lead others to think that he is perhaps weak. The fixation on respect (and fixation is an appropriate description in some prisons) is particularly pronounced among African American gang members in prison. This is also glorified in much of the contemporary gangsta music that emerged in the 1990s and continues today. Because inmates have little else, their sense of self-respect and the respect that they are able to garner from others is paramount to their own welfare and survival.

In addition, it is usually considered a sign of weakness to take help or assistance from another inmate, at least when one is new to the prison environment. Indeed, inmates will be tested when new to the prison world; they may be offered some type of item (e.g., coffee or cigarettes) or provided some type of service (e.g., getting access to the kitchen), but this is never for free or due to goodwill. Rather, inmate subculture dictates that a debt is thus owed by the *newbie* (a term for inmates who are new to a prison). New inmates may be required or coerced to do "favors" for the inmate who provided them with the good or service. For example, they might be asked to be a "mule" for the inmate or the inmate's gang. A **mule** is a person who smuggles drugs into prison for another inmate, often using his or her own body cavities to hide the drugs from prison authorities. In other cases, the inmate may be forced to become a punk for that inmate or for an entire prison gang.

Mule: A person who smuggles drugs into prison for another inmate.

The Con and the Never-Ending Hustle

Among inmates, there is the constant push and pull between the need to "con" others and, at the same time, avoid being conned. Naturally, this constant and contradictory set of expectations completely impedes the ability for inmates to develop any sort of true trust; they must always remain vigilant for the potential "hustle" within the prison system. **Hustle** refers to any action that is designed to deceive, manipulate, or take advantage of another person. Further, consider that the very term *convict* includes the word *con*, which implies that the individual cannot and should not be trusted. Thus, convicts are, stereotypically, always on the hustle, so to speak.

Hustle: Any action that is designed to deceive, manipulate, or take advantage of another person.

Inmates who are able to "get over" on others and/or "skate" through work or other obligations in the prison system are considered particularly streetwise and savvy among their peers. In fact, some prison systems, such as the Texas prison system, have a term for this concept: *hogging*. **Hogging** implies that a person is using others for some type of gain or benefit—manipulating others into doing work or fulfilling obligations on his or her behalf. When inmates are able to find some means to manipulate others into doing their dirty work, they are active in the art of the con. The process by which they encourage or manipulate a person to provide such a service is all part of the hustle.

Hogging: A term used to imply that a person is using others for some type of gain or benefit.

A classic portrayal of this type of logic, though not a prison example, can be found in *The Adventures of Tom Sawyer* by Mark Twain. In this novel, Tom Sawyer convinces other boys in the area that painting a fence (a chore assigned to young Tom) is a fun activity. So fun is it, according to Tom, that he will not allow anyone to help him unless they pay him to do so. Ultimately, other boys pay Tom for the "opportunity" to paint the fence and join in the fun. Once several other boys pay their fee for the privilege of painting the fence, Tom slips off to spend the money that he has procured from those duped into doing his assigned work. This example demonstrates all the fine points of the con, the hustle, and the act of hogging others.

THE IMPACT OF THE INMATE SUBCULTURE ON CORRECTIONS STAFF

Perhaps one of the most interesting dynamics within prison can be found between the inmates and the prison staff. This area of discussion is both complicated and paradoxical in many respects. The paradox involved with this dynamic is that although inmate subculture restricts inmates from "siding" with officers and officer culture restricts officers from befriending inmates, there is a natural give-and-take that emerges between both groups. In fact, a symbiotic relationship usually emerges between prison security staff and the inmate population. This **symbiotic prison relationship** is a means of developing mutually agreeable and informal negotiations in behavior that are acceptable within the bounds of institutional security and also allow inmates to meet many of their basic human needs. This relationship is grounded in the reality of the day-to-day interactions that prison security staff have with inmates.

Because prison is a very intense environment that has a very strong psychological impact on both inmates and staff, it is only natural that this type of relationship often emerges. While the rules of the institution are generally clearly written, these rules are not always pragmatic for the officers who must enforce them. For example, a rule to restrict inmates from having more than one blanket in their cell may, on the face of it, seem easy enough to enforce. However, consider the following scenarios when considering rule enforcement:

Symbiotic prison relationship: When correctional staff and inmates develop negotiation behavior that is acceptable for institutional security and also meets inmates' basic human needs.

Scenario 1

A veteran officer with many years of experience may find that a given inmate, Inmate X, who upholds the convict code and has respect within the institution, has the flu during the winter. The officer has access to additional blankets, and this is known among the inmates. Inmate X tends to mind his own business and usually does as he is expected when the officer is on duty. In this case, the officer may decide to offer Inmate X an additional blanket and would do so with no expectation that the inmate give something in return. Likewise, the inmate (as well as others watching) would know that the officer's kindness should not be taken for weakness or no further empathy will be shown to convicts.

Scenario 2

This same veteran officer finds that Inmate Y, who does not uphold the convict code and generally has average clout (at best) within the prison culture, also has the flu. Inmate Y sometimes causes problems on the cell block and is sarcastic with officers. The veteran officer, in this case, would likely not give an additional blanket to Inmate Y, even if Inmate Y were to be courteous enough to ask for it. In most cases, Inmate Y would know better than to ask, since he knows that he does not honor the convict code or work within the commonsense bounds of the symbiotic prison relationship. If he were to ask and especially if he were to push the issue, the veteran officer would inform him that "the rules are the rules" and would indicate that he needs to keep quiet and go to sick call when that option is available. Further discourse from Inmate Y would result in comments from the veteran officer that would imply that he is being a troublemaker and that he is not doing his time "like a man," which would result in a loss of respect for Inmate Y among others on the cell block. This would likely shame Inmate Y and cause him to lose status, but the veteran officer would likely gain status among the inmates for being firm, streetwise, and cognizant of subcultural norms. Because he does not give in to Inmate Y, he would be perceived as strong and capable, not subject to manipulation and not an easy mark.

Obviously, Inmate X and Inmate Y are being given different standards of treatment. Regardless of whether it is overtly stated or simply presumed, veteran officers will tend to leave convicts and/or trouble-free inmates unbothered and may, in some cases, even extend some degree of preferential treatment, within acceptable boundaries that allow them to maintain respect on the cell block. However, this does not mean that they will do so for all inmates; they reserve the right to use discretion when divvying out the paltry resources available within the prison. In short, these veteran officers become effective resource and power brokers as a means of gaining compliance and creating an informal system of fairness that is understood among the inmates. Essentially, veteran officers and inmates operate with this understanding:

"I will let you do your time, but you will let me do my time—one shift at a time."

This concept is important because it creates a connection between both groups; they are both in a noxious environment and have a role that they must uphold. Yet at the

PHOTO 7.4 Sergeant Tatum talks with inmates regarding the organization of an upcoming function in a meeting hall of the prison. The means by which officers and inmates talk to each other set the tone for respect or disrespect in the prison.

Robert Hanser

Convict boss/officer: A correctional officer with a keen understanding of convict logic and socialization.

Prisonization: The process of being socialized into the prison culture.

same time, some degree of give-and-take is necessary to avoid extremes in rules that do not, ultimately, create just situations. So long as inmates allow the officer to generally do his time, one shift at a time, he will, in turn, leave them to serve their time without problems. On the other hand, if an inmate does not honor this understanding, he should expect no mercy or consideration from the prison security staff; the rules are the rules, and any sense of discretion will simply cease to be acted upon.

Officers who master these types of negotiations tend to gain respect from inmates and even from other officers. They may sometimes be referred to as "convict bosses" by inmates. The term **convict boss** or **convict officer** is applied to a correctional officer who has developed a keen understanding of convict logic and socialization and uses that knowledge to maximize control over his or her assigned post. This term denotes respect gained from inmates and generally comes with time, experience, sound judgment, and a cunning personality that is not easily deceived or manipulated. The officer is not perceived as weak but is instead thought to possess a good degree of common sense by inmates within the facility.

One important note should be added to this discussion. Students should keep in mind that the examples in the prior scenarios present a veteran officer with several years of experience. The use of this type of discretion by newer officers who do not have sufficient time and experience working with the inmate population will not have the same result. If a newer or younger officer attempted such discretion, he or she would likely be seen as a "sucker" and someone who could be easily marked for future exploitation. This person would not be perceived to understand the fine nuances of discretion with regard to prison rules, norms, and mores. This person would also not likely be trusted by peers who, generally, would expect him or her to stay "by the book" until he or she developed the level of expertise to make fine distinctions in blurry situations. This officer would likely be labeled "weak" among inmates and might even be considered an "inmate lover" by other officers. These labels should be avoided in hardcore institutions because once they are applied, it is very difficult (if not impossible) to be rid of them.

Lastly, the mannerisms that are displayed, by both inmates and officers, often reflect the type of upbringing that one has had and also tend to indicate the value system from which that person operates. Within the prison, this is important because during an inmate or beginning officer's first few weeks of indoctrination to the prison experience, he or she is being "sized up" or appraised by others who observe him or her. Both the officers and the inmates begin to determine if the person is likely to be easily influenced and/or manipulated. This formative period whereby inmates and staff are socialized into the prison culture is important due to the influences of the prison subculture that include the inmate's subculture, the officer's subculture, and the need to master the symbiotic prison relationship between the two.

PRISONIZATION

Prisonization is the process of being socialized into the prison culture. This process occurs over time as the inmate or the correctional officer adapts to the informal rules of prison life. Unlike many other textbook authors, the author of this text thinks that it is important to emphasize that correctional officers also experience a form of prisonization that impacts their worldview and the manner in which they operate within the prison institution. In his text on prisonization, Gillespie (2002) makes the following introductory statement: Prison is a context that exerts its influence upon the social relations of those who enter its domain (p. 1).

The reason that this sentence is set off in such a conspicuous manner is because it has profound meaning and truly captures the essence of prisonization. Students should understand that the influence of the prison environment extends to *all* persons who enter its domain, particularly if they do so over a prolonged period of time. Thus, prisonization impacts both inmates and staff within the facility. While the total experience will, of course, not be the same for staff as it is for inmates, it is silly to presume that staff routinely exposed to aberrant human behavior will not be impacted by that behavior.

With respect to inmates, Gillespie (2002) found that both the individual characteristics of inmates and institutional qualities affect prisonization and misconduct. However, he found that individual-level antecedents explained prisonization better than did prison-level variables. This means that the experiences of inmates prior to being imprisoned are central to determining how well they will adapt to the prison experience. For this text, this contention will also be extended to prison guards; their prior experiences and their individual personality development prior to correctional employment will dictate how well they adapt to both the formal and informal exchanges that occur within the prison.

THE GUARD SUBCULTURE

This area of discussion is controversial and has led to a great deal of debate. One reason for developing this text and providing a discussion on this particular topic is to provide students with a realistic and no-nonsense appraisal of the world of corrections, particularly as practiced in the prison environment. In providing a glimpse of the guard subculture (and this text contends that a guard subculture *does* exist), it is important to keep in mind that the specific characteristics of this subculture vary from prison system to prison system and even from prison to prison within the same state system. The reasons for this are manifold but are mostly due to the fact that, unlike inmates, guards are not forced to remain within the prison environment 24/7. Rather, guards have the benefit of time away from the institution, and they can (and sometimes do) transfer from facility to facility, depending on their career path.

Further, since guards are routinely exposed to external society (contact with family, friends, the general public, the media, etc.), they are able to mitigate many of the debilitating effects of the prison environment. Likewise, their integration into society mitigates the degree to which prison socialization will impact them, personally and professionally. Thus, there is greater variability in the required adaptations of prison guards when compared to inmates. The type of institution that they find themselves working within can also impact this socialization. A guard who works at a maximum-security institution or one for particularly violent offenders will likely experience a different type of socialization than a guard assigned to a minimum-security dormitory. All of these factors can impact how the prison culture affects individual officers and the degree to which they become enmeshed into the guard subculture.

The discussion that follows is intended to address guard subculture in maximum-security prisons or those institutions that have histories of violence among inmates. Larger facilities that have more challenging circumstances tend to breed the type of subculture that will be presented here. Though modern-day correctional agencies seek to circumvent and eliminate these subcultural dimensions, they nonetheless still exist in various facilities.

The popular Hollywood image of prison guards is that they are brutal and uncaring and that their relationships with inmates are hostile, violent, and abusive. However, this is a very simplistic and inaccurate view of prison guards that makes for good entertainment but does not reflect reality. For many, corrections work is a stable job available to persons in rural areas where few other jobs exist, and it produces a liveable wage for the effort.

PRISON TOUR VIDEO: Prisonization affects both inmates and correctional staff. Go to the IEB to watch an interview with prisoners about prison culture and behavior.

For others, prison work may be a stepping-stone to further their career, particularly if they are interested in criminal justice employment. Indeed, the author of this text worked at Eastham Unit in Texas while attending school at a state university in the area, and this was a common practice among many students of criminal justice or criminology studies. This means that, at least in this case, many of the prison guards employed in the region actually had an above-average education and most likely possessed depth and purpose that exceeded the Hollywood stereotype.

The author of this text would like to acknowledge the work of Kelsey Kauffman (1988) in explaining the overall processes behind prison guard socialization and the development of prison guard subcultures. Like Kauffman, the author of this text encountered a similar transitional experience where, over time, the aloof and distant feeling between himself and his fellow coworkers grew into a feeling of camaraderie and close connection in identity. To this day, this author considers himself, first and foremost, a prison guard at heart. However, it is Kauffman who so eloquently and correctly penned the formation and description of the guard subculture, and it is her work that will be used as the primary reference for this section.

According to Kauffman (1988), the guard subculture does not develop due to any single contextual characteristic, such as prisonization, indigenous forms of influence, or the importation of values. Rather, the culture is a product "of a complex interaction of importation, socialization, deportation, and cultural evolution" (p. 167). Kauffman notes that prison guards have a distinct and identifiable subculture that separates them from other professionals. The central norms of this subculture dictate how they proceed with the daily performance of their duties, such as with the example scenarios provided earlier when discussing the impact of the inmate subculture on custodial staff. In describing the prison guard subculture, Kauffman produced a list of the basic tenets behind it. These same tenets are presented in this text due to the author's perception that they are reflective of those encountered in most prisons throughout the United States. They include the following:

1. **Always go to the aid of an officer in distress.** This is the foundation for cohesion among custodial staff. This tenet also can, in times of emergency, provide justification for violating norms within the bureaucratic system. This tenet applies to all guards, regardless of how well accepted the officer in distress may or may not be. This norm is key to officer safety and is fundamental. If an officer fails to uphold this norm, he or she will likely be ostracized from the group and will be treated as an outsider.

2. **Do not traffic drugs.** This is also considered fundamental because of the danger that it can create as inmates fight for power over the trade of these substances. In addition, the use of drugs is illegal and does not reflect well on officers, who are supposed to keep drug offenders behind bars. If an officer violates this tenet, it is considered justified within the subculture to inform authorities, but most will not do so due to the reluctance to betray a fellow officer. However, it would not be uncommon for guards to take it upon themselves to put pressure on the officer who violates this norm through threats, intimidation, and coercion. In addition, officers will likely isolate the officer from interactions and will not invite him or her to functions outside of work. The officer will be treated as persona non grata.

3. **Do not be a snitch.** In many respects, this is a carryover from the inmate subculture. This dictate comes in two forms. First, officers should never give information to inmates that could get another officer in trouble. Generally speaking, officers are expected not to discuss other officers, their business, or their personal lives with inmates. The second prohibition applies to investigative authorities of the prison system. Officers are expected to stay silent and not divulge information that will "burn" another officer, particularly when the Internal Affairs Division (IAD) is investigating an incident. While it is expected that officers will not knowingly place their coworkers in legally compromising situations where they must lie for their coworkers, it is still equally expected that coworkers will not snitch on their fellow officer.

This tenet is perhaps one of the most difficult to follow because, in some cases, it puts officers in a position where they must lie to cover for their coworkers, even when they were not directly involved. This can occur during investigations and even if officers are brought to court in a lawsuit. Officers who comply with institutional rules still cannot be assured that they will be safe from liability if they have covered for a fellow officer. This may be the case regardless of whether the officer initiating the situation was acting responsibly or not.

4. **Never disrespect another officer in front of inmates.** This tenet reflects the importance of respect and the need to maintain "face" within the prison culture. Officers who are ridiculed or made to look weak in front of inmates have their authority subject to question by inmates since the word will get around that the officer is not respected (and therefore not well supported) by his peers. This sets the officer up for potential manipulation in the future.

5. **Always support an officer who is in a dispute with an inmate.** This applies to all types of instances ranging from verbal arguments with inmates to actual physical altercations. Simply put, one's coworker is always right, and the inmate is always wrong. However, behind the scenes, officers may not get along and may disagree on issues related to the management of inmates. Indeed, one officer may write up an inmate for a disciplinary infraction but another may overtly object when in the office out of earshot of the inmate population. The reasons for this may be many, but generally older, more seasoned officers will be adept at informally addressing inmate infractions whereas junior officers will tend to rely on official processes. However, given the threat of employee discipline that exists within the system and the need for control of the inmate population, most officers will ultimately maintain loyalty to each other during the final stages where their official support is necessary.

6. **Do not be friends with an inmate.** This is another tenet that has complicated shades and distinctions. For veteran officers, this tenet is not much of a concern. They have already proven themselves to be reliable and/or are known to not be snitches. Further, most veteran officers are capable of enforcing the rules, regardless of their prior conversations with an inmate. However, it is not uncommon for veteran officers (and even supervisors) to have one or two inmates whom they talk with, at least on a topical level. Though they may not consider themselves friends with the inmate, they may allow that inmate some privileges and opportunities that others would not, simply because they have developed a symbiotic prison relationship with that inmate that has existed for a long period of time. In return, these inmates may do the officer small favors, such as reserving higher-quality food from the kitchen for that officer or even, in prisons where the subculture has truly created permeable boundaries, letting the officer know when supervisors or others are watching him or her while on duty. This allows the officer to operate his or her cell block in a more leisurely manner, and, as such, the entire cell block benefits from the officer's laid-back approach.

7. **Maintain cohesion against all outside groups.** In this tenet, *outside groups* applies to members of the supervisory ranks, the outside public, the media, and even one's own family. This tenet is based on the belief that the general public does not understand the pressures placed upon officers and that the media tend to be sympathetic to the plight of the inmate, not the officer. Officers do not wish to implicate their family members and also do not want them to fear for their safety; thus, details are seldom disclosed. Further, the administration is not seen as trustworthy but instead as being politically driven. Administrators care only about their careers and moving up the corporate ladder and are too far removed from the rank-and-file to still understand the complexities of the officer's daily concerns. It is therefore better that officers not talk about what goes on in the institution to persons not within their ranks.

These tenets perhaps most clearly summarize the prison guard subculture. Again, these "guidelines" may not occur exactly as presented in all prisons, but in most larger and older facilities, at least some of them will be familiar to correctional staff.

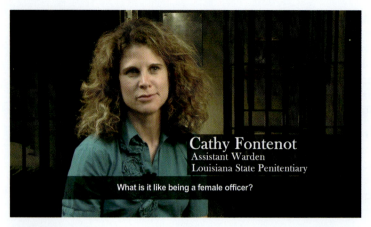

Cathy Fontenot
Assistant Warden
Louisiana State Penitentiary

What is it like being a female officer?

PRISON TOUR VIDEO: Women working in corrections bring key strengths to the role. Go to the IEB to watch an interview with a female correctional officer.

As we have seen in prior chapters, numerous lawsuits emerged during the 1960s, 1970s, and 1980s, and their aftermath greatly impacted the field of corrections in the 1990s as well as the current millennium. Prison systems had to modify and adjust their operations to be considered constitutional, and this required that these systems institute organizational change among their prison staff. An emphasis on professionalism emerged throughout the nation, and, as the War on Drugs resulted in a swelling inmate population, so too swelled the number of prison guards who were hired. In addition, the elimination of building tender and trusty supervision schemes used in many southern states necessitated the recruiting and hiring of more prison security staff. During the 1990s, the term *prison guard* became outdated and was replaced with the official job classification of correctional officer in many state prison systems.

Female Correctional Officers

The correctional field has traditionally been stereotyped as a male-dominated area of work. In later chapters, students will read more about women in the correctional field; an entire chapter is devoted to female offenders in correctional systems (Chapter 8). A general subculture also exists within women's prison facilities, but it is separate and distinct from the male prison subculture. Likewise, the issues that confront women who work in corrections tend to be different as well. We will explore the various aspects related to both women inmates and correctional workers in the next chapter.

For now, it is simply important to note that women are increasingly represented within the field of corrections. While women have had a long history of conducting prison work, they have typically been placed in clerical positions, teaching roles, support services, or the guarding of female offenders. They have not historically worked in direct supervision of male offenders. It was not until the late 1970s and early 1980s that women were routinely assigned to supervise male inmates (Pollock, 1986). The introduction of women into the security ranks has greatly impacted the organizational culture of many prison facilities and the subculture within them.

Women generally do not have the same aggressive social skills that men in prison tend to exhibit. Further, the prison environment tends to emphasize the desire to "be a man" and denigrates women as inferior. This means that women were not widely accepted among correctional officers and/or inmates. Since women have become integrated into the correctional industry, the male-oriented subculture has been weakened. The introduction of women into the security ranks, along with the inclusion of diverse minority groups, the professionalization of corrections, and the proliferation of prison gangs, has eroded the influence of the male-dominated and male-oriented convict code. While the convict code still exists and has its adherents, it is no longer considered a primary standard of behavior in many prison facilities but instead has become more of an ideal.

THE IMPACT OF GANGS UPON PRISON SUBCULTURE

Gang members are another group that tends to not adhere strictly to the tenets of the convict code (Mobley, 2011). This is particularly true among Latino and African American gangs. These gangs, which represent the majority racial lineages among prison gangs, typically view prison stints as just another part of the criminal lifestyle. As such, they have no true use for the convict code since it is their gang family who will protect them, not their reputation according to the code. Their alliances and their allegiances are tied to outside street gangs with members who sometimes get locked up and therefore find themselves operating within prison walls as well (Mobley, 2011). Many young inner-city African American and Latino males who have been incarcerated are able to find home-boys or hombres in nearly every correctional facility within their state (Mobley, 2011). Thus, the young gang member does not need to trouble himself with adapting to the prison subculture (Mobley, 2011).

Gangsters comprise a distinct subculture whether on the street or in prison (Mobley, 2011). They "look out for" one another and protect each other, living in a nearly familial lifestyle. Few African American gang members speak to inmates outside of their gang "set," at least about anything of substance. Though most would claim that they do not snitch to "the man," and most would say that they just wish to do their own time, their true loyalty is to their gang family. Gang members "run with their road dogs" from "the hood" and meet up with each other in prison, forming bonds and making plans for when they reunite in their respective communities, the turf for their street gang activity (Mobley, 2011). This constant cycle, in and out of prison, creates a seamless form of support for many gang members.

The Impact of Cross-Pollination: Reciprocal Relationships Between Street Gangs and Prison Gangs

It is perhaps the emergence of gang life that has been the most significant development within prison subcultures throughout various state systems. In many texts on prisons and/or the world of corrections, there is a section on prison gangs. In most cases, these texts tend to present gang membership as isolated to the prison environment, with little emphasis on the notion that gang membership is permeable, found inside and outside the prison. But members of prison gangs do not simply discard membership once their sentence is served or when they are paroled out into society. Rather, their membership continues, and they will often continue to answer to gang leaders who may still be locked up in prison. In other cases, they may be required to report to other leaders outside of the prison walls and will continue criminal work on behalf of the gang, plying their criminal trade on the streets and in broader society.

Many prison gang members were prior street gang members. An offender may engage in street gang criminal activity for a number of years, with frequent short stints in jail. As noted earlier in this chapter, few inmates in state prison systems are locked up with long-term sentences for their first offense; rather, they have typically committed several "priors" before that point, some of which might not be known to law enforcement. During their activity on the streets, these offenders will develop a reputation, particularly within their gang or their area of operation (if in an urban or a suburban setting) and will develop associations with other gang members. Once they finally do end up with a long-term sentence in a state facility, they have usually already embedded themselves within the gang structure on the outside, which includes members who have been locked up inside the prison system.

In some cases, those doing state time may be the upper leadership of the street gang; these members will tend to direct prison gang activities internally and also "call the shots" for members on the outside. The term *shot caller* refers to those inmates and/or gang members who dictate what members will do within the gang hierarchy. The point to all of this is that a gang's membership does not begin or end with the prison walls. Rather, prison is simply a feature that modern-day gangs must contend with—an obstacle that increases the overhead to conducting criminal enterprise.

Because gang membership is porous in nature, social researchers can only vaguely determine likely gang growth both inside and outside the prison. In 2012, there were an estimated 850,000 gang members across the nation (National Gang Center, 2013). A large proportion of these gang members had served time behind bars at one point or

PHOTO 7.5 This member of the Aryan Brotherhood has clear markings of his gang affiliation for all to see. Members with these bold markings usually remain in the gang for life.

Andrew Lichtenstein/Corbis Premium Historical/Getty Images

another throughout their criminal careers. In fact, some prison systems, such as Texas's, were nearly controlled by gangs; the gangs even controlled many of the staff in these systems through various tactics like friendships or occasional intimidation.

As this shows, prison gangs in some state systems can be both persuasive and very powerful. Potential recruits for existing prison gangs enter prison with natural feelings of anxiety and quickly learn the value of having some form of affiliation. Indeed, inmates without the protection of affiliation are likely to be the target of other inmates who *are* members of a gang. Affiliations tend to be based along racial allegiances—in fact, most prison gang membership is strictly defined by the race of the member.

Historically speaking, the main distinction between prison gangs and street gangs has been the internal structure and the leadership style of the gang (Fleisher & Rison, 1999). However, over time this distinction has become so blurred as to be meaningless in the offender world (Fleisher & Rison, 1999). In the correctional environment of today, prison gangs such as the Mexican Mafia and the Texas Syndicate are just as influential and powerful as classic street gangs like the Gangster Disciples and Latin Kings (Fleisher & Rison, 1999). More telling is the fact that the Mexican Mafia, the Aryan Brotherhood, and even emerging local groups such as the Barrio Aztecas have become just as formidable in their own respective ethnic and/or culturally based neighborhoods or regions as they are in prisons. Indeed, it is sometimes common for leaders of a gang to be incarcerated but still give orders to members outside the prison operating within the community. For this text, we will refer to **gang cross-pollination**, which means that a gang has developed such power and influence as to be equally effective regardless of whether its leadership is inside or outside of the prison walls.

When discussing gang cross-pollination, the term **security threat group (STG)** will be used to describe a gang that possesses the following high-functioning group and organizational characteristics:

1. Prison and street affiliation is based on race, ethnicity, geography, ideology, or any combination of these or other similar factors (Fleisher, 2008, p. 356).

2. Members seek protection from other gang members inside and outside the prison, as well as insulation from law enforcement detection (use of safe houses when wanted).

3. Members will mutually take care of an incarcerated member's family, at least minimally, while the member is locked up since this is an expected overhead cost in the organization.

4. The group's mission integrates an economic objective and uses some form of illicit industry, such as drug trafficking, to fulfill the economic necessities to carry forward other stated objectives (Fleisher, 2008). The use of violence or the threat of violence is a common tool in meeting these economic objectives.

Other characteristics common to prison gangs go beyond racial lines of membership. These characteristics are common to most any gang within jail and/or prison, though they are not necessarily common to those gangs based primarily on the street. First, prison gangs tend to have highly formal rules and a written constitution that are adhered to by all members who value their affiliation, and sanctions are taken against those who violate the rules. Second, prison gangs tend to be structured along a semi-military organizational scheme. Thus, authority and responsibility are very clearly defined within these groups. Third, membership in a prison gang is usually for life. This tendency has often been referred to, as noted earlier, as *blood in—blood out* among the popular subculture. This is one of the root causes of parolees continuing their affiliation beyond the prison walls, and this lifelong membership rule is enforced if someone attempts to exit the prison gang. Thus, when gang members leave the prison environment, they are expected to perform various "favors" for the members who are still incarcerated. Lastly, as members circulate in and out of prison, they are involved in gang activities both inside and outside of the penal institution. Thus, the criminal enterprise continues to be an active business, and prison simply becomes part of the overhead involved in running that business.

Gang cross-pollination: Occurs when a gang has developed such power and influence as to be equally effective regardless of whether its leadership is inside or outside of prison walls.

Security threat group (STG): A high-functioning, organized gang that uses an illegal industry to fund their objectives.

MAJOR PRISON GANGS IN THE UNITED STATES

During the 1950s and 1960s, there was a substantial amount of racial and ethnic bias in prisons. This was true in most all state prison systems but was particularly pronounced in the southern United States and in the state of California. During the late 1950s, a Chicano gang formed known as the Mexican Mafia (National Gang Intelligence Center, 2009). Its members were drawn from street gangs in various neighborhoods of Los Angeles. While many of its members were in San Quentin, they began to exercise power over the gambling rackets within that prison. Other gangs soon began to form as a means of opposing the Mexican Mafia, including the Black Guerilla Family, the Aryan Brotherhood, La Nuestra Familia, and the Texas Syndicate.

This section provides a brief overview of some of the major prison gangs found throughout the nation. These gangs are presented in a manner that is as accurate and historically correct as possible, with attention paid to the basic feeling of the time and context during each gang's development (see Table 7.1). Much of the information presented has been obtained from a recent document titled the *National Gang Threat Assessment 2009*. The following pages provide an overview of 13 of the most prevalent prison gangs in the United States.

The Mexican Mafia prison gang, also known as La Eme (Spanish for the letter *M*), was formed in the late 1950s within the California Department of Corrections. It is loosely structured and has strict rules that must be followed by its 200 members. Most members are Mexican American males who previously belonged to a southern California street gang. The Mexican Mafia is primarily active in the southwestern and Pacific regions of the United States, and its power base is in California. The gang's main source of income is extorting drug distributors outside prison and distributing methamphetamine, cocaine, heroin, and marijuana within prison systems and on the streets. Some members have direct links to Mexican drug traffickers outside of the prison walls. The Mexican Mafia also is involved in other criminal activities, including gambling and homosexual prostitution in prison.

The Black Guerrilla Family (BGF), originally called Black Family or Black Vanguard, is a prison gang founded in San Quentin State Prison, California, in 1966. The gang is highly organized along paramilitary lines, with a supreme leader and central committee. The BGF has an established national charter, code of ethics, and oath of allegiance. BGF members operate primarily in California and Maryland.

The Aryan Brotherhood (AB; see Figure 7.1) was originally formed in San Quentin in 1967. The AB is highly structured with two factions—one in the California Department of Corrections and the other in the Federal Bureau of Prisons. Most members are Caucasian males, and the gang is active primarily in the southwestern and Pacific regions. Its main source of income is the distribution of cocaine, heroin, marijuana, and methamphetamine within prison systems and on the streets. Some AB members have business relationships with Mexican drug traffickers who smuggle illegal drugs into California for AB distribution. The AB is notoriously violent and is often involved in murder for hire. The AB still maintains a strong presence in the nation's prison systems, albeit a less active one in recent years.

The Crips are a collection of structured and unstructured gangs that have adopted a common gang culture. The Crips emerged as a major gang presence during the early 1970s. Crips membership is estimated at 30,000 to 35,000; most members are African American males from the Los Angeles metropolitan area. Large, national-level Crips gangs include 107 Hoover Crips, Insane Gangster Crips, and Rolling 60s Crips. The Crips operate in 221 cities in 41 states and can be found in several state prison systems.

The Bloods are an association of structured and unstructured gangs that have adopted a single-gang culture. The original Bloods were formed in the early 1970s to provide protection from the Crips street gang in Los Angeles, California. Large, national-level Bloods gangs include Bounty

FIGURE 7.1

Symbol of the Aryan Brotherhood

TABLE 7.1

Timeline History of Prison Gang Development in the United States

YEAR FORMED	JURISDICTION	NAME OF GANG
1950	Washington	Gypsy Jokers
1957	California	Mexican Mafia
1958	California	Texas Syndicate
1965	California	La Nuestra Familia
1966	California	Black Guerrilla Family
1967	California	Aryan Brotherhood
Mid-1970s	Arizona	Arizona Aryan Brotherhood
1976	Puerto Rico	Ñeta
1977	Arizona	Arizona Old Mexican Mafia
1978	California	Nazi Low Riders
1980	New Mexico	New Mexico Syndicate
Early 1980s	Texas	Aryan Brotherhood of Texas
Early 1980s	Texas	Texas Mafia
Mid-1980s	California	MS-13
Mid-1980s	California	Bulldogs
1984	Arizona	Arizona's New Mexican Mafia
1984	Texas	Mexikanemi
1984	Texas	Mandingo Warriors
1985	Federal system	Dirty White Boys
1985	California	415s
1986	Texas	Hermanos de Pistoleros Latinos
1988	Texas	Tango Blast
1990	Connecticut	Los Solidos
1993	New York	United Blood Nation

Sources: Orlando-Morningstar, D. (1999). *Prison gangs.* Washington, DC: Federal Judicial Center; Valdez, A. (1999). *Nazi Low Riders.* Police: The Law Enforcement Magazine, 23(3), 46–48.

Hunter Bloods and Crenshaw Mafia Gangsters. Bloods membership is estimated to be 7,000 to 30,000 nationwide; most members are African American males. Bloods gangs are active in 123 cities in 33 states, and they can be found in several state prison systems.

Ñeta is a prison gang that was established in Puerto Rico in the early 1970s and spread to the United States. Ñeta is one of the largest and most violent prison gangs, with about 7,000 members in Puerto Rico and 5,000 in the United States. Ñeta chapters in Puerto Rico exist exclusively inside prisons; once members are released from prison, they are no longer considered part of the gang. In the United States, Ñeta chapters exist inside and outside prisons in 36 cities in nine states, primarily in the Northeast.

The Texas Syndicate (see Figure 7.2) originated in Folsom Prison during the early 1970s. The Texas Syndicate was formed in response to other prison gangs in the California Department of Corrections, such as the Mexican Mafia and Aryan Brotherhood, which were attempting to prey on native Texas inmates. This gang is composed of predominantly Mexican American inmates in the Texas Department of Criminal Justice (TDCJ). Though this gang has a rule to only accept members who are Latino, it does accept Caucasians into its ranks. The Texas Syndicate has a formal organizational structure and a set of written rules for its members. Since the time of its formation, the Texas Syndicate has grown considerably, particularly in Texas.

The Mexikanemi prison gang (also known as Texas Mexican Mafia or Emi; see Figure 7.3) was formed in the early 1980s within the TDCJ. The gang is highly structured and is estimated to have 2,000 members, most of whom are Mexican nationals or Mexican American males living in Texas at the time of incarceration. Mexikanemi poses a significant drug trafficking threat to communities in the southwestern United States, particularly in Texas. Gang members reportedly traffic multi-kilogram quantities of powder cocaine, heroin, and methamphetamine; multi-ton quantities of marijuana; and thousand-tablet quantities of MDMA from Mexico into the United States for distribution inside and outside prison.

The Nazi Low Riders (NLR) evolved in the California Youth Authority, the state agency responsible for the incarceration and parole supervision of juvenile and young adult offenders, in the late 1970s or early 1980s as a gang for Caucasian inmates. As prison officials successfully suppressed Aryan Brotherhood activities, the AB appealed to young incarcerated skinheads, the NLR in particular, to act as middlemen for their criminal operations, allowing the AB to keep control of criminal undertakings while adult members were serving time in administrative segregation. The NLR maintains strong ties to the AB and, like the older gang, has become a source of violence and criminal activity in prison. The NLR has become a major force, viewing itself as superior to all other Caucasian gangs and deferring only to the AB. Like the AB, the NLR engages in drug trafficking, extortion, and attacks on inmates and corrections staff.

Barrio Azteca was organized in 1986 in the Coffield Unit of TDCJ by five street gang members from El Paso, Texas. This gang tends to recruit from prior street gang members

FIGURE 7.2

Symbol of the Texas Syndicate

Source: Orlando-Morningstar, D. (1997). *Prison gangs. Special needs offenders bulletin.* Washington, DC: Federal Judicial Center.

FIGURE 7.3

Constitution of the Mexikanemi

1. Membership is for life ("blood in—blood out").
2. Every member must be prepared to sacrifice his life or take a life at any time.
3. To achieve discipline within the Mexikanemi brotherhood, every member shall strive to overcome his weakness.
4. Members must never let the Mexikanemi down.
5. The sponsoring member is totally responsible for the behavior of a new recruit. If the new recruit turns out to be a traitor, it is the sponsoring member's responsibility to eliminate the recruit.
6. When insulted by a stranger or group, all members of the Mexikanemi will unite to destroy the person or other group completely.
7. Members must always maintain a high level of integrity.
8. Members must never relate Mexikanemi business to others.
9. Every member has the right to express opinions, ideas, contradictions, and constructive criticism.
10. Every member has the right to organize, educate, arm, and defend the Mexikanemi.
11. Every member has the right to wear tattoo of the Mexikanemi symbol.
12. The Mexikanemi is a criminal organization and therefore will participate in all activities of criminal interest for monetary benefits.

Source: Orlando-Morningstar, D. (1997). *Prison gangs. Special needs offenders bulletin.* Washington, DC: Federal Judicial Center.

FOCUS TOPIC 7.2

MS-13 Gang Member Trafficked Drugs to New Jersey From Inside His Prison Cell in California

A member of MS-13 admitted trafficking methamphetamine, heroin, and cocaine to New Jersey from inside a California state prison. The guilty plea follows an investigation by U.S. Immigration and Customs Enforcement's (ICE) Homeland Security Investigations (HSI).

Luis Calderon, 32, aka "Lagrima," of Los Angeles, pleaded guilty before U.S. Magistrate Judge Joseph A. Dickson in Newark federal court to an indictment charging him with conspiracy to distribute, and to possess with intent to distribute, methamphetamine, heroin, and cocaine.

Between August 2015 and November 2015, Calderon was incarcerated at the Calipatria State Prison in California. However, Calderon had access to multiple contraband cellular telephones, which he used to communicate with conspirators outside the prison.

Law enforcement officers lawfully recorded numerous telephone conversations between Calderon and an MS-13 member based in New Jersey, identified in the indictment as "Individual-1." Among other topics, Calderon and Individual-1 discussed plans to distribute crystal methamphetamine, heroin, and cocaine in the New Jersey area. Calderon and Individual-1 ultimately settled on that plan that involved Calderon and others

outside the prison sending a package containing controlled substances to a business center in Edison, New Jersey.

Shortly before the package arrived, Calderon informed Individual-1 by telephone that he was sending Individual-1 a package containing four ounces each of heroin and cocaine. Calderon stated that the package would also likely contain two ounces or more of crystal methamphetamine. Calderon told Individual-1 that the total cost for the heroin, cocaine, and crystal methamphetamine was $9,000, and stated that Individual-1 could keep the proceeds made from selling the drugs once Individual-1 paid Calderon for the shipment. Calderon subsequently gave Individual-1 the names that would appear on the package and the tracking number.

On Nov. 4, 2015, federal agents lawfully intercepted and searched the package. The search revealed approximately 95.5 grams of heroin, 54.7 grams of cocaine, and 52.4 grams of methamphetamine hidden inside a box of Little Debbie Swiss Rolls. ●

Source: U.S. Immigrations and Customs Enforcement. (2018). *MS-13 member admits trafficking drugs to New Jersey from inside a California prison.* Washington, DC: Author. Retrieved from https://www.ice.gov/news/releases/ms-13-member-admits-trafficking-drugs-new-jersey-inside-california-prison

and is most active in the southwestern region, primarily in correctional facilities in Texas and on the streets of southwestern Texas and southeastern New Mexico. The gang is highly structured and has an estimated membership of 2,000. Most members are Mexican national or Mexican American males.

Hermanos de Pistoleros Latinos (HPL) is a Hispanic prison gang formed in the TDCJ in the late 1980s. It operates in most prisons and in many communities in Texas, particularly Laredo. HPL is also active in several cities in Mexico, and its largest contingent in that country is in Nuevo Laredo. The gang is structured and is estimated to have 1,000 members. Members maintain close ties to several Mexican drug trafficking organizations and are involved in trafficking quantities of cocaine and marijuana from Mexico into the United States for distribution.

Tango Blast is one of largest prison/street criminal gangs operating in Texas. Tango Blast's criminal activities include drug trafficking, extortion, kidnapping, sexual assault, and murder. In the late 1990s, Hispanic men incarcerated in federal, state, and local prisons founded Tango Blast for personal protection against violence from traditional prison gangs such as the Aryan Brotherhood, Texas Syndicate, and Texas Mexican Mafia. Tango Blast originally had four city-based chapters in Houston, Austin, Dallas, and Fort Worth. These founding four chapters are collectively known as Puro Tango Blast or the Four Horsemen. In addition to the original four chapters, former Texas inmates established new chapters in El Paso, San Antonio, Corpus Christi, and the Rio Grande Valley. In June 2008, the Houston Police Department estimated that more than 14,000 Tango Blast members were incarcerated in Texas. Tango Blast is difficult to monitor. The gang does not conform to either traditional prison/street gang hierarchical organization or gang rules. Tango Blast is laterally organized, and leaders are elected sporadically to represent the gang in prisons and to lead street gang cells. The significance of Tango Blast is exemplified by corrections officials reporting that rival traditional prison gangs are now forming alliances to defend themselves against Tango Blast's growing power.

United Blood Nation is a universal term that is used to identify both West Coast Bloods and the United Blood Nation (UBN). The UBN started in 1993 in Rikers Island George Mochen Detention Center (GMDC) to form protection from the threat posed by Latin Kings and Ñetas, who dominated the prison. While these groups are traditionally distinct entities, both identify themselves by "Blood," often making it hard for law enforcement to distinguish between them. The UBN is a loose confederation of street gangs, or sets, that once were predominantly African American. Membership is estimated to be between 7,000 and 15,000 along the U.S. eastern corridor.

The Mara Salvatrucha (MS-13) gang originated on the streets of Los Angeles during the 1980s when numerous Salvadoran immigrants came to the United States to escape civil war and conflict that was widespread in El Salvador at that time (Wolf, 2012). The original mandate of the Mara Salvatrucha was to simply protect Salvadoran immigrants from other gangs in Los Angeles, particularly Mexican gangs that had been established on the streets and in prisons decades ago (Wolf, 2012). In fighting the effects of this gang, the United States often deported many MS members when they were arrested. Due to these deportations, members of MS-13 often recruited new members in El Salvador and other surrounding countries. As a result, MS-13 has grown due to this cycle of deportation, recruitment in El Salvador, reentry into the United States, recruitment in the United States, followed by future deportation. Further still, this gang, thought to number around 8,000 to 10,000 in the United States, has migrated to the east coast of the United States, in some part, due to vigorous antigang initiatives in the southwest. Indeed, jail and prison facilities in Virginia and Maryland have seen increases in their MS-13 population, of nearly 30% in just the past 2 years alone. Naturally, with this, there has been an increase of street crime at the hands of MS-13 as well (Hagstrom, 2018; see Focus Topic 7.2).

Gang Management in Corrections

Gang management requires a comprehensive policy that specifies legal precedents, procedures, and guidelines, including the verification of gang members. Over the years, most state systems have developed gang intelligence units and have trained correctional staff on gangs and gang activity. In modern times, state and federal corrections refer to gangs as security threat groups, or STGs, as noted earlier. Students may notice that most of the 14 gangs listed in the prior subsection have links to outside society where they engage in criminal activities that usually have an economic objective. This means that these gangs are all STGs because they operate inside and outside the prison and possess all the other characteristics discussed in previous subsections of this chapter.

The technical aspects of combating STGs in prison, such as the paper classification and procedures needed to investigate gang members, are fairly straightforward.

PHOTO 7.6 This collection of prison contraband consists of an array of small weapons that were confiscated during routine "shakedowns" within a prison facility.

ROBERT GALBRAITH/ REUTERS/Newscom

However, the human element is what makes the fight against STGs much more difficult. In correctional facilities that do not emphasize professionalism or do not encourage open communication among security staff, if there is a strong underlying prison subculture (both inmate and correctional officer), STGs are likely to proliferate. When selecting correctional staff to serve on gang task forces, the prison administrator must exercise care and remain vigilant. In some cases, an inmate's sibling, cousin, girlfriend, former wife, or friend may be employed within the facility. This is, of course, a common tactic used by gangs who seek to infiltrate the correctional system.

Gang Control, Management, and Administrative Segregation

In addition to the physical security of the facility, there are many psychological tactics that can be used to control gang activity. For instance, the immediate tendency of corrections officials may be to restrict privileges for gang-related inmates. But as Fleisher (2008) notes, withdrawing incentives or placing these inmates in long-term states of restricted movement can have financial and social consequences for the prison facility. For instance, in Texas, many gang-related inmates are kept in administrative segregation. Administrative segregation is a security status that is intended to keep the assigned inmate from having contact with the general population. It is not punitive in nature, like solitary confinement. This custody status is intended to protect the general population from the inmate in segregation. However, this form of custody is very expensive.

Further, there is a tendency for prison systems that use administrative segregation to house inmates of the same gang in the same area. This prevents them from coming into contact with enemy gang members and cuts costs that would ensue if they were kept on different cell blocks or dormitories. But doing this replicates the street gang culture, as they are all together but geographically isolated. In other words, on the street it would be one neighborhood, one gang, and here it is one cell block, one gang. This can build solidarity for the group. It can also lead to problematic situations where inmates exercise power (through the gang rank structure) over a cell block or dormitory, encouraging security to work with those in power because the gang leaders can help them maintain authority over the other lower-ranking inmates of that gang. Naturally, this should be avoided because it validates the gang's power and undermines the security staff. Fleisher (2008) notes that other undesirable behaviors can also emerge. For instance, gangs may attempt to run cell blocks, "sell" cells on the cell block, or "own" territory on the recreation yard. These behaviors reinforce their feeling that they have power over the institution and should be avoided.

CONCLUSION

This chapter has provided students with a glimpse of the behind-the-scenes aspects of the prison environment. The notion of a prison subculture, complete with its own norms and standards that are counter to those of the outside world, has been presented to give students an idea of the values and principles that impact the day-to-day operation of many prisons. There has been substantial debate as to whether this informal prison subculture is the product of adjustments and adaptations to prison life or if it is more the product of norms brought in from the outside world—from the street life.

The informal subculture within prisons tends to be largely driven by inmates. The convict code has typically been presented as the "gold standard" of behavior among inmates. This code represents perceptions from older eras, and its guidelines are now out of date and not in sync with the modern era. The effects of professionalism within the correctional officer ranks, the diversity of correctional staff, and the difference in the mindset of this generation of inmates all have led to the near demise of the honor-bound convict code. Within the informal subculture, correctional staff also have some of their own unwritten standards and expectations. The work of Kauffman (1988) has provided very good insight on the behavior of correctional staff in prisons.

The norms associated with both the subculture of prison staff and the convict code of inmates are in a state of flux and decline. The subculture among correctional staff has been impacted by the emphasis on professionalism and the diversification in staff recruiting. These two factors have changed the face of corrections over time and have also undermined the tenets of the prison subculture. What has resulted is a state of ambiguity where inmates may pay lip service to tenets of respectable behavior (according to prison logic) but often break these rules when put under pressure. Simply put, there is truly no honor among thieves.

Gangs have emerged as a major force in state prison systems. The first recorded prison gangs began to emerge in the late 1950s and the early 1960s, primarily in the California prison system. Since that time, gangs have proliferated around the United States and have exerted substantial influence over the prison subculture and even dynamics in prison operations. In this chapter, we have covered 14 of the larger and more well-known prison gangs in the nation. From the coverage of these gangs, it is clear that they have networks that extend beyond the prison walls, and this, in turn, increases their power and influence within the prison walls. Indeed, when offenders in a street gang enter prison, they do not forfeit their membership in their gang. Likewise, when these inmates leave prison, they again do not leave their gang obligations behind. Rather, gangs exist both inside and outside the prison walls.

The methods used to control gang activity inside prisons have been discussed in this chapter. Prison gang intelligence units must have effective means of investigating potential membership and collecting data on gang members. Utilizing electronic equipment for identification, storage, and retrieval is essential to maintaining an effective antigang strategy. The ability to share data with other agencies enhances the public safety in the region surrounding the prison and also improves security within the facility. Whether we like it or not, the prison has an impact on outside society due to the manner in which the inmate population cycles into and out of prison. Thus, it is clear that prisons impact society, both in terms of keeping dangerous persons locked up and in terms of the learned prison behaviors of those persons once they are released back into society.

PHOTO 7.7 An inmate holds another inmate hostage with a homemade shank.

AP Photo/Dirceu Portugal

Want a Better Grade?

Get the tools you need to sharpen your study skills. Access practice quizzes, eFlashcards, video, and multimedia at **edge.sagepub.com/brief**

Interactive eBook

Visit the interactive eBook to watch SAGE premium videos. Learn more at **edge.sagepub.com/hanserbrief/access**.

 Career Video 7.1: Division Chief

 Criminal Justice in Practice Video 7.1: Gang-Involved Offender

 Prison Tour Video 7.1: An Inmate's Experience, Relationships, and Typical Day

 Prison Tour Video 7.2: Officers' Experience and Reputation, Inmates' Feedback Influence, and Grievances

 Prison Tour Video 7.3: Prison Culture

 Prison Tour Video 7.4: Female Correctional Officers

DISCUSSION QUESTIONS

Test your understanding of chapter content. Take the practice quiz at edge.sagepub.com/hanserbrief.

1. How are the prison subcultures for inmates and correctional officers often interrelated?

2. Compare importation theory with exportation theory. Which one do you believe is the stronger influence on prison subculture, and why?

3. Explain some of the common outlooks and views of prison subculture. How does this contrast with the conventional logic of outside society?

4. How have professionalization and the diversification of correctional staff impacted the prison subculture?

5. Explain what a "hyena" is. What are some examples of how these inmates affect group behavior in prison facilities?

6. Identify at least three prison gangs, and explain how they have impacted corrections in their respective jurisdictions. Note their allies and adversaries in the prison, and explain how this affects prison operations.

7. Explain what prison systems can or should do to control gang problems that occur in their facilities.

8. Within the prison subculture, how is labeling theory related to prison rape?

KEY TERMS

Review key terms with eFlashcards at edge.sagepub.com/hanserbrief.

Blood in—blood out, 180

Convict, 185

Convict boss/officer, 188

Convict code, 181

Gang cross-pollination, 194

Hogging, 186

Hustle, 186

Importation theory, 179

Mule, 186

Pains of imprisonment, 180

Prisonization, 188

Punk, 182

Respect, 186

Security threat group (STG), 194

Snitch, 181

Symbiotic prison relationship, 187

APPLIED EXERCISE 7.1

You are the assistant warden of a large, medium-security facility within a state prison system in the southeastern United States. Your facility has a disproportionate number of African American inmates (common in prison systems throughout the United States) and a disproportionate number of Caucasian officers.

In response to concerns from the Grievance Department regarding inmate allegations of racism from officers inside the facility, the warden has asked you to develop a comprehensive diversity training program for the security staff in your facility. Currently, your facility holds an annual 1-day "refresher" course for staff. This course is actually only a 4-hour block of instruction provided at the state training facility. Everyone throughout the system knows that this instruction is not taken seriously and is simply offered as a means of documenting that the state has made the training available.

Your warden desires to change this within your facility for two reasons. First, it is just a good and ethical practice to take diversity seriously. Second, the dollars spent to resolve grievances and other allegations are getting costly enough to make diversity training a fiscally sound alternative to potential litigation.

With this in mind, you are given the following guidelines for the training program that you are to implement:

1. The program should be 1.5 days in length and be given once annually.

2. The program should address numerous areas of concern, including race, gender, age, and religion.

Students should keep in mind that workplace diversity has two components. First, it involves fair treatment and the removal of barriers. Second, it addresses past imbalances through the implementation of special measures to accelerate the achievement of a representative workforce. Further, workplace diversity recognizes and utilizes the diversity available in the workplace and the community it serves.

Workplace diversity should be viewed as a means to attaining the organizational objectives of the correctional facility—not as an end in itself. The link between the agency objectives and the day-to-day processes that occur among staff in the facility is crucial to the success of workplace diversity initiatives.

For this assignment, outline the content that you would recommend for the training session, and explain the rationale behind your recommendation. Also, explain how you will "sell" these ideas to prison staff to ensure that they take the training seriously. Lastly, explain how this training, if successful, can improve security and safety in the institution. Your submission should be between 500 and 1,000 words.

WHAT WOULD YOU DO?

You are a caseworker in a state facility and work closely with the institution's classification department on a routine basis. You have one inmate on your caseload, Jeff, who has presented with a number of challenges. Jeff is a 35-year-old male who is an inmate in your maximum-security facility. He has recently been transferred to your facility from another facility, largely for protective reasons. Jeff has come to you because he is very, very worried. Jeff is a sex offender, and he has been in prison for nearly 11 years on a 15-year sentence. He is expected to gain an early release due to his excellent progress and behavior in prison and due to prison overcrowding problems. He has been in treatment, and, as you look through his case notes, you can see that he has done very well.

But a powerful inmate gang at Jeff's prior prison facility did not want to see him get paroled. Jeff had received "protection" from this gang in exchange for providing sexual favors to a select trio of its members. Jeff discloses that he had to humiliate himself in this way to survive in the prison subculture, particularly since he was a labeled sex offender. The gang knew this, of course, and used it as leverage to ensure that Jeff was compliant. In fact, the gang never even had to use any physical force to gain Jeff's compliance. Jeff notes that this now bothers him, and he doubts his own sense of masculinity. Jeff also discloses that he has had suicidal ideations as a result of his experiences.

Thus, while Jeff has performed well in his treatment for sex offenders, he has also been adversely affected by noxious sexual experiences inside the prison. You are the first person that he has disclosed this to. As you listen to his plight, you begin to wonder if his issues with sexuality are actually now more unstable than they were before he entered prison. Though his treatment notes seem convincing, this is common among sex offenders, and the other therapist did not know that Jeff had engaged in undesired sexual activity while incarcerated. This activity has created a huge rift in Jeff's masculine identity. Will this affect his likelihood for relapse on the outside? Does Jeff need to resolve his issues with consensual versus forced homosexual activity?

As you listen to Jeff, you realize that if you make mention of his experiences in your report, then the classification system is not likely to release him, and this condemns him to more of the same type of exploitation. Putting him in protective custody is not an option because Jeff adamantly refuses this custody level. He fears that the gang would think he was giving evidence against them and would then seek to kill him. However, if you do not mention any of this information and thereby allow a person with a highly questionable prognosis to be released, you run the risk of putting the public's safety at risk.

What would you do?

Stockbyte/Thinkstock

8 Female Offenders in Correctional Systems

Learning Objectives

1. Discuss some of the characteristics of female offenders and compare rates of incarceration and terms of sentencing between male and female offenders.

2. Discuss the history of corrections with relation to female offenders.

3. Describe some of the issues associated with modern-day female offenders.

4. Identify some of the considerations associated with effective correctional operations when considering female offenders.

5. Discuss the legal issues involved with female offenders in the correctional facility.

6. Identify various treatment implications related to female offenders.

A Mom Behind Bars

Maria flopped down on her bunk and looked around the dormitory. She thought to herself, *I wonder how many of these women are not mothers?* With a sigh she addressed her "neighbor," Chelsea. "Chelz, tell me, why do we get involved with these a-holes again?"

Chelsea grinned. "Because we're fools for love . . . and they think we're a bunch of *putas*!"

"Maybe you're right, but he is the baby's daddy, and it seems like there's no getting around that . . ." Maria trailed off and muttered under her breath, ". . . even though he never really helps out much."

She thought about her daughter, Ariana, who was 3 years old. Ariana stayed with Maria's parents, and they had recently told Maria that Ariana was asking about her mommy. The only contact Maria had with Ariana was by phone and a few infrequent visits when Maria's parents would bring Ariana to the prison visitation building. They lived far from the prison, however, and could only come once a month or so, if that.

Maria felt terrible that her parents were raising her daughter while she did time, but they knew the circumstances. She had fallen in love with the wrong man—Juan, a good-looking young man who had lots of "respect" on the streets. He was *cholo* (Spanish describing lower class and/or street gang culture), but he was a dreamer. When he met Maria, he would talk about wanting to get out of the gang life, wanting to go to school, maybe be a school teacher. But, unfortunately, he was into drugs.

They "dated" for about 3 months or so, going to block parties or all-hours night clubs where his *familia* (gang) were considered exclusive members. He seemed to be popular every-where he went. Then she found out that she was pregnant. She remembered that day because she was so afraid to tell Juan. To her surprise, Juan was excited and told all of his friends that he was going to be a father. He told her, "We gotta get engaged, and I'll marry you. . . . I hope your father approves."

Eventually, they moved in together. One day when Maria was about 5 months pregnant, Juan had asked her to hold on to some of the black tar heroin that he had recently obtained from contacts in Guadalajara, Mexico. Maria never used heroin—she smoked pot, sure, and drank a little Patron from time to time, but she absolutely stayed away from other drugs. Naturally, while pregnant, she did not smoke or drink at all.

Juan needed her to walk to an area on the outskirts of the city and leave the heroin in a trash can outside of a local store. Later, some guy who worked at the store would "empty the trash" and obtain the drugs; no one would contact anyone, and nothing would be exchanged in person. Since they had moved in together, Maria had learned more about Juan. He was very macho, expected her to do all of the house work, and would sometimes yell at her when he was drinking or coming off of his drugs. He was not physically abusive, but did have a mean streak.

Maria was hesitant, but Juan said, "Look, one or two more deals like this and I can get out of the street life. I will have the money to quit. Besides, our *familia* is very careful, and we do not use the same location twice in a row. *Nobody ever gets caught*, but I can't do it this time; our rules require that we break it up a bit when doing drop-offs. Come on Maria, *this is our way out!*"

Maria agreed and went through with the drop-off. All seemed well until a few days later. She had been under observation, as Juan and his *familia* were the subject of a large-scale undercover narcotics investigation. When all of the various gang members and their associates were rounded up, numerous people were sent to prison. Even worse, many of those arrested snitched on one another to cut deals with the police. Juan did not sell out Maria, but he did not cop to any of the charges, either. He had enough money to afford a legal defense and got out of jail quickly since no evidence was attached to him. Maria was lucky—she only got a 4-year sentence with the likelihood of making parole in 18 months or so.

Now she had received word from the parole commission that she would be out in approximately 3 months. She had just finished a year in the female correctional center. As her time grew shorter, letters from Juan were arriving more frequently. She was not sure if she really wanted to be with him after she got out because she had heard he was still active in the gang and she knew he was using drugs. She wondered what she would do once she was out again.

INTRODUCTION

This chapter will familiarize the student with the common problems associated with the female offending population. While the majority of this chapter will address issues regarding women in prison, some discussion regarding community-based sanctions will also be included. As will be seen, women offenders have several physiological and psychological characteristics that set them apart from male offenders. Some of these characteristics are common also to the female population in broader society, but others are unique to women who find themselves involved in the criminal process. Of the offending population, only about 7% are women (Carson, 2015). However, female jail, prison, and probation populations grew at a faster rate than the male populations in the years between 2000 and 2010 (Herberman & Bonczar, 2015; Minton & Zeng, 2015). This growth rate ensures that issues regarding the female offending population will be of increasing importance for the police, courts, and corrections system for some time to come.

FEMALE OFFENDERS BEHIND BARS: A DETAILED LOOK AT PERCENTAGES AND RATES

While the total number of female offenders incarcerated increased steadily over the past few decades, recently there has been a slight downturn in this growth. The reasons for this are not clear but may be due to a combination of factors related to a greater emphasis on reentry, treatment, and community-based sanctions.

Despite the general increase in the female prison population, women offenders are, as noted above, a small proportion of the overall prison population. To ensure that the numbers are kept in perspective, consider that at the end of 2017, U.S. prisons held 1,378,003 men compared with 111,360 women. The rate of incarceration for men was 829 out of every 100,000, whereas the rate for women was 63 out of 100,000 (Carson, 2017). These numbers reflect lower overall rates of prison population growth, largely due to states reducing their prison populations as a means of saving money in a tight economy. Because of the War on Drugs and the extended sentences associated with that era, many women were incarcerated for lengthier sentences, and this helped fuel the growth in the female inmate population. Given the rates of drug use among female offenders, their increased number behind bars is not surprising.

Currently, the United States incarcerates more women than any other country in the world. When considering the female inmate population, it is clear that a sizeable portion of that population is concentrated in three states: (1) Texas, which in 2017 held 13,958 female inmates in its prisons and jails; (2) Florida, which held 6,725; and (3) California, which held 5,859 (Carson, 2017). Students should refer to Table 8.1 for a comparison of female offenders incarcerated in 2016 and 2017 in these three and several other selected states.

TABLE 8.1

Female Offenders in Selected States: A Comparison Between 2016 and 2017

STATE JURISDICTION	2016	2017	PERCENTAGE CHANGE 2016–2017
Arizona	3,997	4,059	0.6
California	5,886	5,859	−0.5
Florida	6,836	6,725	−2.0
Georgia	3,788	3,828	1.1
Illinois	2,613	2,281	−12.7
Missouri	3,337	3,396	1.8
New York	2,274	2,277	0.1
Ohio	4,594	4,426	−3.7
Pennsylvania	2,863	2,851	−0.4
Texas	14,335	13,958	−2.6
Virginia	3,109	3,154	1.4

Source: Carson, E. A. (2018). *Prisoners, 2016*. Washington, DC: Bureau of Justice Statistics.

HISTORY OF WOMEN BEHIND BARS

Prior to the 1800s, women were generally imprisoned in the same facilities with men. As students may recall, this was true even in England, where women, children, and men were imprisoned together on floating gaols in the Thames. John Howard's work in the late 1770s helped to draw attention to the plight of women, and, with the developments in prisons in Pennsylvania, the issue of women inmates was squarely addressed. Though men and women were separated during this time, the living conditions were equally unhealthy. Like the men, women suffered from filthy conditions, overcrowding, and harsh treatment.

In 1838 in the New York City Jail, for instance, there were 42 one-person cells for 70 women. In the 1920s at Auburn Prison in New York, there were no separate cells for the 25 or so women serving sentences up to 14 years. They were all lodged together in a one-room attic, the windows sealed to prevent communication with men (Rafter, 1985). Further, sexual abuse was reportedly a common occurrence at this time, with male staff raping women in prison. Consider that in 1826, a female inmate named Rachel Welch became pregnant while in solitary confinement. Examples abound where women in prisons were routinely abused sexually, and, in the state of Indiana, there were accounts that prostitution of these women was widespread (Rafter, 1985).

The Work of Elizabeth Fry

One activist who fought for women in prison during the early 1800s stands out. Elizabeth Fry (1780–1845) was a Quaker prison reform activist and an advocate for women who were incarcerated during these early years. Though most of Fry's work was done in England, her beliefs became known around the world. Fry's work is noteworthy for several reasons. First, she was one of the first persons in the United States and Europe to truly highlight the plight of women in prison. Second, she was a Quaker, and, as we have seen in earlier chapters of this text, the Quakers were instrumental in effecting positive change in American prisons. Third, Fry was so influential as to be invited to share her thoughts across both the North American and European continents; seldom were women held in such high regard during this time.

PHOTO 8.1 Elizabeth Fry reads to female inmates at Old Newgate Prison in 1823.

Department of Education, Arts & Libraries London Borough of Barking & Dagenham.

Chivalry hypothesis: Contends that there is a bias in the criminal justice system against giving women harsh punishments.

Victorian Era: Viewed women from a lens of inflexible femininity where women were to be considered pious and naïve of the evils of the world.

In the early 1800s, Fry visited Old Newgate Prison in England. This prison was notorious for the squalid conditions that it provided for inmates. For female inmates, the circumstances were worse: Approximately 300 women and children were collectively housed in two large cells. Cooking, the elimination of waste, sleeping, and all other activities were conducted in those cells. No fresh linens were provided, nor were nightclothes available. Fry would bring food, clothing, and straw bedding to those unfortunate enough to be at Newgate (Hatton, 2006).

Eventually, in 1817, Fry organized a group of women activists into the Association for the Improvement of the Female Prisoners in Newgate (Samuel, 2001). These women created a school curriculum for female inmates at Newgate and provided materials so inmates could sew, knit, and make crafts that could be sold for income. Naturally, there was a religious orientation to these services, and the women at Newgate were given regular Bible studies to help them form a constructive and spiritually therapeutic outlook on their plight. Fry's concern for women in prison, and for the field of corrections in general, culminated in the completion of a book, *Observations on the Visiting, Superintendence, and Government of Female Prisoners* (Samuel, 2001), in which Fry discusses the need for prison reform extensively.

Prior to the 1800s and even during the early 1800s, judges in the United States were hesitant to sentence women for serious crimes unless they were habitual offenders (Feeley & Little, 1991). Indeed, it would appear that the chivalry hypothesis was the guiding notion behind the view of women and criminality. The **chivalry hypothesis** contends that there is a bias in the criminal justice system against giving women harsh punishments. This is true so long as the offenses that they commit are considered "gender appropriate," or consistent with the stereotyped role that women are expected to maintain.

This type of thinking is typical of the **Victorian Era**, during which women were viewed through a lens of inflexible femininity and considered to be pious and naïve of the evils of the world. Because this was how women were seen, the criminal courts were often hesitant to punish women whom norms defined as pure, passive, and childlike in their understanding and primarily driven by emotion rather than reason. Further, Pollak (1950) has noted that during this era, men were expected to act in a fatherly and/or protective manner. Thus, the male-dominated criminal justice system (including police and judges) tended to downplay female crime, making it less likely to be detected, reported, prosecuted, and sentenced.

The stereotypes regarding appropriate conduct for women come from ideas about their "proper" place in society (Brennan & Vandenberg, 2009). Women who fell outside of these expectations were considered abnormal. Those who fell well outside of these expectations, such as female criminals who committed serious crimes, were considered evil or abhorrent. Thus, if a woman committed a crime that fell outside of what seemed understandable based on gendered stereotypes, such as murder or assault, her punishment was actually likely to be much harsher that what might be given to a man. On the other hand, when a woman committed a crime more gender based, such as prostitution, substance abuse, or theft, she was often given a light sentence or simply avoided official action.

Female Criminality From 1850 Onward

Records of criminal convictions and imprisonment for women increased during and after the Civil War. In speculating why this was the case, it may be concluded that the absence of men (who were off to war) and the industrialization that occurred in the North impacted views of women by creating circumstances where women were more visible within society (Kurshan, 1996). Further, an increase in crimes occurred throughout the nation among both men and women. This meant that by mid-century there were enough women inmates to necessitate the emergence of separate women's quarters. The separation of women from men in prison facilities led to other changes in later years.

In 1869, Sarah Smith and Rhoda Coffin, two Quaker activists, were appointed to inspect correctional facilities, both in Indiana (the state in which they lived) and in other areas of the nation. They concluded that the state of morals and integrity among staff in female prisons was deplorable (Rafter, 1985), and they spearheaded a social campaign against the sexual abuse that they had discovered in Indiana prisons. Their work led to the first completely separate female prison in 1874. By 1940, 23 states had separate women's prisons (Kurshan, 1996).

The Evolution of Separate Custodial Prisons for Women and Further Evidence of the Chivalry Hypothesis

Eventually, as central penitentiaries were built or rebuilt, many women were shipped there from prison farms because they were considered "dead hands" as compared with the men. At first, the most common form of custodial confinement was attachment to male prisons; eventually independent women's prisons evolved out of these male institutions. These separate women's prisons were established largely for administrative convenience, not reform. Female matrons worked there, but they took their orders from men.

FIGURE 8.1

Most Common Offenses for Female Offenders in 2017

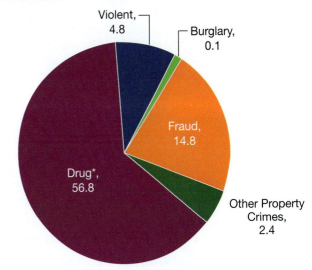

Violent, 4.8
Burglary, 0.1
Fraud, 14.8
Drug*, 56.8
Other Property Crimes, 2.4

*Includes trafficking, possession, and other drug offenses. More than 99% of federal drug offenders are sentenced for trafficking.

Source: Carson, E. A. (2018). *Prisoners, 2016.* Washington, DC: Bureau of Justice Statistics.

Women in custodial prisons were frequently convicted of felony charges, most commonly for crimes against property, often petty theft. Only about a third of female felons were serving time for violent crimes. The rates for both property crimes and violent crimes were much higher than for the women at the reformatories. On the other hand, relatively fewer women were incarcerated for public order offenses (fornication, adultery, drunkenness, etc.), which were the most common in the reformatories. This was especially true in the South, where these so-called morality offenses by African Americans were generally ignored, and where authorities were reluctant to imprison Caucasian women at all.

The reasons for women committing crime, especially violent crime, tend to be quite different from those usually observed among male offenders. Data from New York's Auburn Prison on homicide between 1909 and 1933 reveal the special nature of women's "violent" crime (Freedman, 1981, p. 12). Most of the victims of murder by women were adult men. For many feminist researchers, this would not be surprising; the crime would be taken more seriously if it was against the male social order. Further, as we know from more current research today, most homicides committed by women tend to be domestic in nature. In many cases, issues related to long-term spouse abuse are at the heart of what ultimately turns out to be an instance of victim-turned-perpetrator—the female partner killing her abuser in reaction to years of mistreatment. During the 1800s, issues related to spouse abuse were not identified as they are nowadays, and this means in most cases concepts such as *battered woman's syndrome*—the term for when abused women show identifiable symptoms of trauma associated with domestic battering and psychological abuse—would have been alien in courthouses. Thus, the murders of men, particularly adult men, likely resulted from dynamics that were not acknowledged by courtrooms of that time. In fact, laws often gave husbands the right to use force in their household, including force against their wives. Women, on the other hand, had no such right; they did not even have the right to vote at this time.

Other evidence exists that the chivalry hypothesis and/or stereotyped expectations of women may have been a partial explanation for female criminality. For instance, the

Women's Prison Association of New York, which was active in the social purity movement, declared in 1906 that women who committed crimes were often immigrants who were immoral menaces to sons and daughters throughout society, polluting American values with promiscuous behavior (Rafter, 1985).

This demonstrates how, even in the 1900s, social views related to women and criminality were tied to moralistic outlooks regarding their sexuality and identity as mothers. Consider also that during this time, performing an "illegal" abortion was classified as a violent crime that carried a sentence of imprisonment (Freedman, 1981, p. 12). Naturally, these abortions were illegal in every circumstance because there was no such thing as a "legal" abortion. Since abortion was considered murder, women who exercised autonomy over their own body and wished to avoid unwanted pregnancies had no choice but to commit the ultimate crime, regardless of the circumstances that led to their pregnancy. Because sexual abuse of women went largely unaddressed, many women (especially young girls) who were victims of rape and/or molestation were expected to endure this experience and to birth the offspring of their victimization. A failure to do so would be seen as contradictory to the nurturing and caring image that women were expected to maintain and would, therefore, be demonized by society and by the criminal justice system.

Minority Female Offenders Compared to Caucasian Female Offenders in American History

Despite our discussion of female offending and incarceration throughout past generations, one key point to understanding how women have been sentenced has not yet been addressed: racial disparities that have existed and continue to exist among the female population, particularly among those who are incarcerated. This is a very important point that deserves elaboration. Kurshan (1996) provides excellent insight into this issue, and much of her work has been adapted in this chapter to elaborate on the inherent racism that existed in the justice system when processing female offenders.

Consider that prison camps emerged in the South after 1870, and the overwhelming majority of women in these camps were African American; the few Caucasian women there were imprisoned for much more serious offenses, yet experienced better conditions of confinement. For instance, at Bowden Farm in Texas, the majority of women were African American, incarcerated for property offenses, and they worked in the field (Freedman, 1981). The few Caucasian women there had been convicted of homicide and served as domestics. As the techniques of slavery were applied to the penal system, some states forced women to work on the state-owned penal plantations and also leased women to local farms, mines, and railroads.

An 1880 census indicated that in Alabama, Louisiana, Mississippi, North Carolina, Tennessee, and Texas, 37% of the 220 imprisoned African American women were leased out but the same was true for only one of the 40 imprisoned Caucasian women (Kurshan, 1991, p. 3). Testimony in an 1870 Georgia investigation revealed that, in one instance,

there were no white women there. One started there, and I heard Mr. Alexander (the lessee) say he turned her loose. He was talking to the guard; I was working in the cut. He said his wife was a white woman, and he could not stand it to see a white woman worked in such places. (Freedman, 1981, p. 151)

Acknowledgment of the disparity between minority women and Caucasian women is important for two reasons. First, it is seldom addressed in most textbooks on corrections or correctional systems and practices. Second, we continue to see disparity between minority women and Caucasian women in today's society. In generations past, much of the disparity was residual from the prior slave era, particularly in the South. The reasons for such disparity are perhaps no longer the same, yet they are real and

PHOTO 8.2 These female inmates are participating in a graduation ceremony in an prison education program.

Marmaduke St. John/Alamy Stock Photo

still exist. It can be seen that women who commit crimes have been demonized throughout history, and, for those who were double minorities (particularly African American women), the treatment and type of prison experience received were significantly different. Female offenders were not (and still are not) all cast from one mold—they are diverse, and the experiences for minority women are even more noxious than they tend to be for Caucasian women. This is not to minimize the impact of incarceration upon Caucasian women but is meant to highlight even more the demographics and conditions of confinement for women throughout American correctional history. This provides a more historically correct view of how circumstances have evolved for women in the correctional system and also demonstrates that the disparity that we see today is simply an extension of past dynamics (see Focus Topic 8.1).

FOCUS TOPIC 8.1

Disproportionate Sentencing and Incarceration of Minority Women

To further highlight the disparity between minority women inmates and Caucasian women inmates, consider that in 2015 African American females were between 1.6 and 4.1 times more likely to be imprisoned than Caucasian females of any age group (Carson, 2015). Further, research by Carson (2015) also shows that for female inmates ages 30 to 34, the highest imprisonment rate was among African Americans (264 per 100,000), then Latina inmates (174 per 100,000), then Caucasians (163 per 100,000). In addition, African American females ages 18 to 19 (33 inmates per 100,000) were almost 5 times more likely to be imprisoned than Caucasian females (7 inmates per 100,000), according to Carson (2014). ●

Sources: Carson, E. A. (2015). *Prisoners, 2014.* Washington, DC: Bureau of Justice Statistics; Carson, E. A. (2014). *Prisoners, 2013.* Washington, DC: Bureau of Justice Statistics.

ISSUES RELATED TO THE MODERN-DAY FEMALE OFFENDER POPULATION

When discussing female offenders, it becomes clear that most are minority members with few options and limited economic resources. Thus, many of these offenders are excluded in multiple ways from the access to success and stability commonly attributed to broader society. Of those women who are incarcerated, approximately 44% have no high school diploma or GED, and 61% were unemployed at the point of incarceration (Bloom, Owen, & Covington, 2003). In addition, roughly 47% were single prior to incarceration, while approximately 65% to 70% were the primary caretakers of minor children at the point that they were incarcerated (Bloom et al., 2003). Lastly, over one third of those incarcerated can be found within the jurisdictions of the federal prison system or the state prison systems of Texas and California (Harrison & Karberg, 2004). Students should go to Figure 8.2 for further information on the overall characteristics of the female offender population.

Female offenders are less likely than men to have committed violent offenses and more likely to have been convicted of crimes involving drugs or property. Often, their property offenses are economically driven, motivated by poverty and by the abuse of alcohol and other drugs. Women face life circumstances that tend to be specific to their gender, such as sexual abuse, sexual assault, domestic violence, and the responsibility of being the primary caregiver for dependent children. Approximately 1.3 million minor children have a mother who is under criminal justice supervision, and approximately 65% of women in state prisons and 59% of women in federal prisons have an average of two minor children.

Women involved in the criminal justice system thus represent a population marginalized by race, class, and gender. African American women are overrepresented in correctional populations. Indeed, the imprisonment rate for African American females (96 per 100,000 African American female residents) was almost double that for Caucasian females (49 per 100,000 Caucasian female residents). Further, among female offenders ages 18 to 19, African American females were 3.1 times more likely than Caucasian

FIGURE 8.2

National Profile of Women Offenders

A profile based on national data for women offenders reveals the following characteristics:

- Disproportionately women of color

- In their early to mid-30s

- Most likely to have been convicted of a drug-related offense

- From fragmented families (that include other family members who also have been involved with the criminal justice system)

- Survivors of physical and/or sexual abuse as children and adults

- Individuals with significant substance abuse problems

- Unmarried mothers of minor children

- Individuals with a high school or general equivalency diploma (GED) but limited vocational training and sporadic work histories

Sources: Carson, A. E. (2015). *Prisoners, 2014*. Washington, DC: Bureau of Justice Statistics; Carson, A. E. (2014). *Prisoners, 2013*. Washington, DC: Bureau of Justice Statistics.

FIGURE 8.3

Prevalence of ACE Among Juvenile Offenders

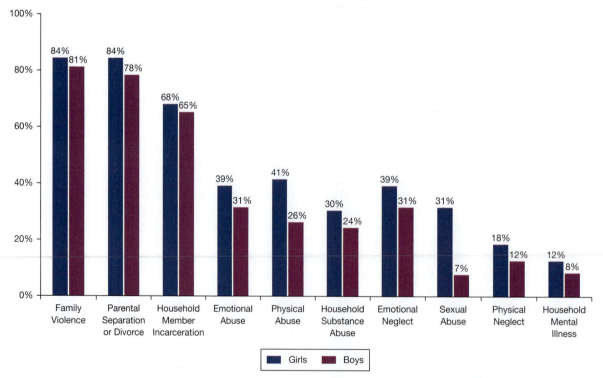

Prevalence of ACE Indicators by Gender

Source: Figure 1 from "The Prevalence of Adverse Childhood Experiences (ACE) in the Lives of Juvenile Offenders," Michael T. Baglivio et al., *Journal of Juvenile Justice*, Vol. 3, Issue 2, Spring 2014.

females and 2.2 times more likely than Latina females to be in prison in 2016 (Carson, 2018). The age and racial characteristics of women who are incarcerated are summarized in Table 8.2 for students wishing to see a more detailed breakdown.

TABLE 8.2

Rate of Female Inmates in State and Federal Facilities by Age and Race in 2017

AGE GROUP	ALL FEMALE[a]	WHITE[b]	BLACK[b]	HISPANIC	OTHER[b]
18–19	11	6	26	14	16
20–24	80	57	129	85	144
25–29	165	135	200	163	250
30–34	185	156	224	187	300
35–39	167	143	208	161	276
40–44	126	108	171	108	197
45–49	100	82	148	91	158
50–54	74	56	121	68	126
55–59	41	29	70	41	84
60–64	20	13	35	26	37
65 or older	5	4	9	7	9
Total Number of Sentenced Inmates	105,000	49,100	19,600	19,400	17,000

Sources: Bureau of Justice Statistics, National Prisoner Statistics, 2007–2017; Federal Justice Statistics Program, 2017 (preliminary); National Corrections Reporting Program, 2016; and Survey of Prison Inmates, 2016.

Note: Counts based on prisoners with sentences of more than 1 year under the jurisdiction of state or federal correctional officials. Imprisonment rate is the number of prisoners under state or federal jurisdiction with a sentence of more than 1 year per 100,000 U.S. residents of corresponding sex, race, Hispanic origin, and age. Resident population estimates are from the U.S. Census Bureau for January 1, 2017. Includes imputed counts for North Dakota and Oregon, which did not submit 2017 NPS data.

[a]Includes American Indians and Alaska Natives; Asians, Native Hawaiians, and Other Pacific Islanders; and persons of two or more races.

[b]Excludes persons of Hispanic or Latino origin.

[c]Includes persons age 17 or younger.

[d]Race and Hispanic origin totals are rounded to the nearest 100 to accommodate differences in data collection techniques between jurisdictions.

The Female Inmate Subculture and Coping in Prison

The conditioning of women in the United States regarding the traditional social role of motherhood and the forced separation from their family during incarceration has a considerable effect on women in prison. A significant aspect of the female coping mechanism inside the prisons is their development of family-like environments with the other female prisoners. These fictional family atmospheres or kinship structures enable the women to create a type of caring, nurturing environment inside the prison (Engelbert, 2001). Associating with a prison family provides a woman with a feeling of belonging and social identity. Many of these relationships are formed as friendships and develop into the companionship roles of "sister-to-sister" and "mother-to-daughter" bonds. These relationships can become intimate and include touching and hugging without having sexual overtones.

However, sexual relationships in prison do exist and are not uncommon. Some women who are in prison for an extended period of time may become involved with another female inmate in order to fill their need for love and companionship. In many of these cases, one will act in a male role and take on male mannerisms. She may walk and talk like a man, cut or shave her hair, and attempt to dress in a manly fashion. The other partner will typically behave in a traditional feminine manner. Some of these women become involved in these relationships only while in prison, and therefore do not consider themselves to be lesbian in the strictest sense of the word. Similar to male prison

PHOTO 8.5
A correctional officer maintains security over female inmates in a prison chow hall.

Jim West imageBROKER/ Newscom

substance abuse, trauma, mental illness, parenting responsibilities, and employment histories; and represent different levels of risk within both the institution and the community. To successfully develop and deliver services, supervision, and treatment for women offenders, we must first acknowledge these gender differences.

Research from a range of disciplines (e.g., health, mental health, and substance abuse) has shown that safety, respect, and dignity are fundamental to behavioral change. To improve behavioral outcomes for women, it is critical to provide a safe and supportive setting for supervision. A profile of women in most correctional facilities indicates that many have grown up in less than optimal family and community environments. In their interactions with women offenders, criminal justice professionals must be aware of the significant pattern of emotional, physical, and sexual abuse that many of these women have experienced. Every precaution must be taken to ensure that the criminal justice setting does not reenact female offenders' patterns of earlier life experiences.

Because of their lower levels of violent crime and their low risk to public safety, women offenders should, whenever possible, be supervised with the minimal restrictions required to meet public safety interests. Understanding the role of relationships in women's lives is fundamental because the theme of connections and relationships threads throughout the lives of female offenders. When an understanding of the role of relationships is incorporated into policies, practices, and programs, the effectiveness of the system or agency is enhanced. This concept is critical when addressing the following:

1. Reasons why women commit crimes

2. Impact of interpersonal violence on women's lives

3. Importance of children in the lives of female offenders

4. Relationships between women in an institutional setting

5. Process of women's psychological growth and development

6. Environmental context needed for programming

7. Challenges involved in reentering the community

Attention to the above issues is crucial to the promotion of successful outcomes for women in the criminal justice system. Substance abuse, trauma, and mental health are three critical, interrelated issues in the lives of women offenders. These issues have a major impact on a woman's experience of community correctional supervision, incarceration, and transition to the community in terms of both programming needs and successful reentry. Although they are therapeutically linked, these issues historically have been treated separately. One of the most important developments in health care over the past several decades is the recognition that a substantial proportion of women have a history of serious traumatic experiences that play a vital and often unrecognized role in the evolution of their physical and mental health problems.

Further, it should be obvious that programs will have to address both the social and material realities of female offenders, just as is true with the male inmate population. This is an important aspect of correctional intervention, particularly at the point of reentry. The female offender's life is shaped by her socioeconomic status, her experience with trauma and substance abuse, and her relationships with partners, children, and family. Most female offenders are disadvantaged economically and socially, and this is compounded by their trauma and substance abuse histories. Improving outcomes for these women requires preparing them through education and training so they can support themselves and their children.

The Prison Rape Elimination Act of 2003 Revisited

Earlier in this text, discussion regarding the Prison Rape Elimination Act of 2003 (PREA) was provided, particularly with regard to findings with male institutions. With this noted, it should be added that the PREA applies to all correctional facilities, including prisons, jails, juvenile facilities, military and Native American tribal facilities, and Immigration and Customs Enforcement (ICE) facilities. Due to the sensitive nature of violent victimization and potential reluctance to report sexual assault, estimates of the prevalence of such acts do not rely on a single measure. The act requires the Bureau of Justice Statistics (BJS) to carry out, for each calendar year, a comprehensive statistical review and analysis of the incidence and effects of prison rape. The act further specifies that the review and analysis shall be based on a random sample or other scientifically appropriate sample of not less than 10% of all prisons and a representative sample of municipal prisons. In 2014, more than 7,600 prisons, jails, community-based facilities, and juvenile correctional facilities nationwide were covered by the PREA. Among the findings of the 2014 data collection and analyses were the following:

1. Administrators of adult correctional facilities reported 8,763 allegations of sexual victimization in 2011, a statistically significant increase over the 8,404 allegations reported in 2010 and 7,855 in 2009.

2. The number of allegations has risen since 2005, largely due to increases in prisons, where allegations increased from 4,791 allegations to 6,660 in 2011 (up 39%).

3. In 2011, 902 allegations of sexual victimization (10%) were substantiated (i.e., determined to have occurred upon investigation).

4. State prison administrators reported 537 substantiated incidents of sexual victimization in 2011, up 17% from 459 in 2005.

5. About 52% of substantiated incidents of sexual victimization in 2011 involved only inmates, while 48% of substantiated incidents involved staff with inmates.

6. Injuries were reported in about 18% of incidents of inmate-on-inmate sexual victimization and in less than 1% of incidents of staff sexual victimizations.

7. Females committed more than half of all substantiated incidents of staff sexual misconduct and a quarter of all incidents of staff sexual harassment.

Clearly, administrators must be concerned with issues related to sexual assault in prisons and jails. As can be seen from the 2014 data, these assaults are completed by both women and men, and they are also completed by both inmates and staff, in nearly equal numbers. During the 1990s, significant attention was focused on female sexual abuse in custodial environments. Much of the reporting from groups such as Human Rights Watch and the news media, coupled with civil case litigation, led to the enactment of the PREA in 2003.

Since that time, the PREA has shed light on sexual misconduct in jails and prisons and, along the way, the dynamics have been found to be much more diverse than many practitioners may have expected. Much of the sexual assault on women behind bars is at the hands of custodial staff, though not necessarily forcible in nature (i.e., a staff member agreeing to assist a female inmate in obtaining free-world goods in exchange for sex), whereas most of the sexual assaults on men are inmate-on-inmate. However, in some cases, female correctional staff will become involved with male inmates, and there are instances where female staff have sexually exploited female inmates, though this is somewhat rare. Lastly, prisons do have to be alert to the fact that in male institutions, some male inmates are assaulted by male prison staff. Thus, the various circumstances associated with custodial sexual assault can be quite varied, but it is the female offender population that seems to be most targeted by corrupt prison staff.

FEMALE OFFENDERS AND TREATMENT IMPLICATIONS

In a study of 110 programs that deal with female offenders, it was found that programs conducive to treatment success of female offenders used female role models and paid particular attention to gender-specific concerns not common to male offenders (Bloom et al., 2003). Treatment for female offenders requires a heightened need to respond to expression of emotions and the ability to communicate openly with offenders.

Based on the research presented and the specific needs of female offenders, it is obvious treatment considerations for female offenders can be quite complicated. Indeed, it is plausible that a female offender could have the sundry challenges outlined in this chapter as well as numerous others included in other chapters of this text. Thus, many female offenders can be viewed as "special needs plus" when considering the myriad issues that may be present. With this in mind, specific recommendations regarding treatment programs for female offenders are outlined as follows:

1. Treatment plans must be individualized in structure:
   a. Clear and measurable goals
   b. Intensive programming with effective duration
   c. Appropriate screening and assessment

2. Female offenders must be able to acquire needed life skills:
   a. Parenting and life skills are taught—these are both *critical*.
   b. Anger management must be addressed.
   c. Marketable job skills are important because female offenders typically have few job skills and, unlike male offenders, have more difficulty obtaining jobs in the manual labor sectors that pay higher (construction, plant work, etc.).

3. Must address victimization issues:
   a. Programs should address self-esteem, which is typically tied into previous abuse issues, which in turn increase likelihood of substance abuse and prostitution, two main segments of female crime.
   b. Programs must address domestic violence issues. These are highly common among female offenders. The violence may have come from the family of origin and/or from previous boyfriends and/or spouses; often this is intergenerationally transmitted.

Dolan, Kolthoff, Schreck, Smilanch, and Todd (2003) discuss the importance of having gender-specific treatment programs for correctional clients with co-occurring disorders. Their insights are important for a couple of key reasons that should be emphasized. First, it is becoming increasingly clear within the treatment literature that therapists and caseworkers and the curriculum that they use must be able to address diverse populations. Many individuals do not consider that women, the elderly, and the disabled should be part of most diversity programs, just like ethnic and racial groups. Correctional programs must deal with this issue. Second, it is important that co-occurring disorders be addressed by correctional systems in order to not only lower recidivism rates but also improve the mental health and overall functioning of those offenders who return back to the community. Just because they may not commit more crime, they will still have issues with stress, depression, anxiety, trauma, and so forth that can impair their ability to successfully function in society.

In Iowa, the First Judicial District Department of Correctional Services established a community-based treatment program in its correctional facility (incidentally, this demonstrates how both correctional facilities and community supervision programs can successfully interface with one another). This program established a gender-specific female program to provide integration of treatment services designed to focus on dual diagnosis of female offenders. This is important since female offenders may be at heightened risk of having dual diagnoses given their frequent prior abuse trauma, maternal concerns for their children, and high rate of substance abuse.

The program addresses physical and sexual abuse issues, substance abuse, mental disorders, family-based counseling, and parenting issues. It is founded on the notion that most of its clients have grown up in dysfunctional families so it is difficult for these clients to even conceive of how a functional family operates on a routine basis. This program specifically strives to include the children in the treatment process since it has been found that this increases likelihood of client program completion and aids in the client's recovery from various other issues. Thus, treatment is optimized because the offender's role as a parent is used as a therapeutic tool to enhance the relevance of the treatment to the client.

PHOTO 8.6 These female deputies work at the Louisiana Transitional Center for Women (LTCW) in Tallulah, Louisiana. LTCW is a reentry facility for female offenders who are leaving prison and will be returning to the community. These officers are shown here at North Delta Regional Training Academy while taking correctional officer training to meet certification standards set by the state.

Robert Hanser

Specific issues pertaining to physical health care, adult sexuality, preventative pregnancy education, and education on sexually transmitted diseases are addressed within this program, as are feelings of grief or loss pertaining to the offender's role as a parent. A consortium of individuals and agencies are involved, including staff from the local Planned Parenthood, the local police force, community supervision, private therapists, and so forth. The topics covered by these individual providers are also discussed in the group therapy sessions to reinforce the learning process. Dolan et al. (2003) note that treatment for a woman with a dual diagnosis and a history of violence is optimized when it does the following:

1. Focuses on the woman's strengths

2. Acknowledges the woman's role as a parent

3. Improves interactions between the parent and child

4. Provides comprehensive, coordinated services for the mother and her children

The Iowa program is an excellent example of how most programs for female offenders should be structured because it is comprehensive and demonstrates how the multitude of issues pertinent to female offenders can be addressed within a single facility while utilizing a wide array of services within the community. This program falls well within the theme of this text, demonstrating how community resources and public and private agencies can combine to address complicated social ills with cost-effective methods. Because of this, it is suggested that other areas of the nation should look to this program when designing treatments for their female offender populations.

CONCLUSION

While female offenders are a small proportion of the offending population, they are a rapidly growing group, and it is increasingly clear that institutional and community intervention programs are inadequately serving these offenders. There is a need to address specific social ills that are fairly unique to female offenders, such as domestic violence, sexual abuse, drug use, prostitution, sexually transmitted diseases, and child custody issues. Many of the problems associated with female offenders have hidden costs that affect the rest of society in a multifaceted manner. Any failure to improve services to this offending population will simply ensure that future generations likewise adopt criminogenic patterns of social coping. This is specifically the case given the "collateral damage" that emanates from the impairment of children whose mothers (and likely sole caretakers) are incarcerated. When thinking long term, it becomes both socially and economically sound practice to work to improve services and ensure that accommodations are made for the special needs of the female offender.

Numerous legal considerations emerge with female inmates. These legal issues can range from the provision of services to pregnancy issues behind bars. The potential for staff misconduct is also a serious problem that resulted in a number of scandals during the 1990s. In addition, liability issues can emerge when employees conduct functions (such as strip searches and pat searches) where cross-gender interactions take place. Agency administrators must be sure to have clear and specific policies in place and see their staff are appropriately trained in supervising female inmates.

PRACTITIONER'S PERSPECTIVE

"Hearing and growing up and seeing those social and inhumane injustices as a child just further fueled my fascination."

Interested in a career in corrections or criminal justice? Watch Ashley Fundack's video on her career as a legal assistant.

Ashley Fundack
Legal Assistant

In order to prepare female inmates for reentry and to make their time in prison more constructive, agencies should follow several key guiding principles. First, agencies must recognize that gender does indeed matter, and agencies should provide an environment that fosters safety, respect, and dignity for women behind bars. Agencies should also develop policies and practices that are relational in nature and that promote good relationships with children, family, significant others, and the community. Agencies must address substance abuse, trauma, and mental health issues in a manner that is culturally competent. In addition, as with male inmates, agencies should provide female inmates with job and life skills that will allow them to change their socioeconomic circumstances. Lastly, effective and collaborative reentry processes must be utilized if agencies wish to realistically lower the likelihood of recidivism for female offenders.

 **SAGE edge**™

Want a Better Grade?

Get the tools you need to sharpen your study skills. Access practice quizzes, eFlashcards, video, and multimedia at **edge.sagepub.com/hanserbrief**

Interactive eBook

Visit the interactive eBook to watch SAGE premium videos. Learn more at **edge.sagepub.com/hanserbrief/access**.

 Career Video 8.1: Legal Assistant

Test your understanding of chapter content. Take the practice quiz at edge.sagepub.com/hanserbrief.

1. What are some of the demographic and statistical characteristics of the female offender population?

2. Compare and contrast the rates of incarceration and the terms of sentencing for male and female offenders.

3. Who was Elizabeth Fry? What did she do to assist female offenders, and why was this unique during her time in history?

4. What is **patriarchy**? How might this concept be associated with the dynamics of women who are incarcerated?

5. How does feminist criminology help to explain criminal behavior among women?

6. Which states are the most punitive for female offenders? Why (based on your prior readings) might this be the case?

7. What have been some common legal problems associated with female offenders in the correctional facility? What has been done in corrections to address these issues?

8. Identify and discuss at least three guiding principles associated with effective correctional operations when considering female offenders.

9. Discuss the various treatment considerations related to female offenders.

Patriarchy: A male-oriented and male-dominated social structure that defers to men and sees women in a subservient position to men.

Review key terms with eFlashcards at edge.sagepub.com/hanserbrief.

Chivalry hypothesis, 208

Collateral damage, 219

Domestic violence, 214

HIV/AIDS, 215

Little Hoover Commission, 219

patriarchy, 225

Victorian Era, 208

For this exercise, students should consider how professionalism in correctional officer training has gained support during the past couple of decades. This trend was discussed in Chapter 7, and students were required to develop a training program on workforce diversity in Applied Exercise 7.1. While that training program was to be comprehensive and address multiple facets of diversity, such programs cannot always go into the depth required when facilities are designed to meet the specific needs of a singly identified group, such as female offenders.

Consider your assignment in Applied Exercise 7.1. With that in mind, you are again asked to assume the role of the assistant warden who was tasked with developing the diversity training program. You have done so, as your warden requested, and the program has been highly successful in reducing grievances based on racial differences. In addition, the overall sense of professionalism possessed by officers in your facility has improved. In fact, your warden made a point to praise your work, even letting the regional director know about your program and the positive impact that it has had on your facility.

It appears that you have done very well indeed. The regional director ultimately created a region-wide position, the director of the new Office of Diversity and Specialized Needs. You were given this new job along with a pay raise. In short, you have been promoted. However, with every promotion comes added responsibilities and duties. Your case is no different. The regional director has asked that you go to the state's largest female prison, which is located in a remote area. He would like you to accomplish two tasks.

First, develop a training program for officers that is 1.5 days in length and addresses *gender-specific* considerations in security, safety, and human services. You will need to develop an outline of the content as well as an explanation for that content. This training will be implemented within the facility during the next fiscal year, when budgeting allows for additional training.

Second, you are to write a brief memo that outlines the top five issues that you believe are important to meeting the special needs of female offenders in your facility. One of these concerns might be the remote location of the current facility, away from friends and family who can visit.

For this assignment, students should create a brief outline of the training session on gender-specific issues in security, safety, and human services. This aspect of the assignment will require anywhere from 250 to 750 words. In addition, the memo, addressed to the regional director of state prison operations, should be approximately 500 to 750 words in length. A brief listing of the five recommendations and the reasons for those recommendations should be included.

WHAT WOULD YOU DO?

You are a correctional officer in a minimum-security facility that houses female inmates. While on your shift you are required to do a "shakedown" of the various inmates' living quarters throughout the day. You are given a list of inmates whose property you are to search. One of these inmates is Lorianna Marisol Vasquez. Her cubicle is ready to be searched, and she is standing close by in case you need to talk with her while conducting the search.

While searching the cubicle, you find a letter that Vasquez has written, which appears to be addressed to one of the male guards at the facility who works the evening shift (you work the day shift). The letter explains that Vasquez is in need of several hygiene products and also desires some cigarettes and sundry food products. Further, the letter seems to imply that Vasquez will make it well worth the officer's time to obtain these items.

However, the letter is written in a vague and comical manner, as if the entire idea were just a game or for sport, rather than being a serious proposition. In addition, you cannot tell from the letter if this has been an ongoing arrangement or if this would be the initial proposition. You look, but find no letters from the officer to Vasquez. However, you have heard rumors that this officer frequently talks with Vasquez for extended periods of time, and you have even detected a degree of jealousy among the other female inmates in the dorm. As it turns out, this officer is a good-looking fellow, and he is fond of being excessively polite to the women on the dorm. It makes you wonder if there is in fact something going on between Vasquez and this officer.

What would you do?

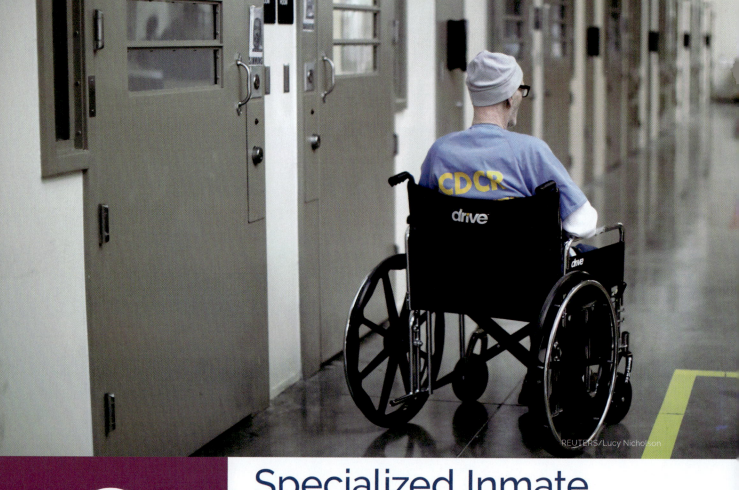
REUTERS/Lucy Nicholson

9

Specialized Inmate Populations and Juvenile Correctional Systems

Learning Objectives

1. Discuss some of the common administrative considerations facing prison systems when housing specialized inmate populations.

2. Explain how the prison subculture impacts, interacts with, and reacts to inmates who are mentally ill and/or are elderly.

3. Discuss the prevalence of mental illness within state prison systems and the difficulties of care and treatment.

4. Describe the difficulties with sex offender management in prisons.

5. Identify the various classifications of elderly inmates within an institution.

6. Identify some of the challenges and issues that correctional agencies face when housing elderly offenders.

Nae-Nae, a Transsexual Behind Bars

Nate, known as "Nae-Nae" by the other inmates on the dorm, sat on her bunk and thought about the situation. It was not an easy decision to make, and she was so close to possibly being granted early release.

Natalie was a transgender woman who had undergone some medical transitions (breasts and cosmetic facial alterations) and was serving a 4-year sentence on a drug charge. In the "free world," she had been an active advocate for gay, lesbian, and transsexual rights.

She was open about her orientation, had a good job in graphic design, and had been in a committed relationship for years. She had taken hormones and completed some cosmetic surgery with the intention of ultimately completely changing her sex.

Natalie's job eventually required that she relocate to another state, one that was more socially conservative, but the position was a promotion to management. Her partner, Karl, had supported the move. Karl was in grad school and between jobs, so it seemed like a perfect time for them both to move as Karl was expected to complete his degree in less than 3 months. Natalie would move to the new location and get settled in, and Karl would wrap up everything at their (then) current residence and follow Natalie to the new location with degree in hand.

Natalie made the move and all seemed well . . . at first. Then she received word that Karl would not be joining her; Karl had been seeing someone else, and it was for this reason that he had been so supportive of Natalie taking the job out of state.

As expected, Natalie found her new location was not nearly as progressive as her old one. The new job was okay, but the managerial challenges required that she alter her mindset. Also, she was emotionally down over losing Karl and being alone. This resulted in her relapse on alcohol and pain killers; the numbing effect helped her to cope with the changes, so she thought.

When Natalie was first booked, she had to complete a strip search. She remembered how the jailers treated her during the process. They were degrading—one in particular made issue of a tattoo that Natalie had on the small of her back. The jailer said, "Hey everyone, look at the tramp stamp that this queenie has on her back!" He pointed at a four-leaf clover with the words "Get Lucky" in a semi-circle above the shamrock design.

From that point onward, the deputies would tease her, usually within earshot of other inmates, asking her if she would show them her tramp stamp or asking if anyone in the dorm had "got lucky" recently.

She went by the name "Natalie" on the outside when involved with Karl, but she was given the nickname "Nae-Nae" by the other inmates inside the institution.

Now, 3 years after her arrest, she was involved with a man named Reggie who was doing time on the same dorm. Reggie was scheduled to be released in 2 days and, of course, would not be seeing Nae-Nae on a daily basis once released. When news of Reggie's parole spread, another man named Jerome stepped forward and announced to everyone, "I claim Nae-Nae as my punk."

Jerome was a huge, strong man who had a prior history of sexual assault. He was currently doing time for armed robbery and was not usually questioned on the dorm. Nobody challenged his claim, and Reggie was not on the dorm at the time.

When Natalie told Reggie about the situation, Reggie said, "Look, you know I've done what I can to keep you protected, but you know how it is. . . . Once our time is up, we gotta put this behind us. I can talk with him, but really, you are just gonna have to do what you got to do to survive."

Whether Nate, Natalie, or Nae-Nae, one thing was a common denominator: Things looked bleak. Jerome would likely be abusive, and there was no way that Natalie would be able to fight the man or prevent the upcoming sexual abuse. She contemplated talking to staff. She only had a few more months to do and figured it would be her safest bet. She did not have to worry about losing honor by being a snitch—as a punk, she had no respect on the dorm anyway. However, she did not like the idea of living in a restrictive protective custody cell during the last few months. Plus, there might be retribution if she talked to authorities. She was also concerned as to whether prison staff would seriously consider her concerns and requests. Natalie sighed. Time was short, and she had to make a decision of some sort.

INTRODUCTION

As we saw in Chapter 6, any modern correctional program relies on being able to reliably and validly classify and determine the needs of offenders. This classification ensures that a facility is able to match treatments and inmates in a manner that optimizes the outcome of the program. From a legal and historical context, the concept of *needs* has been pivotal in designing responses to different types of offenders (juvenile offenders, mentally challenged offenders, etc.). However, when we use the term *need*, this implies that the issue involves something that is absolutely necessary, not something that is a mere desire or preference. Thus, if the individual does not have this need met, he or she will experience some form of serious social, psychological, physical, or emotional impairment in his or her daily functioning. Because correctional institutions are liable for legally defensible standards of care for inmates in their custody, they must be prepared to address specialized needs that may emerge among their prison populations.

Offender with special needs: A specialized offender who has some notable physical, mental, and/or emotional challenge.

For purposes of this text, the term **offender with special needs** refers to specialized offenders who have some notable physical, mental, and/or emotional challenge. These challenges prevent either a subjective or an objective need from being fulfilled for that individual, and the lack of this need impairs the individual's day-to-day ability to function within the confines of the criminal law.

ADMINISTRATIVE CONSIDERATIONS

During the first few hours of being received, an inmate should be given a medical screening or examination. While a full examination is the most effective approach, this may not always be practical. However, screenings can be easily conducted, even when large groups of inmates are being processed into the system simultaneously. The medical screening process is basically a preliminary triage to sort out serious issues from those that are likely to be more routine among other inmates on a cell block or within a dormitory. Likewise, this process can be used to identify medical issues that require immediate attention, thereby alerting staff to the need for a full examination and for additional medical service. During this process, mental health professionals should be available to assist inmates with issues that, while medical in nature, may impact an offender's state of mind.

Access to Program Activities and Availability

Access to specific programs, such as education, addiction treatment, and vocational training, can result in increased satisfaction and improved health while in the prison. Further, participation in these programs can increase the offender's likelihood of receiving early discharge through good-time accumulation and can improve the prognosis of the offender's later integration into the community. Because of these factors, implementing the Americans with Disabilities Act (ADA) requirements may be prudent for correctional administrators for reasons beyond just the legal liabilities involved. Such incorporation may also serve to improve the plight of corrections in the future if these problems are addressed in a proactive manner. As Appel (1999) notes, the ADA is the law, and compliance with the ADA is a classic case of "pay me now or pay me even more later." Thus, the long-term benefits for the wise correctional administrator will likely offset any of the short-term costs paid in the meantime.

PRISON TOUR VIDEO: Offenders with special needs provide unique challenges for correctional practitioners. Go to the IEB to watch a clip about services provided for specialized populations.

Separate Care or Inclusion in General Population

Once the agency has identified special needs offenders, the next important step is to determine specifically how these offenders will be addressed. One primary question is whether these offenders should be kept in contact with the mainstream population of inmates or be segregated. This is an important issue because the welfare of the inmate is at stake and the agency can be held liable for failing to protect inmates from assault or abuse, even when this occurs at the hands of another inmate. Other reasons often given for separating special needs inmates from the general population are as follows:

1. Cost containment: It is generally more efficient and cost-effective if special needs offenders are kept in the same location (both the same facility and the same housing area in that facility), where they can be treated as a group.

2. Managed care: More effective care can be focused on a specialized unit where specifically trained staff can strategically target their skills to populations most in need.

3. Concentration of resources: Relevant staff and resources can be more concentrated if special needs inmates are housed in one central location.

Mainstreaming, or the integration of an inmate with a disability into the general prison population, is a general expectation of the ADA, and this is because it is expected that disabled inmates will be provided full and complete access to prison programs, activities, and services. However, the ADA does allow correctional administrators to remove an inmate from the general population if his or her health condition is a threat to other inmates or if an inmate is at risk of victimization from other inmates due to his or her impairment. What typically occurs is a mix of both separation and inclusion within prison facilities, where, depending on the pertinent circumstances, some inmates are handled separately and apart from the main population while others remain in the general population.

Special Facilities and Housing Accommodations

Many special needs offenders need housing designed to accommodate their specific circumstances. Inmates who are mentally ill or who have cognitive deficits may need to be placed in specialized facilities. When considering facility or housing accommodations,

PRISON TOUR VIDEO: Correctional practitioners have to carefully consider housing for inmates with special needs. Go to the IEB to watch a correctional officer discuss housing for specialized populations.

Universal design: Prison construction design that complies with ADA requirements and that accommodates all inmate needs in a universal fashion.

both physical plant conditions and safety should be considered. It is often the physical plant issues that most concern correctional administrators because they have the potential for exceeding the capacity of already tight budgets. Although the overall spending for corrections has risen along with the increase in inmates throughout the nation, even these larger budgets do not cover the specialized facilities and services required by the elderly and/or disabled. The main problems are with costs of renovating existing facilities; often building a new facility with all of the features required by the ADA is cheaper than modifying an older building. The special equipment and physical modifications needed to bring existing facilities into compliance can result in tight operating budgets being stretched beyond their limits.

Special Facilities and Support

When considering facilities and physical support for special needs inmates, agencies should use a concept that Anderson (2008) calls universal design. **Universal design** refers to prison construction design that complies with ADA requirements and accommodates all inmate needs in a universal fashion, regardless of how varied the needs may be from inmate to inmate. This type of design is ideal for agencies that wish to ensure that future construction efforts meet ADA and other standards of care.

This approach to prison facility construction and to implementing support services within future buildings will likely be the upcoming trend for prison systems. Examples of universal design features might be wider doors, ramps rather than stairs, and heavier building materials that allow safety bars or railings to be added to the construction in later years, if needed (Anderson, 2008). The key issue is that this type of design planning allows for the accommodation of additional needs among the population housed. "While universal design is initially more expensive, it costs less over time as modifications are easier and not as expensive to implement" (Anderson, 2008, p. 364).

PRISON SUBCULTURE AND SPECIAL NEEDS OFFENDERS

In Chapter 7, extensive discussion was given to the prison subculture. This subculture idealizes certain standards of behavior, but, as noted, these norms are in a state of flux due to the emergence of gang life inside and outside of the prison. There are two major points of concern related to the prison subculture and special needs offenders. The first concern has to do with inmate reactions to the increasing representation of special needs inmates within prison facilities. Second, concerns regarding agency staff and their understanding of the issues associated with special needs offenders are important from the perspective of inmate welfare as well as agency liability.

With regard to the inmate population, serious attention should be given to special needs offenders because they are likely to be very vulnerable to exploitation by other inmates. This is true both for elderly offenders, who physically are weaker and may also not be as mentally adept as they were in younger years, and for inmates with a variety of mental health conditions, such as anxiety-based disorders and affective disorders. These inmates are vulnerable to manipulation and are also likely to be victimized by predatory inmates due to their sense of fear (with anxiety-related disorders) or their sense of self-neglect (with affective disorders). Those with serious psychiatric disorders are likely to be taunted, ridiculed, and generally worked into heightened emotional states by other, more stable inmates who may find their reactions amusing or entertaining. Inmates with cognitive impairments are likely to have their property stolen by other inmates, and they are likely to be duped by wiser cons who act as if they wish to befriend them. Sex offenders and inmates with alternative sex identities or preferences are likely to be sexually and physically victimized by other inmates.

As we have seen, respect within the prison subculture is very important. It should be clear to students that special needs inmates would command no respect within the prison environment. In fact, it is likely that regular inmates would be considered weak or soft if they were to associate too closely with special needs inmates, and if those special needs inmates were sex offenders or homosexual offenders, an inmate who

PHOTO 9.1 Inmates participate in a group therapy session as part of a substance abuse program in the Iowa Correctional Institution for Women. Many correction institutions now offer programs to address substance abuse, mental health issues, and even family issues.

AP Photo/Steve Pope

associated with them would likely be ostracized completely from the main inmate power structure. Even with the gang culture that has emerged as a primary shaping mechanism for prison culture, these various types of offenders are not likely to gain respect, with one exception: the elderly inmate.

The elderly inmate who has lived by the convict code or is a veteran member of a gang is likely to have status among his peers. These inmates, for this text's purposes, will be referred to as greyhounds. **Greyhounds** are older inmates who have acquired respect within the offender subculture due to their track record and criminal history both inside and outside the prison; they are career criminals who have earned status through years of hard-won adherence to prison and criminogenic ideals. These inmates are hardly ever likely to be victims within the inmate subculture, and, if they were burned in the past, they generally were able to "even the score" with those who committed a wrong against them. In short, greyhounds are not seen as weak throughout their life of crime or while serving prison time.

Greyhounds: Older inmates who have acquired respect within the offender subculture due to their track record, criminal history, and criminogenic ideals.

Another concern when dealing with special needs inmates is the training and knowledge of correctional staff. Some staff may not understand mental illness and may be ill equipped to deal with inmates who present with odd or maladaptive behaviors. These inmates may have difficulty following institutional rules and may become a source of frustration for security staff. Thus, training is important for staff who will routinely deal with these types of inmates. In addition, in facilities designed for specialized inmates, security staff should be especially well trained in communication techniques, problem-solving approaches, the identification of many mental health illnesses, the best ways to address problematic or challenging inmate populations, sexual issues, communicable diseases, suicidal inmates, and so forth. As one can tell from this long list of criteria, the staff dealing with these populations should be highly specialized.

Lastly, the officer subculture must definitely be professionalized among those who routinely work with various types of special needs inmates. In many cases, correctional officers see themselves as working in a helping capacity as much as a security capacity. However, this may be quite a stretch in job definition for some security staff, particularly those who see their role as more punitive in nature. These offenders are convicted felons and not likely to elicit the sympathy of many correctional staff. This is particularly true for the sex-offending population. The training of correctional officers and organizational change implemented by leadership within correctional agencies is perhaps the best prescription for addressing staff obligations to special needs offenders.

MENTALLY ILL OFFENDERS

Mentally ill inmates are those who have a diagnosable disorder that meets the specific and exact criteria of the ***Diagnostic and Statistical Manual of Mental Disorders (DSM-5)***. This reference tome was formed by gatherings of psychiatrists and psychologists from around the world and sets forth the guidelines in applying a specific diagnosis to a person presenting with a mental disorder. For the purposes of this chapter, **mental illness** will be defined as any diagnosed disorder contained within the *DSM-5*, as

Diagnostic and Statistical Manual of Mental Disorders (DSM-5): A reference manual that sets forth the guidelines in applying a diagnosis of a mental disorder.

Mental illness: Any diagnosed disorder contained within the *DSM-5*.

From the data provided in Table 9.2, it can be seen that in 2006 a full 61% of prison inmates who had a current or past violent offense had some sort of mental health problem. In addition, 74% of these same inmates had some sort of substance dependence or abuse in their histories (see Table 9.3). The statistics among jail inmates were similar in that around three fourths of all jail inmates who had a mental health problem also had a substance abuse problem. These numbers are even more disturbing when one notes that even among prison inmates who did not have a mental health issue, half reported drug use in the month right before arrest, and 42% of jail inmates without a mental disorder also reported the use of substances during the month before arrest.

What these statistics mean is that the vast majority of inmates, whether in prison or in jail, are drug abusers. These statistics also demonstrate that these drug abusers are at an increased likelihood of having mental health problems while in prison. Thus, modern prison and jail systems should all have some sort of effective drug treatment component to mitigate the symptoms of inmates while they serve time behind bars. It behooves correctional agencies to provide these services since this may lower the number of institutional infractions among inmates.

Beyond Screening: Mental Health Assessment

Inmates requiring further assessment should be housed in an area with staff availability and observation appropriate to their needs. The assessment should be assigned to a specific mental health staff member and consist of interviews, a review of prior records and clinical history, a physical examination, observation, and, when necessary, psychological testing. The types of services needed to address these problems are similar throughout each phase of the correctional system's process (e.g., jail detention, imprisonment,

TABLE 9.2

Prevalence of Mental Health Issues Among Prison and Jail Inmates in the United States

SELECTED CHARACTERISTICS	PERCENTAGE OF INMATES IN			
	STATE PRISON		LOCAL JAIL	
	WITH MENTAL PROBLEM	WITHOUT	WITH MENTAL PROBLEM	WITHOUT
Criminal record				
Current or past violent offense	61	56	44	36
Three or more prior incarcerations	25	19	26	20
Substance dependence or abuse	74	56	76	53
Drug use in month before arrest	63	49	62	42
Family background				
Homelessness in year before arrest	13	6	17	9
Past physical or sexual abuse	27	10	24	8
Parents abused alcohol or drugs	39	25	37	19
Charged with violating facility rules*	58	43	19	9
Physical or verbal assault	24	14	8	2
Injured in a fight since admission	20	10	9	3

Source: James, D. J., & Glaze, L. E. (2006). *Mental health problems of prison and jail inmates.* Washington, DC: Bureau of Justice Statistics.

*Includes items not shown.

Note: Most recent data available.

TABLE 9.3

Substance Dependence or Abuse Among Prison and Jail Inmates, by Mental Health Status

SUBSTANCE DEPENDENCE OR ABUSE	PERCENTAGE OF INMATES IN					
	STATE PRISON		FEDERAL PRISON		LOCAL JAIL	
	WITH MENTAL PROBLEM	WITHOUT	WITH MENTAL PROBLEM	WITHOUT	WITH MENTAL PROBLEM	WITHOUT
Any alcohol or drugs	74.1	55.6	63.6	49.5	76.4	53.2
Dependence	53.9	34.5	45.1	27.3	56.3	25.4
Abuse only	20.2	21.1	18.5	22.2	20.1	27.8
Alcohol	50.8	36.0	43.7	30.3	53.4	34.6
Dependence	30.4	17.9	25.1	11.7	29.0	11.8
Abuse only	20.4	18.0	18.6	17.7	24.4	22.8
Drugs	61.9	42.6	53.2	39.2	63.3	36.0
Dependence	43.8	26.1	37.1	22.0	46.0	17.6
Abuse only	18.0	16.5	16.1	17.2	17.3	18.4
No dependence or abuse	25.9	44.4	36.4	50.5	23.6	46.8

Source: James, D. J., & Glaze, L. E. (2006). *Mental health problems of prison and jail inmates.* Washington, DC: Bureau of Justice Statistics.

Note: Most recent data available Substance dependence or abuse was based on criteria specified in the *Diagnostic and Statistical Manual of Mental Disorder (4th ed.) (DSM-IV).* For details, see Karberg, J. C., & James, D. J. (2005). *Substance dependence, abuse and treatment of jail inmates, 2002.* Washington, DC: U.S. Department of Justice.

community supervision). As noted earlier, offenders may enter the system with a certain degree of mental illness that is further aggravated by imprisonment, or they may present new symptomology once incarcerated (Johnson, 1999).

The stress of being involved in the criminal justice process can itself serve as a causal factor for some mental illness. Stressors might include navigating the legal system, separation from community and family support systems, and interactional problems with other offenders within the correctional facility. If mental illness is already present, offenders are often subject to victimization from the remainder of the inmate population (Hanser & Mire, 2010). In fact, those who present with mental illness (particularly along the spectrum of **mood disorders** and/or anxiety disorders) are at an inflated risk of sexual assault within the prison environment (Hanser & Mire, 2010). Further, several documented cases link this victimization with later suicidal ideation and completion among the inmate population (Hanser & Mire, 2010).

Problems with sleeping and eating may be experienced due to limited access to anxiety-reducing activities such as television, exercise, socialization, and smoking (Hanser & Mire, 2010). All of this is compounded by a lack of control over one's environment and the diminished sense of autonomy that is experienced. Thus, inmates arriving at the correctional facility must be closely watched and screened upon initial entry and during their first 90 to 180 days so that their integration into the institution can be appropriately monitored.

The correctional institution has an obligation to maintain security and control over the day-to-day routine for the welfare of staff and inmates involved. This means that inmates who present with psychological problems in coping must be kept in "check" so that they do not create a breach in institutional security. This is important and will likely be the primary (if not only) concern among security staff personnel. When disruptions do occur, it is common practice for security personnel to utilize methods of seclusion and/or restraint that are not necessarily consistent with clinical considerations. This can, to

Mood disorders:
Disorders such as major depressive disorder, bipolar disorder, and dysthymic disorder.

Therapeutic community:
An environment that provides necessary behavior modifiers that allow offenders immediate feedback about their behavior and treatment progress.

Assessment processes will determine whether the offender is amenable to treatment. Most will be provided treatment if they are willing, and those determined to be the most motivated will likely be given a special form of incarceration through a therapeutic community. The **therapeutic community** environment provides necessary behavior modifiers in the form of sanctions and privileges that allow offenders immediate feedback about their behavior and treatment progress. This is important to note because these types of programs occur while offenders are theoretically "behind bars," yet they are utilized to eventually reintegrate the offender into the community.

Basic Sex Offender Management

Institutions must focus on correctly identifying sex offenders as soon as possible after incarceration. For this to realistically occur, assessment and classification (just discussed) must take place. After the classification process, the offender should be separated from the general population of inmates when possible. At a minimum, a separate area for sex offender treatment should exist, and separate housing should be provided if feasible. Much of this has to do with the fact that these inmates are vulnerable to attack from other inmates since, as we have learned, these offenders are not respected within the inmate subculture and are specific targets within that subculture and for gangs active within a facility. Despite the recommendation that sex offenders be given separate housing, this is not often the case, and most are housed in the general population. This is because resources are scarce and because prison facilities are not required to separate sex offenders from the population as long as they are given reasonable protection.

Staff Issues With the Sex Offender Population in Prison

Ingram and Carlson (2008) perhaps say it best in noting that "the most essential step in developing and running a useful program for sex offenders is having trained staff who have the right attitude toward these inmates" (p. 377). Security staff should not view sex offenders as horrible persons who deserve the worst punishment possible. Nor should they be sympathetic to sex offenders to the point that they entertain excuses for the criminal behavior, such as the unfairness of sex laws regarding consent, issues with a troubled home, or misunderstandings regarding appropriate sexual activity (Ingram & Carlson, 2008). Rather, security staff should be realistic and mature. They should view the behavior of the sex offender as inappropriate but should be willing to consider that these offenders can change.

Security staff who will work within sex offender treatment programs should meet high criteria. Most staff who seek to work in this field will already have come to terms with the idea that treatment of sex offenders goes beyond punitive sanctions. However, the reasons that persons may seek these specialized assignments can vary, and some of them may not be healthy. Thus, careful screening and investigation of the person's background should be conducted.

The treatment personnel involved with this type of program should also be well qualified. Typically, a psychologist or psychiatrist is charged with supervising these types of programs, but many treatment providers may be master's-level social workers or counselors. These individuals will often have specialized forms of education or training that is required to work with sex offenders. In fact, most states will have a registry or roster of mental health professionals who are given the legal authority to work with sex offenders. This demonstrates how important it is to ensure that those who work with this population are appropriately trained and credentialed.

Treatment of Sex Offenders

Sex offender programs today most often use a combination of cognitive-behavioral techniques and relapse prevention strategies (Ingram & Carlson, 2008). These programs usually include individual counseling sessions and group therapy sessions to break through offenders' denial of their activity and help them attempt to build empathy for victims of their crime (Hanser, 2010a). When using group therapy, an effective technique to combating denial is to invite offenders who are in an advanced treatment group to challenge offenders in the less experienced group who are in denial. The idea is that more experienced offenders in treatment will be more adept at challenging offenders in denial than others. In other cases, treatment staff may find it more effective to simply create a group of offenders who all exhibit levels of denial. This is called the "deniers group" since it consists of sex offenders who are all in denial. Essentially, this is a "pre-group" that lasts from 11 to 16 weeks as a means of getting the offender in denial primed for the actual group therapy process. According to the Center for Sex Offender Management (2008), treatment providers who employ this method report that the great majority of offenders are able to come out of their denial. This approach targets two major issues:

1. Eliminating cognitive distortions, which, if left intact, allow offenders to continue denying or minimizing

2. Developing victimization awareness, which can allow offenders to understand the physical and psychological harm they inflict and, thus, render them more reluctant to commit future assaults

One primary concern of group members is often related to confidentiality, particularly regarding information divulged by other group members. One way to address this concern is to ask the group participants to come to an agreement among themselves about their own confidentiality. In virtually every instance, the agreement they make is that what is discussed in group does not get discussed outside of group. Typically, it is best to have the group members come to this agreement themselves rather than imposing this rule on them for two reasons. First, it is a simple way to get them involved in a discussion they are likely to understand and be interested in without addressing any threatening content, such as sex offending. This provides practice for what will be occurring in the group. Second, it requires that the group build cohesiveness and trust among its members, at least about this issue. Building trust among themselves can be a useful exercise because it leads to group members sensing that they can be helpful to each other.

Prison Subcultural Reactions and Treatment

Chapter 7 included extensive discussion on the prison subculture. In this discussion, it was made clear that sex offenders tend to have no status behind bars, and, depending on the type of sex offender, they are likely to be victimized. This is particularly true for child molesters and other similar types of sex offenders. However, as with the outside community, social forces have changed some of the beliefs within the prison environment. Nowadays, it is fairly common for younger inmates to have committed some type of sex offense, but this is usually some type of rape of an adult female rather than the molestation of a child. While rape of women tended to also carry negative connotations within the older prison culture prior to the 1960s, the stigma of this is not as great in today's modern prison environment.

There are many reasons for the shift in how the prison subculture responds to different types of sex offenders. First, the gangsta movement during the 1990s consisted of rap music and attitudes that overtly denigrated women and referred to them in terms that implied they were to be exploited. This created an air of tolerance for female victimization among many younger offenders, particularly gang offenders. Second, youth now tend to be very sexually active, and male juveniles continue to commit acts of violence; the shocking aspect of sex crimes has, therefore, diminished. Thus, a culture has bred where women are considered lowly in status, and younger offenders no longer have incentives to be "gentlemen" but instead gain status points for "sexing" young girls and women into gang membership. Further, many gangs may engage in different types of rape as a preferred choice of criminal behaviors. These types of sex offenders will not be stigmatized in the prison culture but instead will be considered part of their gang family.

However, this is not the case for other sex offenders (again, particularly those who have molested children), inmates with sexual identity issues, or inmates who succumb to pressure for same-sex activity inside the prison. These individuals will be stigmatized and very likely sexually victimized in many of the older facilities and systems where the prison subculture predominates. In any event, these inmates will be afforded no respect by other inmates in the facility.

SAFETY, SECURITY, AND ASSISTANCE FOR LGBTI INMATES

Lesbian, gay, bisexual, transgender, intersex (LGBTI) and gender-nonconforming inmates are a highly vulnerable population within the prison environment. This group has distinct medical, safety, and treatment needs that are no longer able to be overlooked by prison staff and administrators. Though many of the concerns for the different categories in this group are similar, it is important to understand that the transgender and gender-nonconforming population is distinct from the gay, lesbian, and bisexual inmate populations (King & Baker, 2014). Because of this, the risk-based classification that a prison will use with the LGBTI population must take into consideration the unique characteristics that will be associated with potential victimization for differing individuals. Specifically, for transgender inmates, this will require individualized decisions regarding gender placement, meaning that classification will need to determine whether the inmate will ultimately be kept in a male or a female facility.

During the past decade or so years, the Prison Rape Elimination Act (PREA) has become increasingly prominent in prison administration. This law was first enacted in 2003, and its purpose was to provide for the analysis of the incidence and effects of prison rape in federal, state, and local institutions (National PREA Resource Center, 2015). This act also created the National Prison Rape Elimination Commission, which was empowered to generate standards and guidelines for correctional facilities to safeguard against prison rape. For now, our discussion of the PREA will be focused on provisions for the LGBTI population. In later chapters, more information will be provided on other provisions and requirements associated with this groundbreaking law.

Since the act's passage, state correctional systems, local jails, and juvenile facilities have worked to demonstrate that they are in compliance with these federal regulations. As correctional administrators work to be in compliance with the act, they do so amid changes and additions to these guidelines, which initially were focused more on heterosexual sexual assaults but have increasingly begun to target the heightened risks that the LGBTI population faces (Schuster, 2015).

This is very important because these changes run completely counter to the old-school thinking that has traditionally been part of the prison subculture. As a result, correctional systems have now found that they must go beyond rape prevention and must address homophobic slurs and other forms of verbal harassment (Schuster, 2015). Further, unlike in the past, correctional systems will be required to discipline and relocate the perpetrators of this harassment rather than isolating or relocating victims "for their own protection." Correctional administrators will now also be required to consider input from transgender inmates as to whether they feel safe or more comfortable living within a male or a female facility; in times past, the inmate was not given any type of documented input on this decision. Schuster (2015) provides an excellent synopsis of these changes by stating that the PREA prohibits

> any hard-line rule about housing these inmates based on their assigned sex at birth. Jails, prisons, and juvenile facilities are now required to determine on a case-by-case basis whether a trans-inmate will be safer housed with men or women. They also must give serious consideration to an inmate's own views regarding his or her safety. Importantly, "transgender" is not defined by whether a person has undergone surgery or hormone treatment to change his or her anatomy or appearance. It is defined solely by a person's internal sense of feeling male or female. (p. 3)

This means that a trans woman (male-to-female) inmate cannot be excluded from being granted protective consideration because she has male genitalia or because prison staff do not believe that she appears to be "female enough" to qualify (Schuster, 2015).

The National Prison Rape Elimination Commission, as a fact-finding body of inquiry, has found that trans women, who often have breasts and a feminine appearance, are disproportionately the target of unnecessary, rough, and sexually oriented frisks and strip searches by male staff within correctional facilities. In addition, these inmates are often the focus of unwanted negative and stigmatizing attention by inmates and staff in shower and recreational facilities. Indeed, in many prison subcultures, the belief among both inmates and staff is that the transgender inmate has invited the derogatory sexual attention by choosing to make these changes in her physical appearance.

PHOTO 9.4 This facility has implemented changes in housing for gay, lesbian, bisexual, and transgender inmates.

AP Photo/Pat Sullivan

The PREA also prohibits the classification of LGBTI inmates as sex offenders based simply on their sexual orientation or identity. While correctional facilities may, of course, prohibit consensual sexual activity between inmates and punish them for engaging in such acts, the PREA requires that this activity be identified as separate from acts of sexual abuse. This is because in times past inmates who engaged in consensual acts of sex would be given punishments that were harsher and more restrictive, such as solitary confinement or even new criminal charges. These practices have been found to disproportionately harm LGBTI inmates.

However, this new requirement has caused difficulties for prison administrators in cases where what appears to be consensual is not. For example, some LGBTI inmates may be the "property" of a prison gang. Due to coercion and out of fear, they may "freely" engage in providing sexual favors in return for protection or commissary supplies. Such sexual activity may appear to be consensual to the outside observer, but it is in fact due to pressure from other inmates. To avoid harm, often the LGBTI inmate will deny that such coercion exists and claim that he or she is engaging in the sexual behavior willingly. In such a case, it will be the responsibility of prison PREA investigators to conduct a thorough investigation to distinguish between this subversive victimization and actual consensual sexual activity.

To further showcase the past treatment of the LGBTI population, consider that many state classification systems had presumed at intake that LGBTI inmates were more likely to be sexual perpetrators and would, therefore, place them in housing usually allocated for sex offenders. This, ironically, put the LGBTI inmate is even worse circumstances, further increasing the likelihood for victimization.

As can be seen, the standards of care for the LGBTI inmate population have been greatly revised in the past decade. These changes are in stark contrast to the old-school prison subculture discussed elsewhere in this text. While these changes are required by administrators, there is still significant resistance among both correctional staff and the inmate population. Just because the laws have been enacted does not mean persons within facilities have changed their own personal views regarding LGBTI offenders. This is, of course, no different than what is seen in outside society in which the population is in conflict regarding views on gay marriage and other issues related to sexual orientation and identity. Nevertheless, just as we have seen significant social change in this area in broader society, similar changes are now permeating the correctional environment as well.

ELDERLY OFFENDERS

The graying of America is a phenomenon that is now fairly common knowledge among most of the public, particularly among those who keep up with recent news and events. However, there is a consequence to the aging phenomenon that stays largely out of view: an elderly inmate population that is burgeoning within state prison systems. There are two factors that primarily contributed to the aging of state inmates between 1993 and 2013. These are (1) a greater proportion of inmates were sentenced to and serving

FIGURE 9.1

Sentenced State Prisoners, by Age

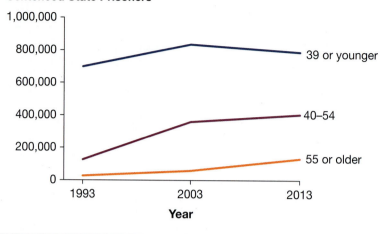

Source: Carson, A. E., & Sabol, W. J. (2016). *Aging of the state prison population, 1993–2013.* Washington, DC: Bureau of Justice Statistics.

Note: Most recent data available.

longer periods in state prison, predominantly for violent offenses; and (2) admissions of older persons increased. The number of persons age 55 or older admitted to state prison increased 308% between 1993 and 2013. Figure 9.1 provides a comparison by age category of inmates in state correctional facilities.

To demonstrate, consider that the number of prisoners age 55 or older sentenced to more than 1 year in state prison increased 400% between 1993 and 2013, from 26,300 (3% of the total state prison population) in 1993 to 131,500 (10% of the total population) in 2013. This ever-growing and ever-aging population tends to have a multiplicity of problems. Indeed, elderly inmates tend to have physiological symptoms that mimic health conditions that are roughly 10 years in advance of their actual chronological age. Thus, a 50-year-old inmate will, on the average, tend to have the physiological health of a 60-year-old in outside society (Carson & Sabol, 2016).

It is a legal fact, indeed a constitutional fact, that prison systems are responsible for the safety and security of the inmates in their charge. Administrators simply cannot run from this responsibility; standards of decency and human treatment require that appropriate medical attention be provided to inmates within their stead. The Eighth Amendment itself ensures that inmates are free from cruel and unusual punishment. Thus, prison administrators cannot inflict harmful conditions upon inmates if these conditions could be considered cruel, malicious, or unusual. In addition, the Supreme Court has ruled in cases like ***Estelle v. Gamble* (1976)** and *Wilson v. Seiter* (1991) that administrators cannot remain deliberately indifferent to the safety and security needs of inmates in their custody. Thus, turning a blind eye to the needs of elderly inmates is not, and will not be, an option for correctional practitioners.

Estelle v. Gamble (1976):
Ruled that deliberate indifference to inmate medical needs constitutes cruel and unusual punishment and is unconstitutional.

Classification of Elderly Offenders

Perhaps the first step in ensuring that any offender's needs are appropriately met is the process of assessment. This assessment then must classify the elderly offender into categories that will ensure that the right standard of care is given. This accomplishes three goals: The process (1) ensures that the inmate receives adequate care, (2) ensures that precious resources are not wasted on inmates not needing more intensive care, and (3) provides a guide when deciding housing and security levels for these offenders to protect them and others from possible harm.

It is also important to note that any classification system for elderly inmates should include a protocol that distinguishes between inmates who entered the prison system before age 50 and those who reached that age while in prison. In essence, the data should clearly and prominently note if the offender falls within the category of an **elderly first-time offender**, a **habitual elderly offender**, or an **offender-turned-elderly-in-prison**. This is an important security consideration that most correctional administrators will find useful when running their institution. Once the elderly offender has been appropriately classified and once any necessary security precautions are resolved, prison staff will then be able to adequately ensure that the needs of these offenders are met.

The rise in numbers of habitual elderly offenders and offenders-turned-elderly-in-prison has to do largely with the advent of three-strikes felony sentencing in many states (Anno, Graham, Lawrence, & Shansky, 2004). These sentences require that third-time felony offenders serve mandatory sentences of 25 years to life. It should be noted that committing a felony does not necessarily mean that the offender is violent in nature. Also adding to these statistics are the punitive sentencing measures associated with the War on Drugs of the 1980s and 1990s. During this period, drug-using offenders were locked up at an all-time high, regardless of whether the crime involved violence or any form of drug trafficking (Anno et al., 2004).

Thus, habitual elderly offenders and offenders-turned-elderly-in-prison are the result of a confluence of social factors and criminal justice polices. These offenders, for various reasons, have been given enhanced penalties that preclude their release in the community. This greatly distinguishes them from the elderly first-time offender, who does not share a similar criminogenic background. Though they may look the same while in inmate clothes in the prison facility, they are usually quite different from one another, and the public would be well served to keep this squarely in mind. States that have abolished parole and other community outlets may want to consider establishing specialized court interventions for this type of offender. It is at this point that we turn our attention to the three typologies of elderly offenders, with each being briefly discussed in the paragraphs that follow.

PHOTO 9.5 Elderly offenders and offenders with disabilities are very costly to correctional facilities.

REUTERS/Lucy Nicholson

Elderly first-time offenders: Those who commit their first offense later in life.

Habitual elderly offenders: Have a long history of crime and also have a prior record of imprisonment throughout their lifetime.

Offenders-turned-elderly-in-prison: Inmates who have grown old in prison who have long histories in the system.

Elderly First-Time Offenders

Elderly first-time offenders are those who commit their first offense later in life. It is estimated that approximately 50% of elderly inmates are first-time offenders, incarcerated when they were 60 or older. For these offenders who commit violent crimes, these are usually crimes of passion rather than premeditated crimes. Conflicts in primary relationships appear to increase as social interactions diminish with age. Older first-time offenders often commit their offenses in a spontaneous manner that shows little planning but is instead an emotional reaction to perceived slights or disloyalties. These offenders do not typically view themselves as criminal, per se, but instead see their situation as unique and isolated from their primary identity. First-time offenders are more likely to be sentenced for violent offenses. These offenses will usually be directed at a family member due to proximity, if nothing else. Some experience a crisis of one sort or another due to disparity regarding the aging process, and this is also thought to instill a sense of abandonment and resentfulness that may lead to aberrant forms of coping. Also, sexual offenses involving children are common among the first-time offending elderly. For persons over the age of 65, aggravated assault is the violent offense most often committed, followed by murder (Aday, 1994).

These offenders are also the most likely to be victimized in prison as their irascible behavior and demeanor are likely to draw the attention of younger inmates (Aday, 1994). These offenders are the most likely to have strong community ties and are therefore usually the most appropriate for community supervision because of this.

Habitual Elderly Offenders

Habitual elderly offenders have a long history of crime and also have a prior record of imprisonment throughout their lifetime. These offenders are usually able to adjust well to prison life because they have been in and out of the environment for a substantial portion of their life. They are a good source of support for first-time offenders and, if administrators are wise enough to implement this, are able to act as mentors. These offenders quite often fit the mold of greyhounds within the prison facility and command a degree of respect within the prison subculture. While they may cope well with prison life, these offenders typically have substance abuse problems and other chronic issues that make coping with life on the outside difficult. Some of these inmates are not considered violent but instead serve several shorter sentences for lesser types of property crimes. This is the group most likely to end up as geriatric inmates who die in prison.

Offenders-Turned-Elderly-in-Prison

Offenders-turned-elderly-in-prison have grown old while incarcerated. They have long histories in the system and are the least likely to be discipline problems. Long-term offenders are very difficult to place upon release because they have few ties in the community and a limited vocational background from which to earn a living. These offenders are often afraid to leave the prison and go back to the outside world because they have become so institutionalized to the predictable schedule of the prison. This also means that these individuals might consider suicide, particularly just prior to release, or even within a short time after release. This phenomenon is no different from that noted in the classic movie *The Shawshank Redemption*, which portrays a released inmate who cannot cope with life on the outside so he chooses to end his life by hanging himself.

Health Care Services and Costs

It should be clear to students that elderly offenders require substantial medical attention. The provision of this medical attention is very expensive and impairs many state prison systems around the nation. The cost of incarcerating an inmate who is 60 years of age or older is around $70,000 a year—roughly three times the cost for the average inmate. These inmates require specialized housing facilities, special programming, and sometimes even hospice or palliative care. Further, prison agencies must remain compliant with standards set by the ADA. All of these various requirements create a serious price tag for prisons, and since the population continues to age, it is likely that these expenses will continue to climb. How states and the federal government will afford these costs is a question often on the mind of many prison wardens and correctional administrators, particularly during times of economic challenge. The answer has yet to emerge, and this represents yet another challenge facing correctional agencies of the modern era.

JUVENILES AND THE CORRECTIONAL SYSTEM

This section is unique from the remainder of this text because it addresses an area of corrections that is in most respects separate and distinct from the other segments of the criminal justice system, including the correctional system. The juvenile system has a different orientation, emphasis, and set of terms and concepts that make it unique from adult corrections. For instance, the use of secure facilities is avoided if at all possible. The vast majority of youth who are processed through the juvenile justice system are placed on some form of community supervision. While this is also true in the adult system, this trend is even more true within the juvenile system. Youth are not considered to be as culpable as adult offenders, and it is presumed that they are more impressionable than adults. Thus, the use of restrictive environments can have negative and counterproductive outcomes.

Processes Involved With Juveniles in Custody

Initially, youth enter the juvenile justice system through some sort of contact with law enforcement. A juvenile can be taken into custody under the laws of arrest or by law enforcement if there are reasonable grounds to believe that the child is in immediate danger from his or her environment or if the youth has run away from home. In addition, court orders may be executed that require that youth be taken into custody. Delinquent youth can be kept in a jail or another type of adult facility when other options are not available. In such a case, the detention space must be a room separate from adults, and the youth can be kept in **detention** only if it seems warranted as a means to ensure public safety. The purpose of these restrictions is to protect youth from exposure to criminals and the negative effects that jails and other secure adult environments may have upon the youth.

Detention: Secure confinement of juvenile offenders.

SCREENING AND CLASSIFICATION OF JUVENILE OFFENDERS

Once it has been determined that a juvenile offender will be charged, the need to classify him or her is important. One instrument used for supervision and treatment decisions when classifying juvenile offenders is the Youth Level of Service/Case Management Inventory (YLS/CMI). This instrument is currently used in a variety of juvenile correctional settings in a number of jurisdictions. This same instrument has been employed to classify youth for judicial disposition decisions, placement into community programs, institutional assignments, and release from institutional custody.

Substance abuse is likewise a very important aspect of juvenile treatment because many young offenders experiment with drugs and alcohol. Further, persons who are under the influence of drugs or alcohol or who have a tendency to use drugs and alcohol typically have a higher rate of suicide when in jails or prisons. As suicide rates tend to be higher than average among teens in general, juvenile offenders are at particular risk for substance abuse–related suicide and/or additional mental health problems (e.g., depression and anxiety) when they are detained or placed in a secure environment.

Regardless, the decisions regarding juvenile justice treatment for thousands of youth in all types of correctional settings are based on some type of screening, assessment, or classification process. This assessment is particularly important with juvenile offenders because of temporal considerations in the youth's development. At this point, the criminal justice system makes its first determination of a juvenile offender, and this determination

---

PRACTITIONER'S PERSPECTIVE

"I fell in love with the adolescent population and, more importantly, kids that have problematic behavior that inevitably end up with criminal prosecutions."

Interested in a career in corrections or criminal justice? Watch Jennine Hall's video on her career as a juvenile court counselor.

is likely to have a very long-range outcome. All types of treatment and intervention given to these offenders will be based entirely upon the results of the assessment(s) given.

Emphasis on Treatment

As we have seen, the primary purpose within the juvenile justice system is to treat and reform youth rather than to punish them. The system has been specifically designed with the mindset that youth should be spared the stigma of a conviction and the trauma of full incarceration, when practical. Indeed, more than 6 in 10 (63%) facilities around the nation have reported that onsite treatment services are provided to youth in their care (Office of Juvenile Justice and Delinquency Prevention [OJJDP], 2016). Students should examine Table 9.4 for further details on the types of treatment programming that tend to be offered. The juvenile justice and mental health fields have increasingly recognized the scope of the mental health needs of justice-involved youth and the inadequacy of services to meet these needs.

One reason that this is important is because youth in juvenile facilities tend to present with affective disorders, adjustment difficulties, and issues related to trauma and stress. For instance, symptoms of depression and anxiety were found to be very common among nationally surveyed youth in treatment facilities, with 51% of the custody population reporting that nervous or worried feelings have kept them from doing what they want to do over the past few months and 52% indicating that they feel lonely for significant periods of time (Sedlak & McPherson, 2010).

Mental health services in the form of evaluation, ongoing therapy, or counseling are nearly universally available in juvenile facilities, with 97% of youth in places that provide one or more of these services either inside or outside the facility. However, despite the relatively high suicide risk in the placement population, individual screening for suicide risk is not common. More than one fourth (26%) of juvenile offenders are in facilities that do not screen all youth for suicide risk, 45% are in facilities that fail to screen all youth within 24 hours, and 26% are in facilities that do not screen any youth at intake or during the first 24 hours (Sedlak & McPherson, 2010).

Suicide is the third leading cause of death among adolescents, and a prior suicide attempt is the single most important risk factor to look for (Wintersteen, Diamond, & Fein, 2007). One fifth of youth in placement admit having two or more recent suicidal feelings. The prevalence of past suicide attempts (22%) of juvenile offenders is more than twice the rate for youth in the general population and nearly quadruple the rate in national samples.

TABLE 9.4

Facilities Providing Treatment Services, by Facility Operation, 2016

	TOTAL	PUBLIC	STATE	LOCAL	PRIVATE
Total number of facilities	1,772	978	365	613	794
Number of facilities providing onsite treatment	1,120	545	292	253	575
Any treatment service	63%	56%	80%	41%	72%
Onsite treatment services					
Mental health	86%	87%	88%	85%	84%
Substance abuse	71%	79%	87%	70%	63%
Sex offender	36%	38%	52%	21%	34%
Violent offender	20%	28%	41%	14%	12%
Arson	9%	12%	20%	2%	6%

Sources: OJJDP Statistical Briefing Book. (2018, March 27). Retrieved from https://www.ojjdp.gov/ojstatbb/corrections/qa08520.asp?qaDate=2016; Office of Juvenile Justice and Delinquency Prevention. (2016). *Juvenile residential facility census 2016* [machine-readable data files]. Washington, DC: OJJDP.

DETENTION VERSUS INCARCERATION

Detention facilities are secure residential facilities for youth and are separate from adult prisons, though they still resemble adult prisons in appearance and have locked doors and barred windows. These facilities are, in theory, supposed to be reserved for the most serious juvenile offenders—those who are violent and require secure confinement as a means of protecting society. Nevertheless, it is not uncommon for other youth, such as runaways, homeless youth, and those who are abandoned, to end up in these types of facilities.

These facilities may also hold youth who are scheduled for trial but have not been sentenced or who may be waiting for the imposition of their sentence. This means that youth of differing custody levels may be kept within these facilities. Other youth may be in violation of their probation and being held for revocation or modification proceedings. In recent years, the juvenile justice system has made efforts to remove status offenders and neglected or abused children from these detention facilities, when and where appropriate. But despite the ongoing effort to limit the use of detention, juveniles are still detained in about 20% of all delinquency cases.

The overarching goal of juvenile corrections is to provide treatment services for youth and to avoid traumatizing or stigmatizing them in the belief that such restrictive measures will orient them toward further criminality. The use of juvenile waiver for serious offenders who are sentenced to periods of incarceration is presented in the following section. But for now, this sentencing option is presented to distinguish it from the use of detention. When youth are sent to prison, they are incarcerated with adult offenders. These youth are usually charged for adult offenses, and their sentencing terms are for several years. The types of incarceration that youth may face, whether in a juvenile detention facility or in prison itself, are discussed in the next few sections.

Juvenile Waiver for Serious Juvenile Offenders

For those child and adolescent offenders who are incarcerated for violent crimes, a special form of incarceration may be employed. This occurs when the juvenile case is transferred to adult court and results in what is commonly known as **juvenile waiver**. Waiver to adult court can theoretically be utilized for any offense, but this process is usually reserved for serious, violent felonies or for property crimes with which juveniles are repeat offenders. Juveniles who are tried in adult criminal court are most often kept in a juvenile facility or separate wing until they are of sufficient age to be transferred to the adult population. Those who are placed in an adult prison are usually separated from the adult population.

Juvenile waiver: Occurs when the juvenile case is transferred to adult court.

The disposition of a juvenile who has been tried and sentenced in an adult court varies from state to state. The usual procedure is that younger offenders who are found guilty are sent to a juvenile detention facility until they are at least 16 to 18 years of age. Their treatment there is basically the same as that of juveniles who were tried under the juvenile justice system. Nevertheless, some youth who are charged as adults may be sent to an adult facility if the crime and circumstances warrant this. The disadvantage to this process is that it sends a message of hopelessness to the youth and society regarding the likelihood of reformation. Further, there is the possibility that this approach will only increase the criminality of juveniles as they are exposed to the adult criminal population. The adult prison system's mission is primarily the punishment of offenders, and this naturally includes those juveniles included within its confines. Perhaps these offenders are best considered to be targeted by a form of legal "selective incapacitation," where the most severe offenders are removed from society for no other reason than to protect society from any further harm from them. However, this automatically translates to an admission from the justice system that it does not have the

PHOTO 9.6 These juveniles listen to a guest speaker on the negative aspects of incarceration in an adult facility.

© ZUMA Press, Inc / Alamy Stock Photo

PHOTO 9.7 It is clear from statistics in juvenile research that African American youth are represented in the juvenile justice system at a proportion that greatly exceeds that of the broader U.S. population.

Antonio Perez/MCT/Newscom

the existence of a delinquent subculture that enhances the opportunity for dominance of the strong over the weak and gives impetus to the exploitation of the unsophisticated by the more knowledgeable. We should not then be surprised when juveniles leave these institutions with more problems than they had prior to incarceration.

Disparate Minority Confinement (DMC) and Factors Associated With DMC

During the late 1980s and early 1990s, the disparate minority representation in juvenile lockdowns became a topic of controversy. Most of the data available on this issue emerged during the late 1990s and during the early parts of the new millennium. The reasons for the disparity in juvenile confinement are many. Most published literature on this issue notes that these statistics are not likely to be due to a racist system. While we cannot possibly answer this question within the scope of one discussion within one section of a single textbook, it should have become clear to students that African American men and women have been disproportionately incarcerated throughout the history of corrections in the United States. Given the various historical precedents associated with the civil rights era and other indicators that our society held minorities in a weakened position, and given the institutional racism that officially existed until the 1960s, it is not unreasonable to presume racism may be part of the explanation in some cases.

Research by Huizinga, Thornberry, Knight, and Lovegrove (2007) investigated the often-stated reason for disparate minority confinement—that it simply reflects the difference in offending rates among different racial/ethnic groups—and found no support for it in their rigorous study examining disparate minority contact with the justice system. Huizinga et al. note that "although self-reported offending is a significant predictor of which individuals are contacted/referred, levels of delinquent offending have only marginal effects on the level of DMC" (p. i). They found these results in terms of both total offending and more focused data that examined violent offenses and property offenses separately. Thus, it would appear that minority youth are no more delinquent than Caucasian youth.

The work of Hsia (2004) is a compilation of surveys and official investigation into the juvenile correctional systems of all 50 states. She notes the following reasons for why such disparities existed in many state systems:

1. **Racial stereotyping and cultural insensitivity:** Eighteen states identified racial stereotyping and cultural insensitivity—both intentional and unintentional—on the part of the police and others in the juvenile justice system (e.g., juvenile court workers and judges) as important factors contributing to higher arrest rates, higher charging rates, and higher rates of detention and confinement of minority youth. The demeanor and attitude of minority youth can contribute to negative treatment and more severe disposition relative to their offenses.

2. **Lack of alternatives to detention and incarceration:** Eight states identified the lack of alternatives to detention and incarceration as a cause of the frequent use of confinement. In some states, detention centers are located in the state's largest cities, where most minority populations reside. With a lack of alternatives to detention, nearby detention centers become "convenient" placements for urban minority youth.

3. **Misuse of discretionary authority in implementing laws and policies:** Five states observed that laws and policies that increase juvenile justice professionals' discretionary authority over youth contribute to harsher treatment of minority youth. One state notes that "bootstrapping" (the practice of stacking offenses on a single incident) is often practiced by police, probation officers, and school system personnel.

4. **Lack of culturally and linguistically appropriate services:** Five states identified the lack of bicultural and bilingual staff and the use of English-only informational materials for the non-English-speaking population as contributing to minorities' misunderstanding of services and court processes and their inability to navigate the system successfully.

Based on the research by Hsia (2004), it is the general contention of this author that much of the reason for disproportionate confinement among minority youth has to do with a confluence of issues that plague minority members of society who have suffered from historical trauma and, generation after generation, have had restricted access to material, educational, and social resources. Indeed, issues such as poverty, substance abuse, few job opportunities, and high crime rates in predominantly minority neighborhoods are placing minority youth at higher risk for delinquent behaviors. Moreover, concerted law enforcement targeting of high-crime areas yields higher numbers of arrests and formal processing of minority youth. At the same time, these communities have fewer positive role models and fewer service programs that function as alternatives to confinement or support positive youth development.

Further, it has been found that a disproportionate number of youth in confinement come from low-income, single-parent households (female-headed households, in particular), and households headed by adults with multiple low-paying jobs or unsteady employment. Family disintegration, diminished traditional family values, parental substance abuse, and insufficient supervision contribute to delinquency development. Poverty reduces minority youths' ability to access alternatives to detention and incarceration as well as competent legal counsel. Thus, all of these factors, associated with historical deprivations over time, have contributed to and culminated in the state of affairs that we now witness among minority juveniles in the United States.

Current Status of Disparity in Juvenile Detention and Incarceration

During the past 5 to 10 years, DMC with the juvenile justice system has received widespread attention throughout the nation. Recent data have shown that the likelihood for detention (the act of holding youth prior to adjudication or disposition in a court hearing, or after disposition when awaiting placement elsewhere) was greatest for African American youth when compared with Asian Americans and Native Americans. Indeed, the rate of detention for African American youth is 1.4 times that of Caucasians, 1.2 times that of Asian Americans, and 1.1 times that of Native Americans. These rates are even higher for African Americans in the case of drug-related cases (Sickmund & Puzzanchera, 2014).

Between 1997 and 2011, the overall number of juvenile offenders placed in residential facilities decreased by nearly 33% (Sickmund & Puzzanchera, 2014). For those youth adjudicated within the juvenile justice system and given sentences that included out-of-home residential placement, options included residential homes, training schools, treatment centers, boot camps, group homes, or drug treatment facilities. From the data available (see Table 9.6), it is clear that both male and female Caucasian youth are less likely than minority youth to be ordered to residential placement (Office of Juvenile Justice & Delinquency Prevention, 2017).

TABLE 9.6

Race/Ethnicity Profile of Juveniles in Residential Placement by Gender, 2015

	CAUCASIAN	TOTAL MINORITY	AFRICAN AMERICAN	LATINO	NATIVE AMERICAN	ASIAN AMERICAN
Male	30%	70%	43%	22%	2%	1%
Female	37%	63%	34%	22%	3%	1%

Source: Office of Juvenile Justice and Delinquency Prevention. (2017, June 1). *Statistical briefing book.* Retrieved from https://www.ojjdp.gov/ojstatbb/corrections/qa08211.asp?qaDate=2015

In addition, security staff must be trained in recognizing and understanding the dynamics of sexual abuse and sexual harassment in juvenile facilities as well as common reactions of juvenile victims of sexual abuse and sexual harassment. Further, security staff must understand how to communicate effectively and professionally with residents, including lesbian, gay, bisexual, transgender, intersex, or gender-nonconforming residents. Lastly, staff must be trained on how to detect and respond to signs of threatened and actual sexual abuse and how to distinguish between consensual sexual contact and sexual abuse between residents (U.S. Department of Justice, 2012). Though other requirements exist, it is clear from the above that juvenile facilities, like adult facilities, must ensure that they are in compliance with PREA standards.

Lastly, as a culture and society, there should be little doubt that the acceptance of same-sex couples is much greater with teens today than in the past. This is particularly true in larger metropolitan areas of the nation, which tend to have greater numbers of LGBTQI youth and where there is less stigma attached to these orientations than in other areas of the United States. Media portrayals and obvious changes in public sentiments toward these sexual orientations reveal that tolerance is much more improved today than even just 20 years ago. Legal decisions and changes in legislative enactments make it all the more clear that these sentiments are becoming codified and serve as official acknowledgment that these individuals are entitled to the rights and privileges extended to any other minority group in the United States.

CONCLUSION

This chapter has provided an overview of specialized offenders, sometimes referred to as special needs offenders. Among the various types of specialized offenders, those who have mental disorders are particularly problematic. These offenders tend to have high rates of institutional infractions and present with more problematic behaviors than do other inmates. Safety of these inmates is a bit more difficult to manage, and security is compromised due to various issues, such as the administration of medications and the supervision of these offenders in the population—a population that may victimize or dupe them when the opportunity presents. Further still, these inmates tend to have substance abuse problems that are comorbid with their mental illness. In many cases, the substance abuse may be their primary disorder, and the presenting mental illness may be co-occurring.

The sex offender population is one category of offenders that was also discussed in this chapter. The nature of their offense and the stigma attached means that these offenders face challenges within the institution from other inmates as well as staff members. The need for effective assessment and treatment programming for sex offenders was discussed in this chapter, as well.

Elderly offenders were also presented as having distinct classifications that are determined by their prison history. The particular history of an inmate, such as whether that offender has engaged in a life of crime or committed a single crime resulting in a long-term sentence, affects the dynamics of his or her experience and the manner in which he or she is perceived within the prison subculture. Elderly inmates are a costly group to house, and, as we have seen, this cost is only going to continue to rise as we see a continued graying of the prison population.

At the other end of the age spectrum, we have also focused on juvenile offenders who are kept in secure environments or incarcerated. Though most youth do get sentences of probation, a demonstration of the issues they face in detention centers and similar facilities warrants specific and separate consideration. It is clear that numerous issues confront juvenile correctional systems, including disparity issues and the need to attend to the specific needs of different offender groups, such as female juvenile offenders.

We have also discussed the use of waiver to transfer youth to adult court. This is often reserved for the most serious juvenile offenders and is an option that has decreased in popularity in many states. It is clear that the primary objective in processing youth is to reform them and, if possible, refrain from stigmatizing them. This is fundamental to juvenile corrections. The vast majority of youth are treated as if they are salvageable, and, in fact, the entire juvenile system is based on the notion that youth should be given intervention-based as opposed to punitive-based programming. Further, these youth report high levels of trauma, substance abuse, and mental illness. This is especially true with female juvenile offenders. Thus, juveniles demonstrate a need for mental health interventions. Current juvenile corrections systems attempt to address these issues to minimize the likelihood that these youth will become further entrenched in crime.

Want a Better Grade?

Get the tools you need to sharpen your study skills. Access practice quizzes, eFlashcards, video, and multimedia at **edge.sagepub.com/hanserbrief**

Interactive eBook

Visit the interactive eBook to watch SAGE premium videos. Learn more at **edge.sagepub.com/hanserbrief/access**.

 Career Video 9.1: Juvenile Court Counselor

 Criminal Justice in Practice Video 9.1: Juvenile Detention

 Prison Tour Video 9.1: Housing for Specialized Populations

 Prison Tour Video 9.2: Health Services, Medicine in Prisons, Work Trustees, and Inmate Organizations That Give Back to Society

Test your understanding of chapter content. Take the practice quiz at edge.sagepub.com/hanserbrief.

DISCUSSION QUESTIONS

1. What are some of the common problems encountered with specialized inmate populations in prison?

2. What is the definition of mental illness? How prevalent is it within state prison systems?

3. What are some of the subcultural issues involved with sex offenders who are incarcerated?

4. Identify and define the various classifications of elderly inmates within an institution.

5. What are some of the specialized requirements or considerations for prisons when housing elderly offenders?

6. How does the prison subculture impact, interact with, and react to inmates who are mentally ill, sex offenders, or elderly?

7. Identify and discuss how racial disparities exist among youth who are placed in confinement. Why do you think that these disparities exist?

8. Identify and discuss the types of facilities that are used to house juvenile offenders.

Review key terms with eFlashcards at edge.sagepub.com/hanserbrief.

KEY TERMS

Co-occurring disorders, 239

Detention, 247

Diagnostic and Statistical Manual of Mental Disorders (DSM-5), 233

Elderly first-time offenders, 245

Four standards of mental health care, 235

Greyhounds, 233

Habitual elderly offenders, 245

Juvenile waiver, 249

Major depressive disorder, 234

Malingering, 238

Mental illness, 233

Mood disorders, 237

Offender-turned-elderly-in-prison, 245

Offender with special needs, 230

Therapeutic community, 240

Universal design, 232

KEY CASES

Estelle v. Gamble **(1976)**, 244

APPLIED EXERCISE 9.1

You are a juvenile caseworker and have been working with Tanya, a 15-year-old at your detention facility. You have worked with Tanya for over a year and have noted that she has made considerable progress in treatment and the other aspects of programming. In fact, Tanya is scheduled to be released to aftercare in the near future. During a session, just 2 weeks prior to her release, Tanya discloses to you that she has been willingly having sexual intercourse with one of the juvenile detention workers at the facility. She notes that she was the one who initiated the relationship with the custodial staff member. She also intends to see this individual when she is released to aftercare. She refuses to disclose the staff person's name as you talk with her. She also notes that she has been sexually active since the age of 8, when her stepfather (who still lives at the domicile) began to molest her. Neither Tanya nor her mother ever reported this. As Tanya nears release, you must consider how to handle this situation.

Consider this case and what you have learned in prior chapters regarding the legal liability of a prison staff member.

Answer the following questions in a 500- to 1,000-word essay:

1. What legal requirements do you have in regard to the current sexual relationship that Tanya has had with the staff person employed at the detention facility?

2. Do you believe that you have a legal responsibility to safeguard Tanya when she is released into aftercare from your detention facility? Why or why not? Be sure to explain your answer.

3. What else do you believe should be done for Tanya?

WHAT WOULD YOU DO?

You work within the administrative segregation unit of a maximum-security prison as a sergeant. You supervise approximately 25 officers on your shift, each of whom works in a different position throughout the five cell blocks that you supervise. One day, while doing your rounds and signing paperwork maintained by your officers, you notice that a pair of officers are on the second level of the cell block standing in front of a cell. You do not reveal your presence, but you listen closely; they are laughing at and ridiculing an inmate in the cell who is known to take a variety of medications due to symptoms of schizophrenia. Their joking is not violent, but it is demeaning, and it is likely to further aggravate the inmate's future mental condition.

This particular inmate attempted to stab an officer 2 months ago with a homemade blade, and thus he is not liked by many of the officers. However, you suspect (based on information from other inmates) that the officer who was attacked had been taunting this inmate and humiliating him for weeks. As their sergeant, you are expected to identify with the other officers and to not befriend inmates. Nor should you, according to the tenets of correctional officer subculture, take the word of inmates over that of other officers.

This situation creates a bit of a dilemma for you, but, if left unresolved, it can escalate to another potential stabbing incident in your department. You consider the options you have as you listen to the laughter from the officers and from other inmates listening in nearby cells.

What would you do?

PRACTICE AND APPLY WHAT YOU'VE LEARNED

▶ edge.sagepub.com/hanserbrief

Jim West imageBROKER/Newscom

10 Correctional Administration and Prison Programming

Learning Objectives

1. Discuss the organization of the Federal Bureau of Prisons.
2. Discuss common organizational features of state and local prison systems.
3. Describe the importance of racial and cultural diversity and professionalism in the corrections workforce.
4. Discuss the effects of educational programming on recidivism.
5. Discuss the effects of vocational programming on recidivism.
6. Identify components of substance abuse treatment in prison.
7. Describe some of the concerns for administrators with different types of recreational programming.
8. Identify some of the legal issues associated with religious programming in prisons.

Sexual Harassment and Prison Culture

Major Turner read through the allegations made by his employee, Officer Kristy Campbell, regarding her treatment by her supervisor, Sergeant Steinbeck. Turner had requested that an inmate by the name of Butch Buchanan be brought to his office.

When Buchanan arrived, Turner said, "Mr. Buchanan, I have to inquire of you as to whether you have overheard any remarks made by officers in the machine shop that could be construed as being sexual in nature."

Buchanan shrugged. "Not really, but then again, I mind my own business."

"Well, as it happens, I have this report on you that was written yesterday by Officer Kristy Campbell, something about you muttering how you'd like to 'take her for a test drive' and making obscene gestures in her direction." The major leaned forward. "What is that all about?"

Buchanan scowled at him. "I did not say that! She's lying!"

"Sergeant Steinbeck was a witness to this, as were some other inmates in the area . . . and just so you know, Campbell has claimed that she felt very threatened by your comments; they seem to have been pretty graphic, like depicting a rape or something."

"C'mon man, that is bullcrap!" Buchanan protested. He started to speak again, then hesitated for a second before asking, "So Steinbeck is witnessing it, huh?"

The major looked at him and said, "Yep," with no additional explanation. After a few seconds, the major continued. "You know, these ain't the old days and all, with all of this PREA stuff, I cannot ignore this stuff or treat it lightly. I think that this case may be one that is threatening-an-officer, which is a higher charge than what Campbell had put here." He paused to ostensibly consult the report in front of him. "She just charged you with being insubordinate to an officer, but I think it could be seen as a threatening-an-officer case . . . just sayin'." The major grinned and gave a little shrug on the "just sayin'."

The major then held up a piece of paper in front of Buchanan's face. It had the signatures of officers on it who worked in the machine shop. "As you can see, Steinbeck is witnessing this, and, if you catch this charge, you are not going to make parole, which is just around the corner for you, isn't it?"

Buchanan held up his hands in surrender. "Okay, look, I did say it, but just because Steinbeck had been egging it on. . . . He's mad at Campbell because she broke up with him. He's setting me up, then siding with her, so that she will be cool with him again. That's what is really going on, and Steinbeck . . ." Buchanan trailed off.

"Yes?"

Buchanan narrowed his eyes at the major. "You gonna' roll me if I explain this, or what?"

The major met his gaze and said, very seriously, "You let me know what the hell is going on in that shop, and I will keep it here. . . . This case has not been processed yet nor logged. I will tear it up and make sure that Campbell knows to let it be handled informally. Nobody will know the difference, especially the parole board."

"Alright, this is the deal. Steinbeck is constantly talking about how Campbell is a tease, and he's always telling us to not listen to her in the shop. If we act all funky with her, then he gives us 'extras'—you know, a few smokes and other goodies—and Campbell finds herself having to ask him for help in keeping us under control. Steinbeck is setting her up and grooming

During the 1990s, the effects of the War on Drugs were being felt, and there was a mass prison-building boom throughout the nation. Politicians heralded "tough on crime" platforms that included three-strikes initiatives, the elimination of parole, and enhanced penalties against drug offenders. During this time, politicians introduced legislation that resulted in the passage of the Violent Crime Control and Law Enforcement Act, which, as discussed above, eliminated the availability of the Pell Grant for inmates and all but killed the offering of higher education in American corrections. However, in 2016 the **Second Chance Pell Pilot Program** (White House, 2016) was announced by the Obama administration. This program, stewarded by the Department of Education, selected 67 colleges and universities who partnered with more than 100 federal and state correctional institutions, enrolling approximately 12,000 incarcerated students in educational and training programs. Selected schools offered classroom-based instruction, online education, or a hybrid of both at corrections facilities; the vast majority of selected schools are public 2- and 4-year institutions. The rationale for providing this funding for inmate educational pursuits developed from a Rand Corporation study that was commissioned by the Department of Justice and found the following:

Second Chance Pell Pilot Program: This program, announced by the Obama administration and stewarded by the Department of Education, selected 67 colleges and universities who partnered with more than 100 federal and state correctional institutions, enrolling approximately 12,000 incarcerated students in educational and training programs.

1. Inmates who participate in correctional education programs had 43% lower odds of recidivating than those who did not.

2. The odds of obtaining employment post-release among inmates who participated in correctional education was 13% higher than the odds for those who did not.

3. Offenders who were provided computer-assisted instruction learned in reading and math in the same amount of time when provided other forms of instruction.

4. For every dollar that is spent on prison-based education programming, there is a clear minimum of $4 saved in future incarceration costs.

Due to the prior president's support of offender-education as a crime fighting tool, some senators began work in 2018 to repeal the ban and make allocations of funds extend beyond the duration intended for the Second Chance Pell Pilot Program. What is interesting is that these monies are made available for general college education rather than being restricted to vocational pursuits. Indeed, one program offered through Ashland University, in Ohio, has proliferated throughout the United States and offers online university-level courses to inmates behind bars. The funding for these course offerings in through a Pell grant process.

General Educational Development (GED): The process of earning the equivalent of your high school diploma.

PRISON TOUR VIDEO: Educational services for inmates have their benefits and controversies. Go to the IEB to watch a clip about educational funding.

Types of Education Programs in Corrections

When discussing types of education programs, of first concern is the need to ensure that inmates have a basic secondary (high school level) education. The two most common routes to achieving this in prison are the completion of the GED or the HiSET. First, the **General Educational Development (GED)** programs generally consist of a program of study with a required minimum number of hours. The individual is usually assessed to determine his or her likely readiness for the test, often through a GED predictive practice test designed to determine if individuals are prepared for the actual exam. When the individual is deemed likely to be prepared for the test, he or she will be administered the GED test. The current GED test assesses four subject areas: reasoning through language arts, mathematical reasoning, science, and social studies. The GED program is, perhaps, the most well-known avenue to obtaining a high school equivalency education in the United States.

Another more recent program is the High School Equivalency Test (HiSET), which covers five basic areas: mathematics, science, social studies, reading language arts, and writing language arts. The primary difference between the HiSET and the GED is that the HiSET is aligned to more modern standards in the common core for secondary education. For persons who are 19 years of age or older, it is recommended that they first score satisfactorily on the Official Practice Test (OPT) before taking the HiSET exam. For persons younger than 19 years of age (i.e., 18-year-olds in adult institutions or youth who are 17 and under), a qualifying score on the HiSET Official Practice Test is required. In some prisons, educational staff are specifically allocated for those who are attempting to meet acceptable scores on the OPT.

College or university courses have continued to exist in some prison systems even after Pell Grant monies were eliminated during the 1990s. The programs were largely through "correspondence," meaning that inmates usually pursued their higher education achievement through the slow and cumbersome process of completing correspondence courses. One primary example would be Adams State University, a regionally accredited university that offers several types of certificates, associate degrees, and bachelor degrees through the mail system. These courses cost somewhere around $500 each with a 12-month timeframe to complete the course. Degrees in business, history, political science, and sociology are some examples of what can be obtained through this program.

Naturally, in some areas of the nation, community colleges and universities may send instructors into the prison itself to teach inmates a variety of courses. These may be vocational or academic in scope. A very good example of such a program is the Windham School District Correctional Education Program in the state of Texas. In 2013, more than 8,000 offenders participated in postsecondary education programs, which resulted in offenders earning a total of 447 associate's degrees, 31 bachelor's degrees, and nine master's degrees (Kuhles, 2013). According to Kuhles (2013), an evaluation of that program found, among other things, the following:

1. The higher the overall education of the offender, the lower their recidivism rate.

2. Offenders who participated in education programs were 1.7 times more likely to be employed than those who did not.

3. The greater the intensity or duration of the program, the more likely it would reduce recidivism.

Lastly, there has been an emergence of programs that offer degrees through the online venue rather than the use of mail correspondence (see Focus Topic 10.2 for more information). These programs tend to use technology that employs secure digital technology that prevents external communication or Internet access beyond the educational platform, is contained within durable hardware, and includes software that tracks and controls equipment usage by education providers and prison administration. Given that online technology has improved so much in the past 20 years and also given that the availability of funds seems likely on the horizon, it may well be that the dissemination of secondary education throughout the incarcerated population will increase at a more rapid pace than ever before in history. Correspondingly, were this to be the case, we would then likewise expect to see lowered recidivism rates and lowered unemployment rates in those states where such programs were offered.

PRISON WORK PROGRAMS

Prison work for inmates tends to consist of two types. The first includes those jobs that maintain the functioning of the prison itself. These types of jobs might include the preparation of meals, working in the laundry, and the cleaning of various prison areas. These tasks do not usually provide inmates with a trade or a skill by which they can earn a living in the outside community. The second type of work is industrial or vocational in orientation. These jobs include production, agriculture, craftsmanship, carpentry, construction, and even clerical positions in prison offices where genuine professional skills might be learned.

in advance that such a diet is a required part of their faith. Legitimate requests based on recognized religious tenets must be accommodated (Van Baalen, 2008). Van Baalen (2008) notes that many religious observances for even traditional faiths may require the consumption of particular foods (e.g., Jewish Passover), and some religions may require that food not be consumed at specific times during the day or night (e.g., Muslim Ramadan). Prisons around the United States routinely accommodate these dietary modifications based on religion. There are even some occasions where a religious group may require the setting of a group meal as part of a ceremony; such types of request should be accommodated so long as appropriate notice is given to prison administrators (Van Baalen, 2008). In these cases, the chaplain or another religious expert will aid security staff in providing the accommodations to ensure that misunderstandings are minimized.

Chaplain Functions and the ACCA

According to the American Correctional Chaplains Association, correctional chaplains are professionals who provide pastoral care to those who are imprisoned as well as to correctional facility staff and their families when requested. In early U.S. prison history, chaplains held positions of relative importance, which is not surprising considering they were part of a system created by religious groups. They were responsible for visiting inmates and providing services and sermons, and they also served as teachers, librarians, and record keepers. At times, the chaplain would also act as an ombudsman for the inmates when issues of maltreatment would arise. During the 1800s and early 1900s, chaplains were often viewed as naïve and easily duped by inmates. Almost in response to this perception, the clinical pastoral education movement emerged during the 1920s and 1930s and promoted the serious study of chaplaincy. This field of study applied the principles, resources, and methods of organized religion to the correctional setting in a structured, disciplined, and professionalized manner. This resulted in the development of competent professional chaplains who were able to meld with the rehabilitation ideas that surfaced from the 1930s through the 1960s.

Correctional chaplains are now recognized as integral professionals in the field of corrections. The **American Correctional Chaplains Association (ACCA)** is a national organization that provides representation and networking opportunities for chaplains who work in various correctional environments. The organization also provides members with research on best practices in the field of correctional chaplaincy. This organization also provides training and certification for correctional chaplains as a means of furthering the standards of service that these professionals provide.

American Correctional Chaplains Association (ACCA): Provides representation and networking for chaplains who work in correctional environments.

PHOTO 10.7 Religious programming is an important element of the daily schedule in many correctional facilities.

Rick Loomis/*Los Angeles Times*/Getty Images

The specific types of religious groups vary from prison to prison and state to state. Nearly all state and federal correctional institutions provide support for at least some of the four traditional faith groups—Catholic, Protestant, Muslim, and Jewish. However, in recent years a number of nontraditional religious groups have been established in prison environments. These include, but are not limited to, Hinduism, Mormonism, Native American religions, Buddhism, Rastafarianism, Hispanic religions (Curanderism, Santeria, Espiritismo), Jehovah's Witnesses, Christian Scientists, and two of the newest faith groups to enter correctional facilities, Witchcraft and Satanism. The religious programs and practices conducted by the different faith groups differ according to the beliefs of the group, inmate interest, amount of time and space available in the prison, competence of the religious staff, and the support of the correctional authorities. It is not uncommon for a large prison to have numerous religious services on a daily basis. As one can see, the role of the chaplain in modern corrections can be quite complicated, and this is, in part, what has led to the various legal challenges noted in this chapter.

Are Inmates Really Motivated by Religion?

There is a good deal of dissent among practitioners in the field of corrections when addressing the question of whether or not inmates are truly motivated by religion. According to Dammer (2002), there is a belief among many who work in prison environments that inmates "find religion" for manipulative reasons (p. 1375). However, though this may sometimes occur, there is evidence that inmates have received positive benefits resulting from their incarceration and religious practice (Dammer, 2002; Johnson, 2004, 2012). Research by Johnson (2012) found that participants in religious programming had significantly fewer infractions while in prison than did inmates who did not participate in such programs. Even more convincing is the finding that inmates who participated frequently in religious programming services were less likely to be arrested or incarcerated when examined 2 years or more after their release. Thus, it would appear that prison religious programs have both short-term and long-term positive effects.

More recent research provides continued empirical evidence indicating that religious programming reduces crime and recidivism among adult offenders (Hercik, 2007). For instance, Johnson and Larson (2003) conducted a preliminary evaluation of the InnerChange Freedom Initiative, a faith-based prisoner reform program. Results show that program graduates were 50% less likely to be rearrested and 60% less likely to be reincarcerated during a 2-year follow-up period. Other research seems to demonstrate the efficacy of faith-based programs, both within the prison facility itself and later when offenders are released into the community. Thus, it appears that most inmates really are motivated by religious programming. With empirical quantitative evidence of positive outcomes, it should be concluded that religious programs in corrections are just as important as other forms of programming.

CONCLUSION

This chapter has provided an overview of the organizational structure of both the federal and state prison systems in the United States. Organization by levels of management and services provided throughout the agency were both addressed, and it is clear that prison agencies are responsible for a disparate array of services and facilities. Students must understand how these systems are organized and how and where rules and regulations originate. It is also important to understand how the lead manager of a facility, the prison warden, must then take these rules and regulations and implement them within his or her own facility. In the process, wardens must emphasize an active form of management, making themselves visible in the facility and communicating routinely with staff and inmates. This informal aspect of prison management is important in dealing with the prison subculture, addressing rumors, and getting a general feel for the climate of the institution.

Topics related to private prison management were covered, and it was noted that such organizations can also operate programs of excellence. Through the use of a unit management system, this type of organization can provide very clear and effective means of governing its employees while also soliciting their input in addressing day-to-day challenges within the organization. Lastly, the reality is that prisons are dangerous institutions, and, as a result, the unexpected can occur and the security and safety of the institution and of the public itself can be jeopardized.

This chapter also provided an extensive overview of many of the typical programs offered to inmates within the prison environment. We have discussed basic services, such as medical services and various types of programming, including educational, vocational, drug treatment, recreational, and religious. Though other types of programs are available to inmates, these types tend to be universal in their implementation, and they all tend to be used by nearly every inmate who serves time in the United States.

Each of these programs work to improve inmate conditions within the prison and that each reduces observed infractions among inmates as well as the likelihood of lawsuits or prison riots. Further, most of these programs have legal requirements that make them at least marginally necessary within the prison environment. Regardless of whether each program is constitutionally required, it is clear that each provides benefits to both the inmates and the prison staff. This is an important observation because it can therefore be said that these programs reduce problems in prison facilities, and, even more encouraging, they reduce recidivism when inmates are released into the community. Thus, prison programming is a smart investment that saves taxpayers more money in the long term than it costs them and enhances public safety in future years when inmates are released from prison. To overlook the importance of prison programming is to be negligent in safeguarding our community's safety.

 **SAGE edge™**

Want a Better Grade?

Get the tools you need to sharpen your study skills. Access practice quizzes, eFlashcards, video, and multimedia at **edge.sagepub.com/hanserbrief**

Interactive eBook

Visit the interactive eBook to watch SAGE premium videos. Learn more at **edge.sagepub.com/hanserbrief/access**.

 Career Video 10.1: Supreme Court Law Librarian

 Prison Tour Video 10.1: Emergency Management

 Prison Tour Video 10.2: Educational Funding

 Prison Tour Video 10.3: Work Programs

 Prison Tour Video 10.4: Recreational Programs

 Prison Tour Video 10.5: Religion Accommodations

DISCUSSION QUESTIONS

Test your understanding of chapter content. Take the practice quiz at edge.sagepub.com/hanserbrief.

1. What are some key features of the organizational structure of the Federal Bureau of Prisons?
2. Identify and discuss the different levels of management in state prison systems.
3. What are the differences between centralized and decentralized forms of management?
4. Identify and discuss the various models of managerial motivation and supervision.
5. What are some of the primary advantages associated with private prisons? Are there any potential problems that may exist for these types of prisons? Explain your answer.
6. Provide an overview of one of the prison work programs presented in this chapter.
7. What are therapeutic communities? How are they utilized in prison facilities?
8. How do substance abuse treatment programs affect potential recidivism?
9. Discuss some of the concerns for administrators with different types of recreational programming (e.g., weight lifting).

KEY TERMS

Review key terms with eFlashcards at edge.sagepub.com/hanserbrief.

Administrator, 263

American Correctional Chaplains Association (ACCA), 290

Armed disturbance control team, 272

Centralized management, 268

Decentralized management, 269

Disturbance control teams, 271

Federal Prison Industries Inc. (FPI), 279

APPLIED EXERCISE 10.1

Students must conduct either a face-to-face or a phone interview with a correctional supervisor who currently works in a prison facility in their state. Students should use the interview to gain the supervisor's insight and perspective on several key questions related to work in the field of institutional corrections. Students must write the practitioner's responses as well as their own analysis of those responses and submit their write-up by the deadline set by their instructor. The submission should be written in the form of an essay that addresses each point below. The total word count should be 1,200 to 2,000 words.

When completing the interview, students should ask the following questions:

1. What are the most rewarding aspects of your job as a supervisor of other prison staff?

2. What are the most stressful aspects of your job?

3. What are some challenges you find in working with the inmate population?

4. Why did you choose to work in this field?

5. What type of training did you receive for this line of work?

6. What are your future plans for your career in institutional corrections?

7. What would you recommend to someone who was interested in pursuing a similar career?

Students are required to provide the following contact information for the person interviewed. While you will probably not need to contact that person, it may become necessary in order to validate the actual completion of the interview.

Name and title of correctional supervisor: _____

Correctional agency: _____

Practitioner's phone number: _____

Practitioner's e-mail address: _____

Name of student: _____

WHAT WOULD YOU DO?

You are a correctional officer assigned to the agricultural section of a minimum-security facility. Your facility is large, with over 2,800 inmates who live on the property, most of them in dorm-like structures. The classification of most of these inmates is either trustee or minimum security. None of them is known to be violent.

You have recently been assigned to a field squad run by several "field bosses," who are correctional officers with experience supervising inmates who work in agricultural settings. Each field boss has his or her own horse, field radio, revolver, and rifle to ensure that security is maintained during the day. The work is hot, and the inmates work very hard.

On your fourth day of work, you notice that the ranking officer, Sergeant Gunderson, allows two inmates, Dooley and Craft, to go off into a wooded section at the edge of the clearing. They emerge a little later and talk with the sergeant and then go back to work.

You watch this same routine continue during much of the month of July. You finally ask one of the other officers about what you've noticed, and he says, "You know, Mack, I just tend to watch over my assigned inmates and don't worry about too much else. So long as they ain't escaping and as long as the sarge is happy, I just stay out of it. Maybe they gotta use the restroom or something . . . like, maybe they got bladder control problems or something."

You ponder this, and after 2 more days, you decide to speak to Sergeant Gunderson himself about the situation.

Sergeant Gunderson eyes you for a moment, pushing his straw-rimmed Stetson hat back a bit on his forehead before speaking. "I wouldn't be worried about them. They're old trustees who never hurt nobody and will likely be here forever. They just wanna ride their time out, and I really don't see the need to rock the boat. Besides, you need to know that they have a long history with one of the assistant wardens here. Believe it or not, those two like to work this detail, but anytime that they want they can pick another detail at the dorm or elsewhere, and they'll get the job switch within 48 hours. They have their reasons, and I have mine, for why things work as they do. You understand?"

You nod. "Yeah, I get it," you say, and leave the issue alone.

However, later that day, you happen to observe the two inmates load a couple of watermelons into the back of a wagon holding work tools and sundry items. They also have several potatoes stacked on the wagon. They cover the produce with a tarp as you ride up and shoot each other worried looks. You ask them what they are doing.

Dooley responds, "We were just gonna see if there were any more work hoes available, but there aren't. Why? What's wrong, boss?"

You stare at him for a beat. "Nuthin'. Just thought I would see what was going on—you guys get back to work." You watch them turn and go toward the main group.

The next day, you wait until Sergeant Gunderson is preoccupied explaining the details of some work assignment to another officer. You quickly guide your horse to the edge of the woods where you have spotted Dooley and Craft disappear so often. You see a faint trail, and when your crane your neck, you spy at its end a small clearing containing a watermelon patch and several potato plants.

Knowing what you know about inmates and the prison subculture, you recognize these two items as being prime ingredients for making homemade alcohol. In fact, some officers who work the dorms have noted that some inmates have appeared drunk on recent occasions.

You look around for Sergeant Gunderson and see that Dooley and Craft have their backs to you, nearly 50 yards away. You gently pull the reins on your grey gelding so as not to turn him too quickly or too obviously. You ride slowly back to a post that provides a full field of vision over the inmates you are supervising. You count them and find that all are accounted for.

You turn over what you have just discovered in your mind. You think, *Those two inmates are growing products to make alcohol, and Gunderson is allowing it.* You wonder how long this has been going on and who else might know about it.

Squinting up at the hot sun, you mutter to yourself.

What would you do?

PRISONER REENTRY RESOURCES

AP Photo/Paul Sancya

11

Parole and Reintegration

Learning Objectives

1. Define parole and basic parolee characteristics.

2. Discuss the historical development of parole.

3. Know and understand the basics regarding state parole, its organization, and its administration.

4. Evaluate the use of parole as a correctional release valve for prisons.

5. Describe the role of parole officers.

6. Explain the common conditions of parole and how parole effectiveness can be refined and adjusted to better meet supervision requirements that are based on the offender's behavior.

7. Be able to discuss the parole selection process, factors influencing parole decisions, and factors considered when granting and denying parole.

8. Describe the process for violations of parole, parole warrants, and parole revocation proceedings.

Making Parole

Rob Hanser drove down the winding, seemingly endless country road, making his way to David Wade Correctional Center for a parole hearing. He arrived at the prison and parked in the visitor's section. As he pulled the keys from his ignition, he glanced over and saw a car parked two spaces away from him. Some people were in the car but did not get out. He thought they looked similar to some photos he had seen before.

Hanser got out of his car, walked over to the other car, and waved through the windshield. Sure enough, it was Ronald's mother and sister.

Hanser leaned down and spoke through the open driver's window. "Ms. Drummer?"

The lady in the car said, "Yes, are you Dr. Hanser?"

"Yes ma'am, I am. It is nice to finally meet!"

The two women, Linda Drummer (Ronald's mother) and Laquitter—pronounced Laquitta—Drummer (Ronald's sister), got out of the car and exchanged introductions and small talk. After these formalities, all three walked toward the entrance to the prison. At the gate, they provided their IDs to the guard and waited.

They were escorted into the building, went through security, and were patted down to ensure that they were not bringing contraband into the prison. Once inside the waiting room, Linda beamed as she saw her son, Ronald "Raúl" Drummer, enter the room. The four sat together at a table.

"How you doing?" Hanser asked Ronald.

"I'm good, just a little nervous, but that's normal, right?"

After about 30 minutes of waiting, Ronald, his family members, and Hanser were in the parole interview room. It was not a large room and, as it turned out, the parole board was in Baton Rouge doing the interview remotely. The board members asked questions of Ronald regarding his job prospects and his place of residence, and they commented on how many courses and programs that he had finished. Then they asked about a letter that had been written.

One of the members said, "I see this letter from the director of your program. . . . Looks like it is from Dr. Hanser. Is he here?"

Hanser spoke up. "Yes, sir, I am here."

The board asked Hanser if he wanted to add any comments, noting that the letter very clearly supported granting Ronald parole and that it had provided some very convincing narrative to that end.

Hanser stated, "Well, first I would like to say that I don't do this very often. I get a lot of guys who ask me to write them letters and such, but I usually don't do it. I only write a letter if I am completely convinced of my appraisal. I certainly don't travel and speak on their behalf unless I'm certain of them. I can say with 100% certainty that Mr. Drummer exceeds all expectations as a mentor in our program. He is intelligent, motivated, and sincere in his efforts. Further, I and several of my colleagues with Freedmen Inc. will keep our eye on him. He will be staying at one of our reentry homes and will be employed with us there. He will also continue his peer support meetings, Bible studies, and other forms of programming in the community. I can ensure that."

After this, the board deliberated a bit and had Ronald, his family, and Hanser step out for a few moments.

When they were summoned back in, the board made its announcement. The head of the panel said, "Mr. Drummer, we are impressed with your record and the fact that you have gone well beyond what you needed to do, minimally, to qualify. You also really seem to have everything in place in the community to help you succeed." The board member paused for a moment, and all that could be heard was the shuffling of some papers. Then he spoke again. "So Mr. Drummer, what I am saying is that we have unanimously decided to award parole. Congratulations."

Linda Drummer let out a shout of excitement, Laquitter clapped her hands in amazement, and Ronald gave the board a wide grin in response.

"Now, Mr. Drummer, you will have some fees, some community service to perform, and, in addition, we're going to require that you commit to volunteering some of your time to speaking to kids at schools or social events on the hazards of drug use and a criminal lifestyle."

"Yes, sir, I understand," said Ronald.

The board allowed for the family to celebrate for a moment, congratulated Ronald again, and then noted that they had a long docket for the day.

Since that time, Ronald Drummer has been the house manager of his reentry home for Freedmen Inc. for over a year. He has also held a full-time job at Planet Fitness for over a year and has received two pay raises. He is studying to become a certified fitness trainer. He has served a leadership role in several community-based intervention programs and has spoken to numerous middle and high school students on the vagaries of the criminal lifestyle. By all accounts, he has been successful and productive while on parole.

PHOTO 11.1 Parole means that offenders will have many more options than might be encountered in a dayroom such as the one shown in this photo.

STAFF/REUTERS/Newscom

INTRODUCTION

This chapter addresses a type of offender outcome that represents a successful end to his or her incarceration experience: the release from prison on parole. This response to offender behavior falls within the field of corrections but comes at the end of the institutional process. This early release from prison reflects the fact that the offender has been well behaved in prison and represents a new beginning for that offender; basically, it is a reward for his or her prosocial behavior in prison. It should be pointed out that the parolee population tends to be quite small within the community-based correctional population when compared with the probation population. Further, the parolee population exists within only a select number of states. This chapter is provided primarily to ensure that this text is complete in its presentation of the correctional system, but it should not be viewed as a fully comprehensive authority on the use of parole.

PAROLE AND PAROLEE CHARACTERISTICS

Parole is a mechanism that has been nearly as controversial as the death penalty. Parole has had a tortured history, resulting in a slow and ongoing effort to eliminate or restrict its use in the federal system, and it has also been eliminated in many states throughout the nation. Nevertheless, a substantial number of inmates are on parole throughout the nation, with some of them still serving the remainder of their sentences under the outdated federal system. **Parole** can be defined as the early release of an offender from a secure facility upon completion of a certain portion of his or her sentence; the remainder of the sentence is served in the community. As of 2016, the nation's parole population had continued to grow to a total of 874,800 offenders. As with prior years, mandatory releases from prison due to good time provisions accounted for around half of those offenders who were released on parole (Kaeble, 2018). Figure 11.1 shows that the number of individuals on parole remained fairly stable from 2010 through 2014, with a slight increase occurring in 2015 and 2016.

Among those on parole, roughly 1 out of every 8 is a female offender. During the past decade, the proportion of female parolees has increased from 10% to 13%, a number that is higher than what has been seen in the past 5 years. The percentage of parolees who are African American tends to be around 38%. The proportion of Caucasian parolees has increased during the past several years, comprising 45% of the overall parole population in 2016. Roughly 15% of all parolees nationwide are Latino American, and another 2% come from other racial categories. Lastly, the majority of parolees were convicted of drug offenses, with 31% of the total parolee population having some drug-related conviction. Another 30% had convictions for violent offenses, and another 21% for property offenses. Figure 11.2 provides an examination of the U.S. parole population by offense.

Lastly, when considering outcomes for parolees around the nation, it has been found that about 27% of all parolees were returned to incarceration in 2016. The 2016 data shows that the rate of parolees going back to prison for violating the terms of their parole was 16%, whereas those who went to prison due to committing a new, separate criminal violation resulting in a new sentence was about 7% (Keable, 2018). Revocations of parole went down substantially from about 2005 to 2015, and the percentage of parolees who commit new criminal violations (thereby getting a new sentence) remains low and appears

Parole: The early release of an offender from a secure facility upon completion of a certain portion of his or her sentence.

FIGURE 11.1

Adults on Parole at Year-End, 2005–2016

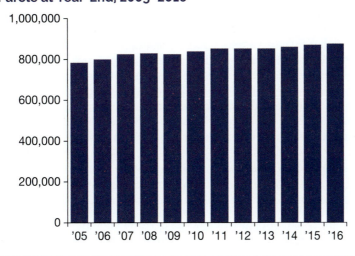

Source: Kaeble, D. (2018). *Probation and parole in the United States, 2016*. Washington, DC: Bureau of Justice Statistics. Retrieved from https://www.bjs.gov/content/pub/pdf/ppus15.pdf; Kaeble, D., & Cowhig, M. (2018). *Correctional populations in the United States, 2016*. Washington, DC: Bureau of Justice Statistics. Retrieved from https://www.bjs.gov/content/pub/pdf/cpus16.pdf

FIGURE 11.2

Parolee Characteristics by Type of Offense in 2016

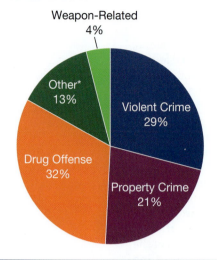

Source: Kaeble, D. (2018). *Probation and parole in the United States, 2016.* Washington, DC: Bureau of Justice Statistics.

*Includes public order offenses.

Ticket of leave: A permit given to a convict in exchange for a certain period of good conduct.

Mark system: A system where the duration of the sentence was determined by the inmate's work habits and righteous conduct.

on track for further decline. Thus, one might conclude that, overall, parole is proving to be a viable option for offender supervision.

THE BEGINNING HISTORY OF PAROLE

The development of parole is attributed to two primary figures: Alexander Maconochie and Sir Walter Crofton. Alexander Maconochie was in charge of the penal colony at Norfolk Island during the 1840s, and Sir Walter Crofton directed the prison system of Ireland in the 1850s. While Maconochie first developed a general scheme for parole, it was Crofton who later refined the idea and created what was referred to as the **ticket of leave**. The ticket of leave was basically a permit that was given to an offender in exchange for a certain period of good conduct. Through this process, the prisoner could instead earn his own wage through his own labor prior to the expiration of his sentence. In addition, other liberties were provided so long as the prisoner's behavior remained within the lawful limits set by the ticket of leave system. This system is therefore often considered the antecedent to the development of parole.

During the 1500s and 1600s, England implemented a form of punishment known as banishment on a widespread scale. During this time, criminals were sent to the American colonies under reprieve and through stays of execution. Thus, the offenders had their lives spared, but this form of mercy was generally only implemented to solve a labor shortage that existed within the American colonies. Essentially, offenders were shipped to the Americas to work as indentured servants under hard labor. However, the American Revolution put an end to this practice until 1788, when the first shipload of prisoners was transported to Australia. Australia became the new dumping ground for offenders, and they were used for labor there just as they had been in the Americas. The labor was hard, and the living conditions were challenging. However, a ticket of leave system was developed on this continent in which different governors had the authority to release offenders who displayed good and stable conduct.

In 1840, Alexander Maconochie, a captain in the Royal Navy, was placed in command over the English penal colony in New South Wales at Norfolk Island, which was nearly 1,000 miles off the eastern coast of Australia. The prisoners at Norfolk Island were the worst of the worst; they had already been shipped to Australia for criminal acts committed in England only to be later shipped to Norfolk Island due to additional criminal acts or forms of misconduct committed while serving time in Australia. The conditions on Norfolk Island were deplorable—so much so that many convicts preferred to be given the death penalty rather than serve time upon the island (Latessa & Allen, 1999).

While serving in this command, Maconochie proposed a system where the duration of the sentence was determined by the inmate's work habits and righteous conduct. Though this was already in use in a crude manner through the ticket of leave process in Australia, Maconochie created a **mark system** in which "marks" were provided to the offender for each day of successful toil. This system was quite well organized and thought out. It was based on five main tenets, as described by Barnes and Teeters (1959, p. 419):

1. Release should be based not on the completing of a sentence for a set period of time but on completion of a determined and specified quantity of labor. In brief, time sentences should be abolished and task sentences substantiated.

2. The quantity of labor a prisoner must perform should be expressed in a number of "marks" that he must earn, by improvement of conduct, frugality of living, and habits of industry, before he can be released.

3. While in prison, he should earn everything he receives. All sustenance and indulgences should be added to his debt of marks.

4. When qualified by discipline to do so, he should work in association with a small number of other prisoners, forming a group of six or seven, and the whole group should be answerable for the conduct of labor of each member.

5. In the final stage, a prisoner, while still obliged to earn his daily tally of marks, should be given a proprietary interest in his own labor, and be subject to a less rigorous discipline, to prepare him for release into society.

Under this plan, as first described in Chapter 1, offenders were given marks and moved through phases of supervision until they finally earned full release. Because of this, Maconochie's system is considered indeterminate in nature, with offenders progressing through his five specific phases of classification: (1) strict incarceration, (2) intense labor in forced work group or chain gang, (3) limited freedom within a pre-scribed area, (4) a ticket of leave, and (5) full freedom. This system was based on the premise that inmates should be gradually prepared for full release. Due to the use of primitive versions of indeterminate and intermediate sanctioning, Maconochie's mark system is perhaps best thought of as a precursor to both parole and the use of classification systems for offenders. Maconochie's system provided a guide to predicting the likelihood of success with an offender, making his process well ahead of its time.

However, Maconochie's system appears to have been *too* far ahead of its time; many government officials, influential persons, and even ordinary citizens in both Australia and England believed this approach was too soft on criminals. (This is not much different from today, when prisons and punitive sanctions are the preferred forms of punishment in the opinion of most Americans.) For his part, Maconochie was fond of criticizing prison operations in England; he believed that confinement ought to be rehabilitative in nature rather than punitive (note that this is consistent with the insights of John Augustus and his views on the use of probation). Due to the unpopularity of his ideas, Maconochie was ultimately dismissed from his post on Norfolk Island for being too lenient with offenders. Nevertheless, Maconochie was persistent and, in 1853, he successfully lobbied for the **English Penal Servitude Act**, which established several rehabilitation programs for offenders.

English Penal Servitude Act: Established several rehabilitation programs for convicts.

The English Penal Servitude Act of 1853 applied to prisons in both England and Ireland. Though Maconochie had spearheaded this act to solidify, legalize, and make permanent the use of ticket of leave systems, the primary reason for this act's success had more to do with the fact that free Australians were becoming ever more resistant to the use of Australia as the location for banished English prisoners. Though this act did not necessarily eliminate the use of banishment in England, it did provide incentive and sugges-

Father of Parole: Alexander Maconochie, so named due to his creation of the mark system, a precursor to modern-day parole.

tions for more extensive use of prisons. This law provided guidelines for the length of time that inmates should serve behind bars before being granted a ticket of leave and served as the basis for a general form of parole. The conditions mentioned in the English Penal Servitude Act of 1853 are also common to today's use of parole in the United States, though, of course, there are now many more technical aspects involved. However, the guidelines clearly stated that the offender's early release was contingent on his or her continued good behavior and avoidance of crime and criminogenic influences. Because of his work spearheading this act, his advocacy for other significant improvements in penal policies in England, and his contributions to early release provisions in England, Maconochie has been dubbed the **Father of Parole**.

PRISON TOUR VIDEO: There are many factors to consider when evaluating an inmate for parole. Go to the IEB to watch a warden discuss the role correctional officers play in parole decisions.

During the 1850s, Sir Walter Crofton was the director of the Irish penal system. Crofton was familiar with Maconochie's ideas, which he drew upon to create a classification system for the Irish prison system. In this system, an inmate's classification level was measured by the number of marks that he had earned for good conduct, work output, and educational achievement. This idea was, quite obviously, borrowed from Maconochie's system on Norfolk Island. It is important to point out that the Irish system developed by Crofton was much more detailed. It provided specific written instructions and guidelines that provided for close supervision and control of the offender and the use of police personnel to supervise released offenders in rural areas. It also called for an inspector of released prisoners in the city of Dublin (Cromwell, del Carmen, & Alarid, 2002).

Release was contingent upon certain conditions—for example, offenders had to submit monthly reports to either a police officer or another designated person and curtail their social involvements. Violations of these conditions could lead to reincarceration. This obviously is similar to modern-day parole programs. In fact, it could be said that contemporary uses of parole in the United States mimic the conditions set forth by Sir Walter Crofton in Ireland.

Parole From 1960 Onward

From 1930 through the 1950s, correctional thought reflected the medical model, which centered on the use of rehabilitation and treatment of offenders (see Chapter 2). The medical model presumed that criminal behavior was caused by social, psychological, or biological deficiencies that were correctable through treatment interventions. The 1950s were particularly given to the ideology of the medical model, with influential states such as Illinois, New York, and California turning to this type of treatment. In general, support for the medical model of corrections began to dissipate during the late 1960s and had all but disappeared by the 1970s.

The reintegration era, which lasted until the late 1970s, advocated for very limited use of incarceration; only a small proportion of offenders were imprisoned, and short periods of incarceration were most commonly recommended. Probation was the preferred sentence, particularly for nonviolent offenders. Indeterminate sentences were utilized, and deinstitutionalization was the theme for this period of corrections. However, this era in corrections was short-lived and received a great deal of criticism. Indeed, the prior medical model of corrections had hardly come to its full conclusion before the reintegration model was also being seriously questioned by skeptics.

The mid- to late 1970s saw a slowly emerging shift in corrections thought due to high crime rates that were primarily perceived as the result of high recidivism among offenders. Skepticism of rehabilitation was brought to its pinnacle by practitioners who cited (often in an inaccurate manner) the work of Robert Martinson. As students may remember from Chapter 2, Martinson conducted a thorough analysis of research programs on behalf of the New York State Governor's Special Committee on Criminal Offenders. Martinson (1974) examined a number of various programs that included educational and vocational assistance, mental health treatment, medical treatment, early release, and so forth. In his report, often referred to as the Martinson Report, he noted that "with few and isolated exceptions, the rehabilitative efforts that have been reported so far have had no appreciable effect on recidivism" (p. 22).

PRACTITIONER'S PERSPECTIVE

"Get in there—do internships!"

Go to the IEB to watch Dan Burke's video on his career as a federal agent.

FIGURE 11.3

Historical Developments in Parole

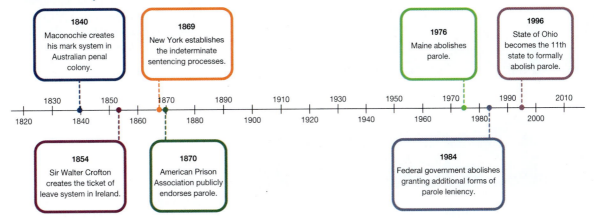

Source: Adapted from Association of Paroling Authorities International. (2001). *Parole board survey.* Retrieved from http://www.apaintl.org/resources/documents/surveys/2001.pdf

From this point forward, there was a clear shift from a community model of corrections to what has been referred to as a crime control model of corrections (see Chapter 1). During the late 1970s and throughout the 1980s, crime was a hotly debated topic that often became intertwined with political agendas and legislative action. The sour view of rehabilitation led many states to abolish the use of parole. Indeed, from 1976 onward, more than 14 states and the federal government abolished the use of parole. The state of Maine abolished parole in 1976, followed by California's elimination of discretionary parole in 1978, and then the full elimination of parole in Arizona, Delaware, Illinois, Indiana, Kansas, Minnesota, Mississippi, New Mexico, North Carolina, Ohio, Oregon, Virginia, and Washington (Sieh, 2006). In addition, the federal system of parole was also phased out over time. Under the Comprehensive Crime Control Act of 1984, the U.S. Parole Commission only retained jurisdiction over offenders who had committed their offense prior to November 1, 1987. The act also provided for the abolition of the Parole Commission over the years that followed, with this phasing-out period extended by the **Parole Commission Phaseout Act of 1996**. This act extended the life of the Parole Commission until November 1, 2002, but only in regard to supervising offenders who were still on parole from previous years. Thus, though the Parole Commission continued to exist, continued use of parole was eliminated, and federal parole offices across the nation were slowly shut down over time (see Figure 11.3 for further information on various developments in parole).

In addition to eliminating parole, many states implemented determinate sentencing laws, truth-in-sentencing laws, and other such innovations that were designed to keep offenders behind bars for longer periods of time. The obvious flavor of corrections in the 1980s was toward crime control through a correctional ideology of incapacitation. This same crime control orientation continued through the 1990s and even through the beginning of the new millennium, with an emphasis on drug offenders and habitual offenders during the 1990s. Also of note were developments in intensive supervision probation (ISP), more stringent bail requirements, and the use of three-strikes penalties. The period during the last half of the 1990s and beyond the year 2000 had a decidedly punitive approach. The costs (both economic and social) have received a great deal of scrutiny even though crime rates lowered during the new millennium. Though there was a dip in crime during this time, it was not necessarily made clear if this was, in actuality, due to the higher rate of imprisonment or due to other demographic factors that impacted the nation. Further, as we have seen during recent years, this led to correctional systems having to implement mass forms of early release of inmate populations that had swollen during the 1990s and early 2000s. This has very serious potential implications for meeting public safety objectives in corrections.

Parole Commission Phaseout Act of 1996: Extended the life of the Parole Commission until November 1, 2002, but only in regard to supervising offenders who were still on parole from previous years.

PHOTO 11.2 This inmate appears before the parole board, answering questions regarding his eligibility for early release.

© Mikael Karlsson/Alamy Stock Photo

History of Federal Parole and Supervised Release

Parole of federal inmates began after enactment of legislation on June 25, 1910. At that time, there were three federal penitentiaries, and parole was granted by a parole board at each institution, with the membership of each parole board consisting of the warden of the institution, the physician of the institution, and the superintendent of prisons. By 1930, a single Board of Parole in Washington, D.C. was established. This board consisted of three full-time members appointed by the attorney general. The Federal Bureau of Prisons performed the administrative functions of the board. In August 1945, the attorney general ordered that the board report directly to him for administrative purposes. In August 1948, due to a postwar increase in prison population, the attorney general appointed two additional members to the board, increasing it to five members (Hoffman, 2003).

Legislation in 1950 saw the board increase to eight members who served 6-year terms. The board was placed in the Department of Justice for administrative purposes. Three of the eight members were designated by the attorney general to serve as the Youth Corrections Division pursuant to the Youth Corrections Act. In October 1972, the board began a pilot reorganization project that eventually included the establishment of five regions, the creation of explicit guidelines for parole release decision making, the provision of written reasons for parole decisions, and an administrative appeal process (Hoffman, 2003).

In May 1976, the **Parole Commission and Reorganization Act** took effect. This act retitled the Board of Parole as the U.S. Parole Commission and established it as an independent agency within the Department of Justice (Hoffman, 2003). The act provided for nine commissioners appointed by the president, with the advice and consent of the Senate, for 6-year terms. These members included a chair, five regional commissioners, and a three-member National Appeals Board. In addition, the act incorporated the major features of the Board of Parole's pilot reorganization project that were listed above.

Parole Commission and Reorganization Act: Established the U.S. Parole Commission as an independent agency within the Department of Justice.

Eight years later, the **Comprehensive Crime Control Act of 1984** created the U.S. Sentencing Commission to establish sentencing guidelines for the federal courts and established a regimen of determinate sentences (Hoffman, 2003). The chair of the Parole Commission is an ex-officio, nonvoting member of the Sentencing Commission. The decision to establish sentencing guidelines was based in large part on the success of the Parole Commission in developing and implementing its parole guidelines. In 1987, the Sentencing Commission submitted to Congress its initial set of sentencing guidelines, which took effect that year. As set forth by the Crime Control Act, offenders whose acts were committed on or after November 1, 1987, serve determinate terms under the sentencing guidelines and are not eligible for parole consideration (Hoffman, 2003).

Comprehensive Crime Control Act of 1984: Created a U.S. Sentencing Commission to establish federal sentencing guidelines favoring determinate sentences.

Per the Sentencing Commission's guidelines, parole for federal inmates was essentially abolished, but the use of supervised release from federal prisons was not entirely eliminated. While official parole and the use of parole boards no longer exist within the federal justice system, the modified version of early release is afforded some federal inmates based on requisites related to sentence completion. This post-release supervision is termed **supervised release** and is provided as a separate part of the sentence under the jurisdiction of the court. This type of early release is administered by the sentencing court for a given inmate, similar to community supervision under probation. In the federal system, the court, not the parole board, has the authority to impose sanctions on released inmates if they violate the terms or conditions of their supervision.

Supervised release: Post-release supervision.

The Comprehensive Crime Control Act of 1984 provided for the official abolition of the Parole Commission on November 1, 1992, 5 years after the sentencing guidelines took effect. This phase-out provision did not adequately provide for persons sentenced under the law in effect prior to November 1, 1987, who had not yet completed their sentences. Elimination of, or reduction in, parole eligibility for such cases raised a serious ex post facto issue. To address this problem, the **Judicial Improvements Act of 1990** extended the life of the Parole Commission until November 1, 1997 (Hoffman, 2003). However, this extension still did not sufficiently address the complexities related to the residual paroled population, and this resulted in yet another act, the Parole Commission Phaseout Act of 1996, which again extended the life of the Parole Commission, this time until 2002. In addition, it required the attorney general to report to Congress annually beginning in 1998 on whether it was more cost-effective for the Parole Commission to continue as a separate agency or for its remaining functions to be transferred elsewhere. It is important to note the U.S. attorney general has reported each year that it is more cost-effective for the Parole Commission to continue as a separate agency. After the Phaseout Act's date expired, the 21st Century Department of Justice Appropriations Authorization Act of 2002 extended the life of the Parole Commission until November 1, 2005.

Currently, the U.S. Sentencing Commission oversees the supervision of offenders who leave federal confinement early due to credit for good behavior. According to the *2007 Federal Guidelines Manual*, the court

> is required to impose a term of supervised release to follow imprisonment if a sentence of imprisonment of more than one year is imposed or if a term of supervised release is required by a specific statute. The court may depart from this guideline and not impose a term of supervised release if it determines that supervised release is neither required by statute nor required for any of the following reasons: (1) to protect the public welfare; (2) to enforce a financial condition; (3) to provide drug or alcohol treatment or testing; (4) to assist the reintegration of the defendant into the community; or (5) to accomplish any other sentencing purpose authorized by statute. (p. 409)

Though the Sentencing Commission oversees the majority of federal offenders released from prison, there is apparent support for the continued use of the Parole Commission. On July 21, 2008, during the meeting of the 110th session of Congress, the **United States Parole Commission Extension Act of 2008** was passed to provide for the continued performance of the Parole Commission (U.S. Congress, 2008). It became Public Law No. 110-312. So strong was support for this act that the initial bill passed in the Senate by unanimous consent and was ultimately signed by President George W. Bush (GovTrack.us, 2010).

From this brief discussion of federal parole, it should be clear that this mechanism is a vestige from the past, but it continues to reemerge as an operational organization. Although it is unknown if federal parole will ever return to its previous prominence, it is clear that the federal system, like the state systems, will continue to need to supervise offenders released from prison, regardless of the agency mechanism that is ultimately used.

PAROLE AT THE STATE LEVEL

With respect to the administration and organization of parole boards, it is clear that there is a great deal of variation in their structure and implementation. Table 11.1 shows the diverse nature of state parole boards with various characteristics such as the term of service for parole board members, the number of persons serving on parole boards, and the use of either part-time or full-time members varying greatly from state to state. The types of sentencing schemes used in a given state also vary greatly. Some states have maintained parole release processes within indeterminate sentencing systems where the judge imposes a maximum sentence while parole boards determine the exact release date for many inmates. In other states that utilize determinate sentencing approaches, parole boards do not usually decide on an offenders' release date. However, they may sometimes use discretionary decision-making powers when considering offenders who were convicted prior to the passage of a given determinate sentencing statute or with offenders who serve life sentences. This demonstrates that there is a great deal of disparity throughout the United States among the top organizationally ranked decision-making bodies that decide on issues related to parole.

Judicial Improvements Act of 1990: Extended the life of the Parole Commission until November 1, 1997.

United States Parole Commission Extension Act of 2008: Provided for the continued performance of the U.S. Parole Commission.

TABLE 11.1

State Parole Board Appointments, Structure, Terms, and Functions From Selected States

	GOVERNOR APPOINTED	TERM YEARS	NUMBER ON THE BOARD	FULL OR PART TIME	INDETERMINATE SENTENCING	DETERMINATE SENTENCING
California	X	3	15	Full		X
Florida	X	6	3	Full		X
Illinois	X with Senate consent	6	13	Full		X
New York	X with Senate confirmation	6	12	Full	X	
Ohio	Appointed by Dir. of Rehab. & Correction	Up to two 6-year terms	Up to 12 members	Full		X
Texas	X with Senate consent	6	7	Full	X	
Georgia	X with Senate confirmation	7	5	Full	X	

Source: Author independent research of each state: Robina Institute. (2018). *Parole boards with indeterminate and determinate sentencing structures.* Minneapolis: University of Minnesota.

The variation in state implementation of the day-to-day supervision of parolees (for example, either as a separate function or combined with probation caseloads) and the variation that exists among parole boards themselves make it clear that the entire organizational structure can be quite complicated. However, there is a great deal of similarity in the types of laws, forms of supervision, and regulations that are required throughout the nation. While each state has the right and ability to administer community supervision functions in a manner that is most suitable for that state, there are as many similarities between probation and parole programs around the nation as there are differences.

The Granting of Parole in State Systems

In most states, offender cases are assigned to various individual parole board members who are tasked with reviewing the case so that they can formulate their initial recommendations. The recommendations that they provide are typically honored and accepted as written. Most states that follow this process hold a formal hearing where parole board members may share their views. When the parole hearing is conducted with the offender seeking parole, all members involved with the decision may be present (as is often the case in television or movie portrayals), but it is also possible that just one member is present. Lastly, when parole hearings are conducted, the board may convene at the facility where the inmate is located (requiring the board to travel), or the inmate may be brought to the board, wherever the board is located (in many cases, the state capital).

The process and guidelines for parole selection vary considerably from state to state. Some states have a minimum amount of time that must be served. Others have stipulations on the types of crimes that the offenders have committed. Some states use both of these criteria as well as others. However, the actual decision by any parole review body is often made by members who have a great deal of discretion. Indeed, it appears that parole boards are influenced by a wide variety of criteria, many of which are not necessarily noted by statute or official agency guidelines. Institutional infractions, the age of the offender, marital status, level of education, and other factors may all weigh into the parole board's decision. Naturally, one of the key concerns with granting parole is the probability of recidivism. To a large extent, the prediction process has been little better than guesswork. For decades, the development of prediction tools has continued in an attempt to standardize risk factors. Psychometric tools and statistical analyses have ultimately rested upon actuarial forms of risk prediction. In most cases, it is the objective use of statistical risk prediction that turns out to be more accurate than that which allows for individual subjectivity. There are, of course, some exceptions since the context surrounding the statistical data may be important and may provide alternative explanations as to why a certain set of numbers or statistical outcomes may have been obtained. However, this often simply results in the overprediction of likely reoffending. Overprediction of offending is costly to prison systems because they will continue to incarcerate persons who are, in reality, at no risk of reoffending. In some states, there may be a need to reduce prison system overcrowding and such overprediction can further exacerbate problems with this overcrowding.

Indeed, parole mechanisms can serve as release valves for prison systems that become overstuffed with offenders. When this occurs, there may be a need for a certain amount of offender releases, and parole boards may have to make tough decisions that do not necessarily comport with the formal risk assessment based on a standardized instrument. This is where the difficulty tends to occur, and this demonstrates why, on the one hand, it is counterproductive for standardized risk assessment instruments to overpredict (as with the Wisconsin Risk Assessment scale), yet, on the other hand, subjective decision making is a necessary evil that is fraught with peril, resulting in incorrect predictions that ultimately lead to serious mistakes in determining an offender's likelihood to recidivate.

Other factors also affect the decision to grant parole. For instance, an inmate may (according to a standardized instrument) have a high likelihood of reoffending. But the type of reoffending may be of a petty nature. In such instances, parole boards may decide to grant parole despite the fact that the offender is not considered a good risk based on a pure analysis of whether he or she will or will not reoffend. Thus, it is clear that the specific type of reoffending is also an important consideration among parole board personnel. As just noted, this may be an especially important consideration when parole boards are aware that the state's prison system is overcrowded and that a certain number of releases will assist prison administrators in maintaining their prison population levels. Therefore, it is better to release a person likely of relapsing on drugs or alcohol or who may commit some form of shoplifting than it is to release someone likely of committing some form of violent crime. It is using this next-best-solution approach that parole boards may be compelled (though not legally required) to make their releasing decisions.

It is important for students to appreciate the problems associated with prison overcrowding. As discussed in past chapters, federal court rulings during the 1970s and 1980s penalized many state prison systems and essentially forced them to honor a variety of civil rights standards when incarcerating inmates. Thus, the issue of overcrowding cannot be taken lightly by prison administrators, and state systems resort to a number of alternatives to alleviate this issue. In states that still utilize parole, this is one method state prison systems use to resolve their overcrowding problems, which means community correctional systems are used to augment and support the states' institutional correctional systems. Thus, parole boards may play a key role in bridging these two components in an effort to ameliorate challenges facing a state correctional system.

The fact that parole boards can play such a role should not be underestimated. They may, in fact, be under some pressure to assist the overall state system. Further, consider that these boards are often constructed by the governor of a given state. In some cases, state politics and state priorities may come into play, affecting the decisions in some parole board cases. This is particularly true when the parole board's administration is consolidated rather than independent in nature (as discussed earlier in this chapter). The point in this discussion is to demonstrate that parole boards do not operate in a complete vacuum. The influences of the surrounding contextual reality are inevitable, and these influences come from a number of directions. Indeed, prison wardens, state offices, victims, the parolee's family, and the public media may all have an impact upon the discretion employed by parole board members, individually and collectively. Students should refer to Table 11.2 for details of parole populations throughout the nation, including the federal system, all states combined, and a few other select states. Note that Table 11.2 includes only the top 10 states with the largest parole populations.

PAROLE AS THE CORRECTIONAL RELEASE VALVE FOR PRISONS

The intent behind parole, at least initially, was to provide an incentive to inmates for exhibiting prosocial behavior while also providing for a gradual process of reintegration into society. This process was intended to be based upon the behavior of the inmates and their progress in work assignments and programming while serving their sentence. The use of parole was not intended to be a mechanism to assist prisons in maintaining their population overflow. In fact, the soundest decisions for parole do not consider prison populations at all but instead are based entirely upon the factors relevant to the inmates and their behavior.

Nevertheless, states around the nation find themselves considering the increased use of parole or early release programs due to problems with prison overcrowding. State correctional systems may find it difficult to house the influx of offenders when their budgets are not increased to accommodate this continual flow of new inmates. Whenever correctional systems use parole with the intent to reduce correctional populations rather than facilitate reintegration of offenders, they are using parole as a **release valve mechanism** for their prison population.

The use of parole as a release valve is becoming increasingly common due to the recent economic challenges that have faced state governments. While reentry efforts such as parole can be useful in reforming offenders and thereby reducing their likelihood of recidivism, such practices should operate regardless of prison population levels. If prison population levels determine the likelihood of release, then release decisions are made due to monetary, not public safety, considerations. This is a dangerous game to play in corrections and puts the public at risk for future criminal victimization.

Release valve mechanism: When correctional systems use parole to reduce correctional populations.

TABLE 11.2

Adults on Parole in Selected States, by Size of Parole Population

ORGANIZATION	NUMBER OF PAROLEES AS OF 12/31/2013	NUMBER OF PAROLEES AS OF 12/31/2015	PERCENTAGE CHANGE
United States Total	853,215	870,526	+2.0
Federal	111,226	114,471	+3.0
Texas	111,302	111,892	+0.5
California	87,532	86,053	−1.7
New York	45,039	44,562	−1.0
Louisiana	28,744	31,187	+8.5
Illinois	29,586	29,146	−1.4
Georgia	26,611	24,130	−9.3
Arkansas	21,709	23,093	+6.3
Wisconsin	20,251	19,453	−3.9
Ohio	16,797	18,284	+8.9
Michigan	18,439	17,909	−2.9

Sources: Herberman, E. J., & Bonzcar, T. P. (2015). *Probation and parole in the United States, 2013.* Washington, DC: Bureau of Justice Statistics; Kaeble, D. (2018). *Probation and parole in the United States, 2016.* Washington, DC: Bureau of Justice Statistics.

THE ROLE OF INSTITUTIONAL PAROLE OFFICERS

Institutional parole officers, often referred to as case managers or caseworkers, will work with the offender and a number of institutional personnel to aid the offender in making the transition from prison life to community supervision while on parole. This professional serves both a security function (assessing suitability for parole) and a reintegration function (providing casework services inside the prison and networks that extend beyond the prison). Much of the information presented in this section regarding prerelease planning and the role of the institutional parole officer draws on information from the state of Oklahoma's Pre-Release Planning and Reentry Process guidelines (Jones, 2007).

During prerelease planning, prison staff work together to provide a bridge of services that connect the offender to the outside world. A great deal of work can go into the planning and preparation process of an inmate's exit from prison. This section will shed some light on the institutional parole officer's function, since these professionals provide a link between the prison world and the outside community.

Upon determining that an inmate is suitable for parole, the institutional parole officer will begin the prerelease planning process that attends to the offender's transition from prison to the community. This process typically begins about 6 months prior to release and involves a shift from institutional case planning to individual community preparedness. The goal of a good reintegration program should be to ensure that the offender has the support, information, and contacts necessary to begin anew prior to exiting the prison. Even small details must be attended to, such as providing offenders with essentials like clothing and shoes that are appropriate for the season, proper identification, and appropriate referrals to community agencies that can assist with other services.

Throughout the process, agency administration will tend to track the offender's progress, keeping a careful eye on the 6 months prior to release. At this point, the offender may experience problems with anxiety due to nervousness over his or her expected freedom, the responsibilities of the outside world, and the effects of prisonization inside the facility. A good prerelease program will address these issues, preparing the inmate psychologically for release. As with their initial entry into prison, this period is often one of the most stressful for offenders since so much of their future is uncertain and they will be held to expectations that they have not had to meet in years.

Interviews, forms, and checklists will be completed during this time as part of the review case plan that documents the offender's approach toward release. These interviews will seek to identify various needs that the offender might have upon release. Needs-based assessment instruments perform this function. Identified needs can be many and varied but often include some program that the offender did not complete while in prison, such as an educational plan or substance abuse program. Other needs may be related to the payment of restitution, transportation, or making provisions for child support.

In addition to these concerns, correctional staff should note any unique circumstances in the prerelease plan that might provide a challenge to the successful reintegration of the offender. This will typically be included in the adjustment review and also in what is often referred to as the Offender Accountability Plan. The **Offender Accountability Plan** addresses the need for restitution, the need to respect the rights and privacy of prior victims, and any particular arrangements that have been made with the victim as well as any necessary provisions to ensure the offender's responsibility to the community at large.

Institutional parole officer: Works to aid the offender in making the transition from prison life to community supervision.

Offender Accountability Plan: Addresses needs for restitution, any particular arrangements that have been made with the victim, and provisions to ensure the offender's responsibility to the community at large.

PHOTO 11.3 The sheet of paper underneath this urinalysis cup clearly identifies the conditions of an offender's parole. In this case, the parole officer is making it clear that the offender is expected to remain drug free by placing the urinalysis cup directly on top of the list of parole conditions.

Robert Hanser

The actual day of release is an important milestone for the offender and is actually critical to his or her successful reintegration. This should be treated as more than a nostalgic moment, and the seriousness of the new challenge ahead should be kept in focus. (This is illustrated in this chapter's opening vignette when the Louisiana Parole Board made note of additional requirements for the parolee, Ronald Drummer. Note that this vignette is based on true events during a hearing that the author actually attended to provide recommendations for the offender's release to the parole board.) Activities should focus on the last few tasks required for a seamless transition to the community. In addition, the offender should be provided a portfolio of the various services available, requirements of parole, and so forth, allowing him or her to keep the information and requirements organized. Organizational skills may be somewhat impaired given the newness of the release experience and the likely excitement that will be experienced.

Lastly, Torres (2005) points out that although release from the prison facility can be a euphoric experience, it can also result in unexpected disappointment and frustration for the offender. Torres provides an insightful description of the psychological challenges associated with offenders' reintegration as they navigate between their life prior to incarceration and the life that they now face:

> The parolee's memories of family, friends, and loved ones represent snapshots frozen in time, but in reality, everyone has changed, moved away, taken a new job, grown up, or perhaps most disappointingly become almost strangers. The attempts to restore old relationships can be very threatening and eventually disappointing. In addition, the presence of almost complete freedom after years of living in a structured, confined prison setting can also add tremendous stress to adjusting to the open community where the offender must now assume major responsibilities of transportation, obtaining a driver's license, finding a job, reporting for drug testing, and so on. If married, with children, the spouse may unrealistically expect the offender to immediately begin providing financial relief to the family that perhaps has endured financial hardships while the breadwinner was away. Other barriers to success include civil disabilities that prohibit the felon from voting . . . and most importantly, from being employed in certain occupations. (p. 1125)

Institutional parole officers are cognizant of the situation that faces upcoming parolees. They must ensure that the offender has the full range of support necessary to face the potentially traumatic adjustment to the outside world. The offender will need to come to grips with issues that most people do not consider, making the experience of release sometimes a bit bittersweet. Offenders may (or may not) themselves realize the full range of emotional experiences that they will have upon release, and it is one job of the institutional parole officer to ensure that appropriate support for coping is provided to the offender who may be disappointed or overwhelmed by the experience.

COMMON CONDITIONS OF PAROLE

The terms and conditions for parolees, in most cases, are identical to many of those for offenders on probation. For instance, the state of Oklahoma requires parole fees of $40 at a minimum and clearly outlines the potential outcomes if an offender violates the terms of his or her parole. These outcomes include additional levels of supervision, reintegration training, the addition of day reporting centers, weekend incarceration, nighttime incarceration, intensive parole, jail time, and incarceration. This is similar to what happens if offenders on probation violate the terms and conditions of their supervision. If possible, the parolee will be kept on supervision (depending on the nature of the violation) but will experience a graduated set of increasingly restrictive sanctions and requirements that will become additional conditions to his or her parole requirements. Figure 11.4 provides an example of some of the general conditions of parole in the state of California. These conditions, for the most part, tend to be very similar from state to state unless the offender happens to be in a specialized category of population typologies (such as with sex offenders), but even in these cases, restrictions tend to be similar to those required of other sex offenders.

REENTRY INITIATIVES

Before we can discuss offender reentry programs, we must understand what constitutes offender reentry. Some observers note that offender reentry is the natural by-product of incarceration because all prisoners who are not sentenced to life in prison and who do not die in prison will reenter the community at some point. According to this school of thought, reentry is not a program or some kind of legal status but rather a process that almost all offenders will undergo. A variant on this approach to reentry is the concept that **offender reentry**, simply defined, includes all activities and programming conducted to prepare ex-offenders to return safely to the community and to live as law-abiding citizens (Nunez-Neto, 2008).

To demonstrate how the reentry effort has become a nationwide priority, consider the Second Chance Act, which was signed into law on April 9, 2008, and designed to improve outcomes for people returning to communities from prisons and jails. This first-of-its-kind legislation authorizes federal grants to government agencies and nonprofit organizations to provide employment assistance, substance abuse treatment, housing, family programming, mentoring, victim support, and other services that can help reduce recidivism.

When considering the Second Chance Act and the grant funding that has stemmed from this act, it is clear that the federal government has provided significant attention to the notion of reentry and, in the process, has developed an excellent overarching paradigm for reentry service delivery, commonly referred to as the Roadmap to Reentry. The **Roadmap to Reentry** identifies five evidence-based principles that guide efforts to improve correctional programs that are developed for those who will reenter society after being incarcerated. Much like standard discharge planning among therapeutic treatment professionals, the federal government operates on the idea that reentry begins on the very first day of incarceration. Likewise, the work of correctional programmers does not end when the individual leaves the prison gates; rather, correctional efforts are expected to continue from intake, to incarceration, and onward through release. As developed by the Department of Justice, the five evidence-based principles to optimal reentry efforts are as follows:

- **Principle I:** Upon incarceration, every inmate should be provided an individualized reentry plan tailored to his or her risk of recidivism and programmatic needs.

- **Principle II:** While incarcerated, each inmate should be provided education, employment training, life skills, substance abuse, mental health, and other programs that target their criminogenic needs and maximize their likelihood of success upon release.

- **Principle III:** While incarcerated, each inmate should be provided the resources and opportunity to build and maintain family relationships, strengthening the support system available to them upon release.

- **Principle IV:** During transition back to the community, halfway houses and supervised release programs should ensure individualized continuity of care for returning citizens.

- **Principle V:** Before leaving custody, every person should be provided comprehensive reentry-related information and access to resources necessary to succeed in the community.

Again, this outlook on reentry is very similar to the continuity-of-care idea that is common among medical and mental health patients. The whole idea is that as the offender continues to progress through the completion of their sentence in a satisfactory manner, their level of supervision should become less intense and less invasive while their standard of living improves in quality and in terms of overall opportunities available. Once released, the offender has been provided effective tools to avoid criminogenic influences and embrace involvement in prosocial lifestyle choices.

The Use of Reentry Councils

In providing this leadership role, many local, regional, and state levels of government have created what are often known as reentry councils. In fact, the author is a member

Offender reentry: Includes all activities and programming conducted to prepare ex-convicts to return safely to the community and to live as law-abiding citizens.

Roadmap to Reentry: Identifies five evidence-based principles that guide efforts to improve correctional programs that are developed for those who will reenter society after being incarcerated.

of the governor's reentry council in the state of Louisiana. Most states have some type of reentry council or task force, and many larger counties also have their own reentry councils. Likewise, many larger cities have found it beneficial to develop their own councils, as well. Regardless, all of these tend to consist of a collaboration of partners who meet routinely throughout the year, so as to pool resources and optimize reentry efforts in their area. However, it is the federal government that has established the most clear and distinct version of this collaborative effort.

According to the U.S. Department of Justice (2016), the Federal Interagency Reentry Council brings together 20 agencies that are tasked with the mission of achieving the following:

- making communities safer by reducing recidivism and victimization

- assisting those who return from prison and jail in becoming productive citizens

- saving taxpayer dollars by lowering the direct and collateral costs of incarceration

---

FOCUS TOPIC 11.1

Freedmen Inc. Halfway House for Offenders Released From Prison

Freedmen Inc. is a faith-based organization that works with a variety of organizations in the community to provide offenders with housing, transportation, employment, job skills, spiritual guidance, mental health, and substance abuse assistance. This organization's board of directors includes numerous people who are active in reentry efforts in their community. The House of Healing, as it is called, is the primary home in which offenders are housed, but there are other homes as well.

It is important to understand that most of the efforts of this organization are funded through donations and church-based collaborations. Naturally, this means that there is a strong biblical basis to much of the programming. While this may be problematic to some people, this program is designed for offenders who desire this type of reentry experience.

Though this program was originally designed for men, there is now a sister program that aids female offenders in reentry. This points toward the growing reentry needs of the community. These women engage in programming that is similar to the programs in which their male counterparts engage; however, they do not stay in the same facility as the male participants.

This organization is one example of how grassroots efforts in communities can provide services that aid persons trying to rebuild their life after incarceration while, at the same time, making the community safer by offering participants alternatives to crime. ●

Source: R. Hanser (2018).

A prime focus of the Federal Interagency Reentry Council is to remove barriers to effective reentry. The idea is to provide sufficient groundwork for motivated individuals who have served their sentence so they can gain employment, have stable housing, be available to support their families, and contribute to the surrounding community. Aside from recidivism reduction, the agencies brought together by the Reentry Council seek to improve public health, child welfare, employment outcomes, housing, and other important aspects to offender reintegration (U.S. Department of Justice, 2016). A partial list of offender needs that are specifically cited by the Federal Interagency Reentry Council (U.S. Department of Justice, 2016) includes the following:

1. Employment

2. Education

3. Housing

4. Collateral Consequences

5. Child Support

6. Women and Reentry

7. Children of Incarcerated Parents

8. Health Care

9. Reservation Communities

10. Veterans

As one can see, most all of the offender needs listed above are those that have been discussed previously throughout this chapter. Naturally, since this is a federal reentry organization, issues related to Native Americans and reservation communities as well as veterans issues were, not surprisingly, included in this list. This also brings to bear the point that regardless of whether the reentry program is federal, state, local government, a faith-based community organization, or a grassroots organization, cultural groups can and should be considered. Likewise, veterans issues related to health care, potential trauma, educational benefits, and so forth should probably also be included, as well as other potential areas of need. It would appear that the list of potential needs and services can be quite encompassing, depending on the offender typology that is served.

VIOLATIONS OF PAROLE, PAROLE WARRANTS, AND PAROLE REVOCATION PROCEEDINGS

No discussion pertaining to parole (and particularly an entire chapter on the subject) would be complete without at least noting some of the issues associated with the revocation of that sentencing option. The revocation process is often a two-stage one that was initially set forth in the Supreme Court ruling of *Morrissey v. Brewer* (1972). The first hearing is held at the time of arrest or detention and is one where the parole board or other decision-making authority will determine if probable cause does, in fact, exist in relation to the allegations against the parolee that are made by the parole officer. The second hearing then is tasked with establishing the guilt or innocence of the parolee. During this hearing, the parolee possesses a modified version of due process; he or she is provided with written notice of the alleged violations, is entitled to the disclosure of evidence to be used against him or her (similar to discovery), has the right to be present during the hearing and to provide his or her own evidence, has the right to confront and cross-examine witnesses, has the right to a neutral and detached decision-making body, and has the right to a written explanation of the rationale for revocation.

In some states, such as South Carolina, a person known as the **parole revocations officer** is primarily tasked with the routine holding of preliminary parole revocation hearings. This officer reviews the allegations made by parole officers against parolees. These hearings are administrative and not nearly as formal as those held by a judge in a true court of law. Typically, these courts are routine in nature, but some rulings and findings of fact may vary. Though most hearings are not complicated, a degree of discretion is required on occasion when determining if the evidence has been presented well or to determine if the violation requires a true revocation of parole or just more restrictive

Parole revocations officer: Primarily tasked with the routine holding of preliminary parole revocation hearings by reviewing allegations made by parole officers against parolees.

PHOTO 11.4 This parole officer is checking on one of his parolees who is at work.

Mikael Karlsson/Alamy Stock Photo

FIGURE 11.4

Example of General Conditions of Parole in the State of California

- Your Notice and Conditions of Parole will give the date that you are released from prison and the maximum length of time you may be on parole.

- You, your residence (where you live or stay) and your possessions can be searched at any time of the day or night, with or without a warrant, and with or without a reason, by any parole agent or police officer.

- You must waive extradition if you are found outside of the state.

- You must report to your parole agent within one day of your release from prison or jail.

- You must always give your parole agent the address where you live and work.

- You must give your parole agent your new address **before** you move.

- You must notify your parole agent **within three** days if the location of your job changes, or if you get a new job.

- You must report to your parole agent whenever you are told to report or a warrant can be issued for your arrest.

- You must follow all of your parole agent's verbal and written instructions.

- You must ask your parole agent for permission to travel more than 50 miles from your residence and you must have your parole agent's approval before you travel.

- You must ask for and get a travel pass from your parole agent before you leave the county for more than two days.

- You must ask for and get a travel pass from your parole agent before you can leave the State, and you must carry your travel pass on your person at all times.

- You must obey ALL laws.

- If you break the law, you can be arrested and incarcerated in a county jail even if you do not have any new criminal charges.

- You must notify your parole agent immediately if you get arrested or get a ticket.

- You must not be around guns, or anything that looks like a real gun, bullets, or any other weapons.

- You must not have a knife with a blade longer than two inches except a kitchen knife. Kitchen knives must be kept in your kitchen.

- Knives you use for work are allowed only when approved by your parole agent but they can only be carried while you are at work or going to and from work. You must ask for a note from your parole agent that approves carrying the knife while going to and from work, and you must carry the note with you at all times.

- You must not own, use, or have access to any weapon that is prohibited by the California Penal Code.

Source: California Department of Corrections and Rehabilitation, Division of Adult Parole Operations (2019).

sanctions. The position of parole revocations officer does not require formal legal training but instead simply requires that the officer know the laws and regulations involved with that state's parole system.

Regardless of whether the decision-making body consists of the parole board itself or a parole revocations officer, there are some situations where the offender may be entitled to some form of legal counsel. In ***Gagnon v. Scarpelli* (1971)**, it was held that parolees do have a limited right to counsel during revocation proceedings, as determined by the decision-making person or body and to be determined on a case-by-case basis. This is, of course, relevant only to those circumstances where the parolee contests the allegations

Gagnon v. Scarpelli **(1971):** Held that a parolee's sentence can only be revoked after a preliminary and final revocation hearing have been provided.

of the parole officer, and the retaining of counsel is done at the parolee's own expense; there is no obligation on the part of the state to provide such representation.

CONCLUSION

This chapter has provided students with a view of parole, the process by which offenders are allowed to leave prison before serving the entirety of their sentence. Parole has often been criticized due to the concern for public safety when inmates are released. Citizens around the nation read news reports of offenders who are released early from prison and commit heinous acts shortly after reentering society. This obviously makes it seem as if our justice system is being soft on criminals and prison authorities are indifferent to the safety of surrounding communities. However, prison authorities largely have their hands tied and must rely on the direction of the central state administrations that generate parole decisions. Prison overcrowding can lead and has led to legal complications as it can pose a violation of constitutional rights held by inmates in confinement. Thus, the only options are to build more prisons, house inmates in some other type of facility, or let them out on early release. Parole is one of the early release mechanisms used by some prison systems and thus has been likened to a release valve that opens when prisons are overstuffed with inmates. These types of "numbers game" release decisions are not safe for society and result in continued crime problems.

The correctional process must often contend with public concern and controversy. It is almost as if the correctional system can never meet the competing demands placed upon it by society. On one extreme is the desire to punish; on the other is the desire to reform. Amid this is the concern for the victim, which has become increasingly important to the field of corrections. What may be in store for the correctional system and its practitioners is a matter of debate, but it is certain that the challenges will never disappear.

PHOTO 11.5 Parole officer and supervisor Pearl Wise (middle) is pictured here with officers David Jackson (left) and Chris Miley (right) in tactical gear. As parole officers, they are required to complete in-service tactical training that includes nonlethal and lethal weapons proficiency. Ms. Wise is well known throughout her community as an active supporter of reentry efforts in Louisiana.

Robert Hanser

PRISON TOUR VIDEO: Reintegrating inmates into society is an enormous challenge for correctional practitioners. Go to the IEB to watch a clip about parole and reentry programs.

Want a Better Grade?

Get the tools you need to sharpen your study skills. Access practice quizzes, eFlashcards, video, and multimedia at **edge.sagepub.com/hanserbrief**

Interactive eBook

Visit the interactive eBook to watch SAGE premium videos. Learn more at **edge.sagepub.com/hanserbrief/ access**.

Career Video 11.1: Federal Agent

 Criminal Justice in Practice Video 11.1: Reentry Into the Community

 Prison Tour Video 11.1: Parole Decisions

 Prison Tour Video 11.2: Counseling as a Profession

 Prison Tour Video 11.3: Reentry Programs

Prison Tour Video 11.4: Drug Testing and Parole Issues

DISCUSSION QUESTIONS

Test your understanding of chapter content. Take the practice quiz at edge.sagepub.com/hanserbrief.

1. Identify and discuss the contributions of Alexander Maconochie to the development of parole.

2. Identify and discuss the contributions of Sir Walter Crofton to the development of parole.

3. What are some basic concepts regarding state parole, its organization, and its administration?

4. How does the parole selection process work, and what are the various factors that influence parole decisions?

5. What is meant by the "release valve" function of parole?

6. How is an effective reentry program a component of any effective crime prevention model through the reduction of recidivism?

7. Discuss Braithwaite's theory on crime, shame, and reintegration, and explain how it is related to the effectiveness of parole.

KEY TERMS

Review key terms with eFlashcards at edge.sagepub.com/hanserbrief.

Comprehensive Crime Control Act of 1984, 304

English Penal Servitude Act, 301

Father of Parole, 301

Institutional parole officer, 309

Judicial Improvements Act of 1990, 305

Mark system, 300

Offender Accountability Plan, 309

Offender reentry, 311

Parole, 299

Parole Commission and Reorganization Act, 304

Parole Commission Phaseout Act of 1996, 303

Parole revocations officer, 313

Release valve mechanism, 308

Restorative justice, 317

Roadmap to Reentry, 311

Supervised release, 304

Ticket of leave, 300

United States Parole Commission Extension Act of 2008, 305

KEY CASE

Gagnon v. Scarpelli (1971), 314

APPLIED EXERCISE 11.1

Students must conduct either a face-to-face or a phone interview with a parole officer or other parole specialist who currently works in a community corrections setting. Students should use the interview to gain the practitioner's insight and perspective on several key questions related to work in his or her field. Students must write the practitioner's responses, provide their own analysis of those responses, and submit their draft by the deadline set by their instructor. Students should complete this application exercise as an essay that addresses each point below. The total word count should be 1,400 to 2,100 words.

When completing the interview, students should ask the following questions:

1. What are the most rewarding aspects of your job in parole?
2. What are the most stressful aspects of your job in parole?
3. What is your view on treatment and reintegration efforts with offenders?
4. What are some challenges that you have in keeping track of your caseload?
5. Why did you choose to work in this field?
6. What type of training have you received for this line of work?
7. What would you recommend to someone who was interested in pursuing a similar career?

Students are required to provide contact information for the parole practitioner. While instructors will probably not need to contact this person, it may become necessary so that they can validate the actual completion of an interview.

Name and title of correctional supervisor: _____

Correctional agency: _____

Practitioner's phone number: _____

Practitioner's e-mail address: _____

Name of student: _____

WHAT WOULD YOU DO?

You are a state parole officer who has been active in various aspects of offender reentry. You currently work and live in a medium-size community. On occasion, your supervisor asks you to serve on community committees and advisory boards in order to increase partnerships in your area and extend the sources and abilities of your own agency. Recently, you have been asked to serve with a group of agencies, some state level, some county level, and many of them private or nonprofit in nature. This group is known as the Community Reentry Initiative (CRI). It has had very good success in creating employment opportunities for parolees in the community, and this has been a great help in reducing recidivism. This group has also had some success in obtaining affordable housing for offenders who do not have a place to stay.

The CRI has decided to add a **restorative justice** component to its efforts. This will require contact with prior victims of crime and will require their consent in participating in the process. This is likely to provide a challenging aspect to the project. However, all victims must be allowed to provide their input in the process as offenders are paroled into the community and integrated into the restorative justice process.

Restorative justice: Interventions that focus on restoring the community and the victim with involvement from the offender.

While at the meeting, it becomes clear that many people look to you as an expert on reentry issues. In fact, several members suggest that a subcommittee be created to begin the development of the restorative justice program, and they would like you to lead this subcommittee. However, there has been a recent backlash in the community against offender reentry initiatives. Some citizens have even gone to city hall to protest the implementation of these initiatives. Due to this, you are a bit uneasy with this responsibility, but you know that your supervisor would be disappointed if you did not agree to help with this task. Your supervisor is very progressive and is fond of saying, "Change is good, so let's have more good by making more change!" So with no real time to consider the implications, you hesitantly agree to accept the position as head of the subcommittee.

At this point, you want to help but do not know exactly what you should do. Your subcommittee consists of two local religious leaders, a police officer assigned to the neighborhood stabilization team, a victim's rights advocate from a local domestic violence facility, a low-ranking person from a local television station, a social services supervisor, a classification specialist who is employed by the regional prison, and a counselor from a local substance abuse treatment facility. Your parole agency supervisor encourages you to help this group and even offers to give you a half day off each week so that you can spend time supporting this initiative.

What would you do?

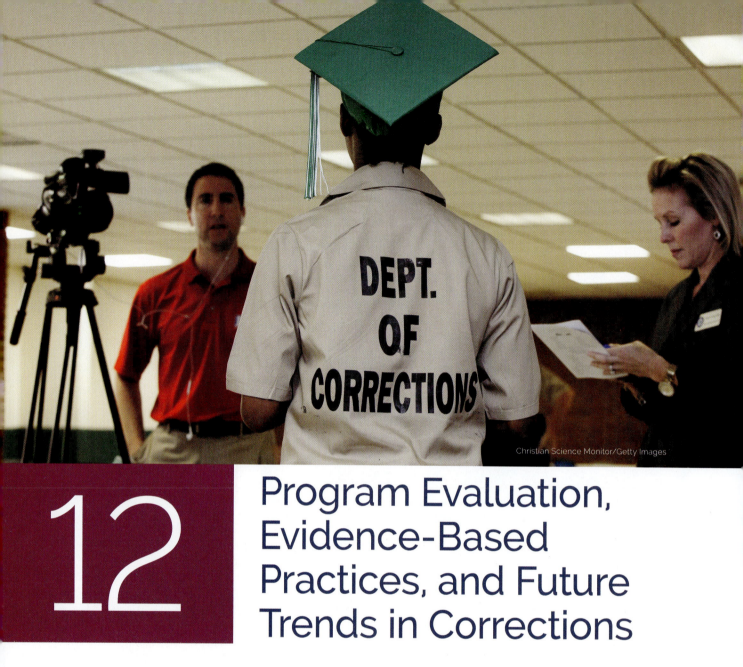

Christian Science Monitor/Getty Images

12

Program Evaluation, Evidence-Based Practices, and Future Trends in Corrections

Learning Objectives

1. Identify the process of evaluative research and distinguish between process and outcome measures.

2. Identify and discuss the use of evidence-based practices in corrections.

3. Be aware of likely future trends in correctional agencies.

Research-Based Funding Decisions for Rehabilitation in Thailand

The date was Monday, August 3, 2015, and Rob Hanser sat on the panel that had been gathered at the Grand Mercure Bangkok Fortune Hotel. Like the other correctional experts surrounding him, he had been invited to lend his expertise on correctional rehabilitation to about 80 professionals within the Kingdom of Thailand's national corrections service.

Thailand was in the process of revamping and restructuring its correctional system to accommodate and emphasize a rehabilitative orientation rather than one that was focused solely on security. Thus, the International Meeting on Offender Rehabilitation had been coordinated to bring experts from other parts of the world together with Thai correctional administrators to exchange ideas and to discuss matters related to rehabilitation.

The project under which this process was implemented had been named Through Care: A Model for Thailand's Coherent Rehabilitation by its two key researchers, Dr. Srisombat Chokprajakchat and Dr. Attapol Kuanliang. The two professors had been awarded a government grant to begin setting the groundwork for developing a rehabilitation model in the country.

The various administrators listened with interest to the panel members, who were from Japan, Singapore, the United Kingdom, and the United States. During his presentation, Hanser noted his involvement with various facets of correctional treatment but, in particular, showcased his work with Freedmen Inc., an organization that operated reentry homes for ex-offenders. After his presentation, the panel broke for coffee.

During the break, Hanser was approached by two people from a regional nongovernmental organization (NGO) that aided offenders in their reentry efforts. The director of the program presented some focused questions to Hanser.

"So, when you do reentry, does the government pay you to do this, or do they just give you permission to engage in the activity?" asked the director.

"Well, really they just partner with us in providing logistical support and access to facilities. We do work in collaboration with our probation and parole office, but, in the end, we are a self-funded program."

The director nodded. "It is the same with us here. It is important that if we are going to discuss rehabilitation, the government must make it a priority. . . . If they are not willing to give any money to this assistance, then I wonder if it is really a priority."

"Yes," Hanser replied, "I have said the very same thing in the United States. As you know, it is difficult to get the community to understand that if the offender is reformed, we all save trauma and expense by not having crimes in the future. The government understands this but will not act on it unless the public pressures them to do so. And the public will not pressure them unless it is educated on the importance of reentry programming and the need to fund such programs."

Both men shook their heads regretfully. The director then invited Hanser to come to his facilities to visit and to share ideas on reentry programming. Hanser thanked him, and both exchanged business cards, agreeing that Hanser would visit later that week.

Once the break ended, the Q & A portion of the panel began. The NGO director who had talked with Hanser during the break posed a question to one of the other members of the panel. He noted that his organization received marginal government funding but that it seemed as if programs from other countries that were being showcased often received government funding. He also pointed out that some of the countries represented at the meeting had correctional populations that were substantially smaller than Thailand's. He asked the panel member to comment on this and whether this was an important factor to consider when writing a reform bill for rehabilitation in Thailand.

The panel expert responded by acknowledging that her country's correctional population was indeed smaller than Thailand's. She noted that this did make the program much more manageable.

Then, a high-ranking administrator from the Thai correctional system rose and stated, "I know that it is true that we do not fund most of our reentry homes, and that is unfortunate. But currently money is not always available, and there are many pushes and pulls on the system."

The NGO reentry director shook his head and acknowledged that matters were complicated. He nonetheless reminded the administrator that NGOs and other partners were a critical component to reducing recidivism among released offenders in Thai society.

Hanser listened intently to the discussion and realized that most all of the challenges that he faced in the United States were being faced by correctional administrators in Thailand as well. This was, of course, the point to the conference—to compare practices, research, and outcomes of various programs so that the Thai could determine the best course of action for rehabilitation with their own correctional population.

As Hanser pondered this, he thought to himself, *It's a small world after all.*

INTRODUCTION

Effective research and evaluation of correctional programs is critical if we are to begin to understand what is likely to produce lasting change in offenders. If correctional systems do not take these important steps, they are resigning their efforts to chance. Mere chance is not acceptable, however, when huge amounts of money and human lives are at stake. Therefore, the question becomes, how do we effectively evaluate correctional programs, and how do we know that those programs are working? In response to this, we begin this chapter with an explanation of the specific function of evaluation research, which can be described as a form of explanatory research. We conclude the chapter with a discussion of trends likely to impact corrections in the future. This last aspect is based on prior research as well as observations of policy and decision making that have been touted among numerous correctional agencies around the nation.

EVALUATION RESEARCH

In the past several years, a massive effort on behalf of the U.S. government has been aimed at enhancing evaluation practices and services of the correctional system. Various documents have been published and placed in the public domain to help the administrators of community corrections programs better understand the impact of various treatment services. Much of the following information is borrowed from the Center for Substance Abuse Treatment (2005), a government agency responsible for implementing and evaluating many treatment programs that attempt to better serve offenders suffering from mental illness and co-occurring disorders.

Research and evaluation is a critical dimension of correctional programs. Evaluations are needed for program monitoring and for decision making by program staff, criminal justice administrators, and policymakers. Evaluations provide accountability, identify strengths and weaknesses, and provide a basis for program revision. In addition, evaluation reports are useful learning tools for others who are interested in developing effective programs. Many treatment programs in the criminal justice system have operated without evaluations for many years only to find out later that essential outcome data were needed to justify program continuation.

PHOTO 12.1 The government-sponsored International Symposium on Offender Rehabilitation was held in Bangkok, Thailand. At this conference, Robert Hanser provided input on rehabilitation programming in the United States.

Robert Hanser

Conducting an adequate evaluation requires one to clearly formulate the treatment model, reasonable program goals, and specific objectives related to client needs. General goals must be translated into measurable outcomes. The evaluator generally works closely with program administrators to translate the evaluation guidelines into operational components. In essence, scientific principles for conducting research should be carefully adhered to in order to enhance the viability of findings.

There are three basic types of evaluation: implementation, process, and outcome. An important note before we discuss these components is that although implementation and process evaluations can begin when the program is initiated, outcome evaluation should not begin until the program has been fully implemented. Outcome evaluations are generally more costly than other types of evaluation and are warranted for programs of longer duration that are aimed at modifying lifestyles (such as therapeutic communities), rather than drug education interventions that are less intensive and less likely to produce long-term effects.

Implementation Evaluation

While programs often look promising in the proposal stage, many do not succeed as planned in the security-oriented correctional environment. Sometimes programs are too rigidly implemented and adjustments are not made for the realities of community corrections; this often renders these programs less effective. **Implementation evaluation** is aimed at identifying both complications and accomplishments during the early phases of program development in order to provide helpful feedback to clinical and administrative staff. Such evaluations involve informal and formal interviews with correctional administrators, staff, and offenders to ascertain their degree of satisfaction with the program and their perceptions of any problems.

Implementation evaluation: Identifying problems and accomplishments during the early phases of program development for feedback to clinical and administrative staff.

Process Evaluation

Traditionally, **process evaluation** refers to the assessment of the effects of a program on clients while they are in the program; this makes it possible to assess the institution's intermediary goals. Process evaluation involves analyzing records related to the following:

Process evaluation: Traditionally refers to assessment of the effects of the program on clients while they are in the program, making it possible to assess the institution's intermediary goals.

1. Type and amount of services provided

2. Attendance and participation in group meetings

3. Number of offenders who are screened, admitted, reviewed, and discharged

4. Percentage of offenders who favorably complete treatment each month

5. Percentage of offenders who have infractions or rule violations

6. Number of offenders who test positive for substances (this can be compared to urinalysis results for the general prison population)

Effective programs produce positive client changes. These changes initially occur during participation in the program and ideally continue upon release into the community. The areas of potential client change that should be assessed include the following:

1. Cognitive understanding (e.g., mastery of program curriculum)

2. Emotional functioning (e.g., anxiety and depression)

3. Attitudes/values (e.g., honesty, responsibility, and concern for others)

4. Education and vocational training progress (e.g., achievement tests)

5. Behavior (e.g., rule infractions and urinalysis results)

Within institutional corrections, it is also important to evaluate program impact on the host prison facility itself. Well-run treatment programs often generate an array of positive developments affecting the morale and functioning of the entire inmate population. Areas to examine include the following:

1. *Offender behavior.* Review the number of rule infractions, the cost of hearings, court litigation expenses, and inmate cooperation in general prison operations.

2. *Staff functioning.* Assess stress levels, which may become manifest in the number of sick days taken and the rate of staff turnover. Generally, the better the program, the lower the stress and the better the attendance, the involvement, and the commitment of staff.

3. *Physical plant.* Examine the physical properties of the program. Assess general vandalism apparent in terms of damage to furniture or windows as well as the presence of graffiti. Assess structural damage to walls and plumbing, for example.

Outcome Evaluation

Outcome evaluation:
Involves quantitative research aimed at assessing the impact of the program on long-term treatment outcomes.

Outcome evaluations are more ambitious and expensive than implementation or process evaluations. An **outcome evaluation** involves quantitative research aimed at assessing the impact of a program on long-term treatment outcomes. Such evaluations are usually carefully designed studies that compare outcomes for a treatment group with outcomes for other, less intensive treatments or a no-treatment control group (i.e., a sample of offenders who meet the program admission criteria but who do not receive treatment). These evaluations involve complex statistical analyses and sophisticated report preparation.

Follow-up data (e.g., drug relapse, recidivism, employment status) are the heart of outcome evaluation. Follow-up data can be collected from criminal justice records and face-to-face interviews with individuals who participated in certain programs. Studies that use agency records are less expensive than those that involve locating participants and conducting follow-up interviews. Outcome evaluations can include cost-effectiveness and cost–benefit information that is important to policymakers. Because outcome research usually involves a relatively large investment of time and money as well as the cooperation of a variety of people and agencies, it must be carefully planned. A research design may be very simple and easy to implement, or it may be more complex. In the case of more complex studies, it is usually advisable to enlist the assistance of an experienced researcher.

Program Quality and Staffing Quality

In addition to outcome and process measures, there are a number of other areas that agencies may wish to evaluate. These other areas may or may not require the input of the offender, and they may or may not be dependent upon the offender population's outcome results. One example of this is when agencies wish to assess the quality of their program, their staff, or their curricula. Each of these three components is very important and may require more than simple outcome evaluation measures. In some cases, such as with program curricula, it may be necessary to examine the general process measures of a program as a whole. It is important when agencies evaluate curricula that they keep this aspect of a program separate and distinct from the effects that staff may have upon the process. Staff members may modify the general process through their own therapeutic slant or means of implementing aspects of a job requirement. In other words, the

FOCUS TOPIC 12.1

Commonly Used Measures of Reentry Program Performance

Process Measures

Substance abuse treatment services received

Employment services received

Housing assistance received

Family intervention and parent training received

Health and mental health services received

Outcome Measures

Rearrest rates

Reincarceration rates

Proportion employed

Rates of drug relapse

Frequency and severity of offenses

Proportion self-sufficient

Participation in self-improvement programs ●

Source: Bureau of Justice Assistance, Center for Program Evaluation. (2007). *Reporting and using evaluation results.* Washington, DC: Author.

individual preferences of different persons employed in the agency may not be what you hope to observe in a curricular assessment; rather, it is the uniform and written procedures that are of interest.

Therefore, it is clear that evaluations can be quite complex and detailed, depending on the approach taken by the agency. The key to an effective and ethical evaluation is evaluative transparency. **Evaluative transparency** is when an agency's evaluative process allows for an outside person (whether an auditor, an evaluator, or the public at large) to have full view of the agency's operations, budgeting, policies, procedures, and outcomes. In transparent agencies, there are no secrets, and confidential information is only authorized when ethical or legal requirements mandate that the information not be transparent. In some cases, information should *not* be available to the general public; such information may include a client's treatment files or a victim's personal identity. In such cases, the intent is a benevolent safeguarding of the client's welfare, not the agency's own welfare.

If one is to evaluate the quality of a program, it stands to reason that the program must be transparent to the evaluator who is tasked with observing that program. Agencies that seek to meet high ethical standards must be transparent. This is a core requisite to ensuring the quality of the program that is implemented. Further, programs of quality are accountable to the public, which is, in part, an element of transparency. Public accountability is a matter of good ethical standing, and this is consistent with the reason that ethical safeguards are put into place—to protect the public consumer. In the case of corrections agencies, the product that is "sold" to the public is community safety, and it is the obligation of the agency to be accountable and transparent to the public when providing this product to its jurisdiction. Thus, the quality of the program should be measured by its ability to deliver ethical, open, and honest services that hold community safety as paramount.

With regard to staffing quality, agencies should also make a point to evaluate the support that they provide to their staff. Naturally, recruitment and hiring standards should be evaluated routinely, and it is also important that agencies examine their own support services for staff. Some examples of necessary support for staff include the existence of an effective human resources division, sufficient budgeting for equipment to effectively do one's job, and the nature of the job design, particularly with regard to caseload. As one might guess, this is also related to the overall quality of the program.

Evaluative transparency: When an agency's evaluative process allows an outside person to have full view of the agency's operations, budgeting, policies, procedures, and outcomes.

Quite naturally, agencies should evaluate their hiring standards and should examine factors such as the number of complaints generated by the community regarding staff functioning. Grievances made by offenders can also be examined if it should turn out that there is some legitimacy to them. Employee standards of conduct are important, as are incidents where employees do not meet standards that are expected by the agency. Evaluators should consult staff as to whether the staff feel prepared for their jobs or whether they consider their work environment to be on par with other agencies. All of this staff-related information provides a richer analysis of agency operations and adds transparency to the day-to-day routines that occur.

Feedback Loops and Continual Improvement

In evaluating correctional agencies, it is important that the information obtained from the evaluation serve some useful purpose. The Bureau of Justice Assistance Center for Program Evaluation (BJA; 2007) elaborates on the need for evaluations to be constructed in a manner that is useful to the stakeholders of the evaluation. **Stakeholders** in corrections evaluations include the agency personnel, the community in which the agency is located, and even the offender population that is being supervised. According to the BJA, it is important for evaluators to be clear on what agency administrators wish to evaluate, and it is also important that evaluators ensure that administrators understand that evaluative efforts are to remain objective and unbiased in nature.

Stakeholders: Agency personnel, the surrounding community, and the offender population.

It is also important that evaluators provide recommendations for agencies based on the outcome of the evaluation (BJA, 2007). It is through the use of these recommendations that agencies can improve their overall services and enhance goal-setting strategies in the future. Indeed, evaluation information can be a powerful tool for a variety of stakeholders (BJA, 2007). Program managers can use the information to make changes in their programs that will enhance their effectiveness (BJA, 2007). Decision makers can ensure that they are funding effective programs. Other authorities can ensure that programs are developed as intended and have sufficient resources to implement activities and meet their goals and objectives (BJA, 2007).

Learning organizations: Have the inherent ability to adapt and change, improving performance through continual revision of goals, objectives, policies, and procedures.

Agencies that are adept at implementing evaluative information and recommendations are sometimes referred to as learning organizations. **Learning organizations** have the inherent ability to adapt and change, improving performance through continual revision of goals, objectives, policies, and procedures. Throughout this process, learning organizations respond to the various pushes and pulls that are placed upon them by utilizing a continual process of data-driven, cyclical, and responsive decision making that results in heightened adaptability of the organization. The ideal community corrections agency is a learning organization—one that can adjust to outside community needs and challenges as well as internal personnel and resource challenges.

Lastly, in its ideal state, evaluation is an essential component in the process of program planning, goal setting, and modification and improvement. The BJA (2007) notes that evaluation findings can be used to revise policies, activities, goals, and objectives (see Focus Topic 12.2) so that community supervision agencies can provide the best possible services to the community to which they are accountable.

This is to demonstrate the importance of policymaking as well as the setting of goals and objectives that guide a community supervision agency into the future. This cyclic pattern of going from assessment to implementation to evaluation demonstrates a continual circle of development that uses past data to better face future challenges. This is the most effective means of utilizing real-world research to tailor programs that can meet the challenges within a jurisdiction. With this in mind, we once again look to the work of Van Keulen (1988), who roughly 20 years ago noted that

goals and objectives also play a critical role in evaluation by providing a standard against which to measure the program's success. If the purpose of the program is to serve as an alternative to jail, the number of jail-bound offenders the program serves would be analyzed. If the program's focus is to provide labor to community agencies, the number of hours worked by offenders would be examined. Last, having a statement of goals and objectives will enhance your program's credibility by showing that careful thought has been given to what you are doing. (p. 1)

Van Keulen (1988) demonstrates the reasons why clarity in the definition and purpose behind a community corrections program is important. Clearly articulated goals not only help to crystallize the agency's philosophical orientation on the supervision process but also provide for more measurable constructs that lend themselves to effective evaluation. Clarity in program goals and objectives allows the agency to perform evaluative research to determine if its efforts are actually successful or if they are in need of improvement. Such clarity then facilitates the ability of the agency to come "full circle" as the planning, implementation, evaluation, and refinement phases of agency operations unfold.

Community Harm With Ineffective Programs, Separating Politics From Science in the Evaluative Process

As we near the close of this text, it is important to reflect on the potential consequences that might be incurred if agencies are allowed to operate ineffectively. Public and institutional safety and security constitute the top priority for correctional agencies. However, one of the best long-term approaches to improving public safety is the use of effective reintegration efforts. Thus, programs that fail to adequately oversee offenders on community supervision run the risk of allowing the community to be harmed. Likewise, programs that fail to implement effective treatment approaches also put the community at risk.

Multiple methods of supervising offenders have been provided throughout this text. All of these supervision interventions—jails and prisons, residential treatment facilities, probation, intermediate sanctions, parole, and even the death penalty—should be implemented so that the intervention chosen best fits the offender and his or her likely level of risk. This comes back to one of the most critical aspects of corrections, the assessment and classification process. This process that sets the stage for determining how an offender is housed and supervised. It is also at this point where correctional resources can be optimized by ensuring the best fit between resources and offender risk as well as offender treatment and needs.

The evaluation of correctional agencies is directly tied to the assessment component that occurs as the offender is first processed. Indeed, the assessment of the offender typically serves as the baseline measure when examining evaluation processes. Both process and outcome evaluations tend to examine data from the initial assessment

---

FOCUS TOPIC 12.2

What Are Policies, Activities, Goals, and Objectives?

Policy: A governing principle pertaining to goals, objectives, and/or activities. It is a decision on an issue not resolved on the basis of facts and logic only. For example, the policy of expediting drug cases in the courts might be adopted as a basis for reducing the average number of days from arraignment to disposition.

Activities: Services or functions carried out by a program (i.e., what the program does). For example, treatment programs may screen clients at intake, complete placement assessments, provide counseling to clients, and so on.

Goals: A desired state of affairs that outlines the ultimate purpose of a program. This is the end toward which program efforts are directed. For example, the goal of many criminal justice programs is a reduction in criminal activity.

Objectives: Specific results or effects of a program's activities that must be achieved in pursuing the program's ultimate goals. For example, a treatment program may expect to change offender attitudes (objective) in order to ultimately reduce recidivism (goal). ●

Source: Bureau of Justice Assistance Center for Program Evaluation. (2007). *Reporting and using evaluation results.* Washington, DC: Author.

against the data received when the offender exits a particular program or sentencing scheme. It is in this manner that the evaluation of correctional supervision programs serves to reinforce the initial assessment process. The initial assessment and classification process will be considered effective if at the end of the offender's involvement in a given supervision program, the evaluation of the program demonstrates that the offender is indeed less likely to recidivate, particularly if this likelihood falls below that experienced at other agencies in the area and throughout the country. Thus, the evaluation process is a feedback loop into the initial assessment process, demonstrating to agencies that their programs are (or are not) working. If a program is found to be in need of improvement, evaluators can then determine if this is due to the initial assessment or to some process issue further within the program's service delivery. Checking the initial assessment and ensuring that this process is adequate follows a "garbage in, garbage out" philosophy.

Assessment-evaluation cycle: Comparing assessment and evaluation data to determine the effectiveness of programs and to find areas for improvement.

This process in which an agency is constantly assessing and evaluating itself is known as the **assessment-evaluation cycle**. This is the process whereby assessment data and evaluation data are compared to determine the effectiveness of programs and to find areas where improvement of agency services is required. Agencies that successfully implement the assessment-evaluation cycle tend to use public resources more effectively and are also less prone to placing the community at risk of future criminal activity. On the other hand, agencies that do not successfully implement the assessment-evaluation cycle will be more likely to waste agency resources and place the community at a level of risk that otherwise would be preventable.

EVIDENCE-BASED PRACTICE

Evidence-based practice is a significant trend throughout all human services that emphasizes outcomes. Interventions within community corrections are considered effective when they reduce offender risk and subsequent recidivism and therefore make a positive long-term contribution to public safety. In this section of the chapter, students are presented with a model or framework based on a set of principles for effective offender interventions within state, local, or private correctional systems. **Evidence-based practice (EBP)** implies that (1) one outcome is desired over others, (2) the outcome is measurable, and (3) the outcome is defined according to practical realities (e.g., public safety) rather than immeasurable moral or value-oriented standards (Colorado Division of Criminal Justice, 2007). Thus, EBP is appropriate for scientific exploration within any human service discipline, including the discipline of corrections.

Evidence-based practice (EBP): A significant trend throughout all human services that emphasizes outcomes.

Research Evaluation for Effectiveness of Evidence-Based Practice

Too often programs or practices are promoted as having research support without any regard for the quality of the research support or the research methods that were employed. Consequently, a research support pyramid (see Figure 12.1) has been included that shows how research support for evidence-based practices might be conducted or implemented.

The highest-quality research support depicted in this schema (gold level) reflects interventions and practices that have been evaluated with experimental/control design and with multiple site replications that concluded significant sustained reductions in recidivism were associated with the intervention. The criteria for the next levels of support progressively decrease in terms of research rigor requirements (silver and bronze), but all the top three levels require that a preponderance of the evidence supports a program's effectiveness. The next rung lower in support (iron) is reserved for programs that have inconclusive support regarding their efficacy. Finally, the lowest designation (dirt) is reserved for those programs that have been subjected to research (utilizing methods and criteria associated with gold and silver levels) but the findings were negative and the programs were determined to be not effective (National Institute of Justice, 2005).

FIGURE 12.1

Research Support Pyramid for Evidence-Based Practice Implementation

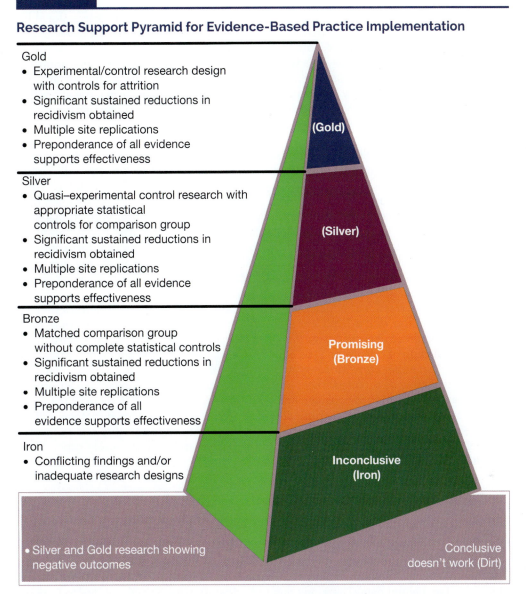

Gold
- Experimental/control research design with controls for attrition
- Significant sustained reductions in recidivism obtained
- Multiple site replications
- Preponderance of all evidence supports effectiveness

Silver
- Quasi–experimental control research with appropriate statistical controls for comparison group
- Significant sustained reductions in recidivism obtained
- Multiple site replications
- Preponderance of all evidence supports effectiveness

Bronze
- Matched comparison group without complete statistical controls
- Significant sustained reductions in recidivism obtained
- Multiple site replications
- Preponderance of all evidence supports effectiveness

Iron
- Conflicting findings and/or inadequate research designs

(Gold)

(Silver)

Promising (Bronze)

Inconclusive (Iron)

- Silver and Gold research showing negative outcomes

Conclusive doesn't work (Dirt)

Source: National Institute of Justice. (2005). *Implementing evidence based practice in corrections.* Washington, DC: U.S. Department of Justice.

THE FUTURE OF CORRECTIONS

It is fitting that this chapter concludes with a section on future trends since it is through the use of the evaluative research discussed so far that we predict future trends in correctional agencies throughout the nation. It is on this note, and with much trepidation, that some speculative predictions regarding future trends in community corrections will be provided. Before doing so, it should be noted that many psychologists, particularly behavioral psychologists, claim that the best predictor of future human behavior is past human behavior. In fact, empirical research has shown this to be generally true. Keeping this in mind, we will also look at prior research from the field of corrections and make some general observations and predictions.

First, it should be noted that this text has been addressing three key trends in corrections that are likely to continue in the future. These trends are (1) the tendency for technology to have more impact on correctional agency operations, (2) the need for continued training of correctional staff, and (3) an increased emphasis on the reentry process for offenders. These trends are likely to continue for a variety of practical rather than theoretical reasons. Evidence of this already exists throughout the correctional literature, and, as this text clearly demonstrates, agencies around the nation have already been incorporating these themes into their future operations.

The use of technology is already commonplace within the correctional industry. However, it is the effectiveness of technology that will determine its usefulness and if it is worth the cost. Technology is becoming increasingly reliable and also more affordable for state agencies. Technology can also save on the amount spent on human resources. Thus, technology will continue to be a driving force that will shape the landscape of correctional services in the future.

Regardless of the role that technology will play in corrections, there will still be a need for trained and educated staff. In fact, the proliferation of technology will require that persons who work in corrections be competent enough to utilize high-tech tools of the trade. No longer will it be sufficient for staff to act as mere turnkeys; rather, they will be required to use sophisticated equipment and tools. In addition, correctional populations are becoming more complex in terms of mental and medical issues as well as cultural factors that must be considered. Due to legal requirements and concerns with liability, staff cannot be negligent or careless. Rather, they must act competently as trained professionals when working with the inmate population—a population that can be deceptive and dangerous.

Lastly, the correctional population will continue to grow, and our nation's prisons are full. Continuing with a "stuff 'em and cuff 'em" mentality will simply not work for a system that is already over capacity. If something cannot be done to alleviate the continued overcrowding of prisons and the already overloaded case management approaches that exist, then either correctional systems will need much larger budgets or offenders will have to simply be set free without any form of supervision. Obviously, the latter possibility is too dangerous to consider, and the former will just result in more of the same problems that already face the criminal justice system. Thus, reentry initiatives will be used with much greater frequency in the future, due both to practicality and economic concerns. With this in mind, a discussion of the likely trends in community supervision programs will follow.

A Continued Emphasis on Reentry—Until the Backlash Pendulum Swings the Other Direction

For at least the next 5 to 10 years, we are very likely to see a continued emphasis on the development of reentry programs. While this may seem like good news, there is an underlying truth to this issue—many state correctional systems are unable to afford keeping offenders behind bars. Therefore, due to budgetary reasons and not due to humanitarian or public safety reasons, we are likely to see a continued emphasis on reentry. While this will fix state budgetary concerns in the short term, it will likely only exacerbate crime problems and response issues in the future.

With this said, it is likely that within another 5 years or so, we will then see a backlash where society again supports more vigorous enforcement and more punitive reactions to crime. However, it is unlikely that this will be effective or that it will remedy problems. Rather, because the circumstances have gone so far and because crime and offending have become such a common feature of society, the final outcome will simply be more offenders and more challenges. Essentially, orange is the new black and, given this, we will likely not see black return with the same formality that it once had, socially.

Privatization in Corrections

The use of privatized operations in corrections carries some degree of controversy. This is largely because some people have a moralistic opposition to the idea that money might be made off the misery of others. In other words, some people may take exception to the idea that in order for one person to fulfill his or her sentence, the pockets of a private corporation must be lined. However, it is probably a more realistic view to understand that inmates must be housed *somewhere* when serving a stint of incarceration, and, so long as their constitutionally protected rights are upheld, the specific ownership of the physical facility is irrelevant. In fact, if the private prison company can maintain security with less overhead to the taxpayers, then this is a good thing for everyone concerned.

The simple reality is that, like it or not, private corrections is here to stay, and the use of private facilities is growing steadily each year. Private companies must ensure that they stay within budget and operate in an efficient and productive manner; to do otherwise results in a loss of revenue. For many business-minded administrators, this is superior to many state systems that are full of inherent waste, red tape, and civil service regulations that paralyze agencies from moving in an adaptive fashion. Private correctional systems, however, adhere to all of the same legal requirements as state institutions but are able to adapt to changes much more quickly and with little need for excessive staggering of new policies and procedures. For the most part, executive-level decision makers can generate a memo or set of protocols and disseminate it in a day, if they so desire; state systems do not have this flexibility.

Also, one must keep in mind that it is the state correctional systems that have had legal problems related to the constitutionality of their operations, not the private systems. It appears then that state systems are not superior and, in some cases, may be more lacking in legal integrity than privately run operations. Thus, state systems may provide worse services than private systems (in some cases) while still being more costly to taxpayers. Also consider that in Texas and the Federal Bureau of Prisons, the vocational or industry operations of these prison systems that produce revenue do so through private corporations (PIE in Texas, UNICOR for the Bureau of Prisons). The point in mentioning this is that it may well be that these types of operations are most appropriate for profit-generating endeavors, and such programs are generally viewed as successful in operation.

In situations where states may be reluctant to resort to complete privatization of industrial or security operations, hybrid versions of privatization may be utilized. The author of this text works as a substance abuse counselor in a private prison (Richwood Correctional Center, owned and operated by LaSalle Corrections) that houses inmates from the Louisiana Department of Corrections. Counselors in the treatment program are hired by either the state or by LaSalle Corrections, depending on budgetary constraints and offender-client population levels. This integration of state and private service delivery allows for services to continue at optimal levels beyond what the state could afford to provide on its own.

The use of private correctional facilities has grown more in some states than in others, but regardless of how common privatization becomes, it is doubtful that state systems will ever completely disappear. States will likely always maintain at least a fraction of facilities since it is unlikely private systems will be appropriate for housing the most dangerous of inmates. But with budgets being reduced throughout the nation, it is unlikely that many states will be able to fund their correctional systems to the level that is necessary; other alternatives must be considered. Thus, while private prisons will probably never completely replace state prisons, it is likely that private facilities will become much more common than they currently are.

Increased Use of Technology

The world as a whole has become more technologically driven, and this has, of course, impacted correctional operations. As time goes on, correctional administrators are relying more and more on technological innovations to aid them in matters of security as well as with peripheral operations such as visitation, food service, or telemedicine services. As this technology continues to develop, there will be a need to assess and evaluate the effectiveness of these innovations. This points to a continued need for future research in corrections as well.

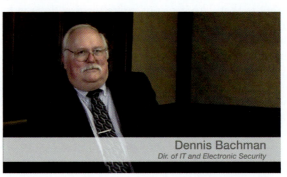

Dennis Bachman
Dir. of IT and Electronic Security

Further, technology will continue to be an issue of security concern, both in terms of the dangers that it presents as well as the benefits that is provides. For example, the use of drones has provided a means for persons on the outside to drop contraband into prison facilities so that inmates can obtain the contraband on the inside. Easy-to-hide cell phones are smuggled into prisons on a routine basis; this item of contraband was unheard of 15 to 20 years ago. Each of these forms of technology has changed the scope of searches and means of fighting the introduction of contraband into the facility. In addition, facilities are finding the use of cell phones for transportation crews to be useful as inmates are transported between facilities. Likewise, drones provide a means of observing inmate actions when sending a person into areas that might prove dangerous (such as during a riot). Thus, these forms of technology have their useful applications for correctional officials.

While it is not clear in most cases whether there are likely drawbacks to relying so heavily on technology, it is clear that facilities are becoming more sophisticated in their operations. This means that many people working in the correctional field will need to develop some degree of technical savvy lest they be unable to utilize basic equipment or security system innovations. Throughout this text, it has been shown that technology can assist in aiding security, improving living standards, enabling safe working conditions, and safeguarding against liability. For all of these reasons and more, the world of corrections will become more technologically driven.

Standards and Accreditation

Over time, an abundance of professional organizations and associations have emerged to represent a wide variety of professions in the field of corrections. This is a positive trend in corrections that will likely continue with increased enthusiasm. The desire to professionalize the workforce is strong since this can add legitimacy to a facility's operations. In addition, when employees adhere to the requirements of various governing bodies, it is presumed that they are at least meeting the nationally accepted standards of competence within these associations.

In addition, when employees meet the standards of external governing bodies, agencies are unlikely to be open to lawsuits for failing to properly train their staff or other types of negligence. This is an effective incentive for most agencies, including private ones. Further, many individual practitioners will seek to obtain credentials that may help them to receive promotions or higher levels of pay. Thus, the incentives for systems and facilities to improve standards and credentials are strong, and this translates to a desire among agencies to obtain accreditation. When agencies seek accreditation, they are making strides to enhance their reputation and to make a public statement that demonstrates their integrity.

Accreditation speaks volumes to the agency's professionalism and sense of ethical responsibility. The transparency of management necessary to obtain and maintain accreditation means that administrators have nothing to hide. This is of benefit when one considers the closed nature of corrections from the broader community. A correctional agency's image within the community is likely to be enhanced when it is accredited, and this is likely to generate community support for correctional personnel. As we have seen in prior chapters, this can benefit the administrators and staff who are the "face" of the agency for which they work. It can be a morale booster for employees and generate a culture of pride in the mission and work of the agency.

PHOTO 12.2 This treatment group participates in a discussion of a controversial issue.

Robert Hanser

An Emphasis on Cultural Competence Will Continue to Be Important

Earlier chapters of this text addressed cultural competence and the increasing representation of minority offenders in the criminal justice system. It is clear that the United States is becoming more racially and culturally diverse. Thus, multilingual skills and knowledge of different religious beliefs, lifestyle orientations, and other matters of diversity will become mandatory for future corrections employees. The continued diversity of inmates, along with other challenges they present (such as mental health needs), means that offender populations are likely to become much more complicated to manage in the future. Naturally, this can add to the already stressful conditions under which correctional staff work, so agencies will need to mitigate these challenges and the stress associated with them.

An Emphasis Will Be Placed on Employment Programs

Research has shown that offenders need jobs if they are to be able to make ends meet while paying restitution, fines, and other obligations to society. Chapter 10 of this text has provided sufficient evidence to demonstrate that recidivism among offenders will decline if they are given the minimal educational and vocational skills necessary to function. Programs such as Project RIO in Texas are proving vocational programs that aid in job placement and are quite effective in reducing recidivism. Also, since employers are able to secure tax incentives for hiring offenders, these programs benefit society as a whole. These programs ensure that society is compensated for property crimes and can often repair damage resulting from nonviolent crimes when used within a restorative justice or victim compensation framework. This is important because the victim, society, and the offender all benefit from such programs. Thus, vocational training will be critical, and the use of work-release and restitution programs will continue to be necessary.

Processing of Geriatric Offenders Will Be Shifted to Community Supervision Schemes

The entire population is graying in the United States, and the nation's prison population is graying at a faster rate than is the general society. This issue has been given

PHOTO 12.3 This elderly inmate has equipment for breathing assistance in his cell. Such accommodations are becoming increasingly necessary with the graying of the prison population.

Robert Galbraith/Reuters/ Newscom

considerable attention in recent years, with Texas, California, Florida, New York, and Louisiana all experiencing a rise in per capita elderly inmates who are incarcerated. As noted previously, the states just mentioned have either one of the largest prison populations or one of the highest rates of incarceration in the country. The costs associated with the elderly inmate are exponentially higher than those associated with the average inmate. This leads administrators to look for possible solutions to this problem, such as the possibility of early release of inmates who are expected to die or the implementation of human caregiver programs like hospice. However, accountability to the public places prison administrators in a dilemma since public safety is the primary concern for all custodial programs. The sobering reality is that society will, one way or another, pay for the expense of keeping elderly inmates.

It is with this point in mind that state-level correctional systems will need to increase their use of community supervision programs with elderly offenders, including those who are chronically ill. This may seem to be an oversimplified recommendation, but it is one that has not truly been implemented by many states. Most states do have programs designed for the early release of elderly inmates, but these programs are not used extensively. The recommendation here is that community supervision be *automatically implemented* when an inmate reaches the age of 60 years old, unless he or she is a pedophile. In the case of pedophiles, the typical risk assessment methods should remain intact since these offenders have such poor prognoses for reform. However, all other elderly inmates should be automatically placed on community supervision since this would reduce costs of upkeep significantly.

Those inmates being automatically released should be given intensive supervision that uses the latest and greatest forms of human supervision and electronic gadgetry. Though this adds to the cost, the outcome is much cheaper than prison alternatives. Through routine weekly contact by probation or parole officials, frequent electronic phone monitoring, the use of GPS tracking devices, house arrest, and other such innovations, the risk to public safety can be greatly minimized. Further, it is a simple fact that recidivism for elderly offenders is very low, and the crimes committed are seldom assaultive in nature. Thus public safety is not compromised. In fact, this is the safest population to place on community supervision, as long as one excludes using such a policy for pedophiles.

Official News Sources, Social Media, and Correctional Operations

Certain groups of offenders, such as sex offenders, have drawn a great deal of public attention and concern. This is understandable, but the public may make erroneous conclusions regarding the offender population if they are not given correct information. This can completely undermine a prison's ability to implement an effective reentry program and cast negative views on the prison if the public perceives that it is just letting offenders out so they can commit crime again. **Media effects** are an important consideration in corrections and refer to the effects that the media have upon the public perception of prison or community supervision programs. How the media report on specific incidents can affect this. When programs are successful or innovative, the media can provide effective coverage of these as well to ensure that the public is getting the most accurate information possible.

In addition, administrators must now be aware of the online community and its impact on public perceptions. Indeed, a simple Google search of many prisons in the

Media effects: The effect that the media has upon the public perception of prison or community supervision programs.

United States will reveal that there is often a blog or other form of online communication that exists among persons who have experiences with that facility. This **prison social media**, in most cases, consists of information regarding visitation, sending necessities to loved ones who are locked up, or other concerns. However, in some instances, different forms of clandestine communications may be included, such as coded messages or communications between gang members outside the prison regarding issues that occur, or will occur, within the prison. In other cases, there is sometimes libelous and embarrassing commentary that appears on these and other sites. This creates problems for administrators because accusation can be made by anonymous individuals to which the administration cannot reply or defend themselves. Further, the online reporting process has nibbled away at the isolation of the prison environment, to some extent. Nowadays, prison operations and incidents that occur therein are disseminated amid the public much more rapidly. Though illegal, some persons may use cell phones and take photos or record incidents within the facility and then post them online for the entire online community to see and hear. This is a much different public image scenario than existed for prison administrators 25 to 30 years ago, and it will continue to be a key source of concern for administrators in the future.

PHOTO 12.4 The use of video equipment to facilitate visitation between inmates and their family has become more common in prison facilities.

AP Photo/David J. Phillip

Prison social media:
Online dialogue that consists of information regarding visitation, rules of the correctional facility, or sending necessities to loved ones in prison.

CONCLUSION

Research and assessment of correctional programs are vital in this day and age. Through this process, we are able to identify program strengths and weaknesses that serve to inform the literature. The ultimate question that should guide research and assessment projects is "So what?" In other words, if we choose to conduct a research project, will the results provide a meaningful contribution to what is known about some phenomena? Or, if we did *not* conduct the research project, would there continue to be a significant gap in the literature, hindering our ability to make optimal decisions regarding community corrections programs?

It is important to keep in mind that the assessment process is an integral component of the evaluation of any agency. This means that what agencies put into their program will be reflected in its final output. Thus, assessment can be seen as a measure of what goes into a program, and evaluation can be seen as a measure of what comes out of the program. The two work hand-in-hand. Because of this, students should be familiar with the assessment-evaluation cycle, since this is the primary means by which community supervision agencies measure their performance and since this is what ultimately determines if an agency is meeting its goals and objectives. For agencies with unfavorable evaluations, an examination of the policies and activities of that agency may be in order, or, in some cases, a reassessment of the goals and objectives may take place.

Lastly, this chapter reflected on some of the themes that have emerged throughout this text. These themes are important to contemporary correctional practices and will also be a focus for corrections agencies in the future. A number of likely trends were noted, and recommendations were given for agencies that will face challenges that loom on the horizon. The field of corrections is both dynamic and demanding; practitioners who work in this field have their work cut out for them. However, the work of correctional personnel is critical and warrants support from the various funding sources as well as the public at large. Without this support, the entire public is likely to pay dearly for such negligence.

Want a Better Grade?

Get the tools you need to sharpen your study skills. Access practice quizzes, eFlashcards, video, and multimedia at **edge.sagepub.com/hanserbrief**

Interactive eBook

Visit the interactive eBook to watch SAGE premium videos. Learn more at **edge.sagepub.com/hanserbrief/ access**.

 Career Video 12.1: Director of IT and Electronic Security

 Prison Tour Video 12.1: Getting Out and Performance Grid

 Prison Tour Video 12.2: Return Rate and Rules

DISCUSSION QUESTIONS

Test your understanding of chapter content. Take the practice quiz at edge.sagepub.com/hanserbrief.

1. How is evaluative research conducted, and why is it important?

2. Compare process and outcome measures. Why is each important to evaluation projects?

3. Explain, in your own words, how the assessment-evaluation cycle works in correctional research.

4. What is meant by the term *evidence-based practice*? Provide at least two examples of how such a practice might be used in a correctional agency.

5. Why does the author note that more emphasis on diversity and multiculturalism will be important in the field of corrections?

KEY TERMS

Review key terms with eFlashcards at edge.sagepub.com/hanserbrief.

Assessment-evaluation cycle, 326

Evaluative transparency, 323

Evidence-based practice (EBP), 326

Implementation evaluation, 321

Learning organizations, 324

Media effects, 333

Outcome evaluation, 322

Prison social media, 333

Process evaluation, 321

Stakeholders, 324

APPLIED EXERCISE 12.1

For this exercise, you will need to consider what you have learned in this chapter as it applies to the information in Chapter 10 on prison programming, and in particular life skills programs for offenders. Your assignment is as follows:

You are a researcher recently hired by the prison system of your state, and you have been asked to design and evaluate a prison life-skills program. Specifically, you are asked to determine if the completion of the life-skills program actually reduces recidivism among the offenders who take it. You will need to provide a clear methodology for testing and evaluating your proposed program, taking into account such factors as the validity and reliability of your study, the validity and reliability of your instruments (if any), the use of control and experimental groups, distinctions between process and outcome measures, and any ethical issues that might be involved with conducting your research.

Your response to each of the questions below must consist of 300 to 700 words. You should complete this application exercise as an essay that explains the program that you develop and then address each question. The total word count should be 900 to 2,100 words.

1. Identify the specific aspects of the life skills program that you would evaluate, and explain why each would be important to your evaluation. Further, identify and distinguish between the process and the outcome measures within your study.

2. Identify and discuss the various research methods considerations that you might wish to employ. Be sure to discuss the likely validity and reliability of your study as well as the validity and reliability of the instruments that you might use. Also, determine if you will be able to use a control group or a comparison group within your study.

3. Be sure to explain how you would use your evaluative study to determine the effects of a life-skills program. (Note that you have a wide degree of latitude with this aspect of the assignment.)

Spelling, grammar, and writing style will be considered. In addition, the paper *must* be within the prescribed range in word count. Lastly, *must* cite and reference work, in APA format, or no credit will be given.

WHAT WOULD YOU DO?

You are a correctional officer working the graveyard shift at a maximum-security prison. One night, during turnout, the lieutenant explains to you and the other staff on your shift (there are about 50 of you) that there will be a researcher who will be coming to each officer and giving you a survey. This will be done intermittently throughout the week, and the lieutenant makes it clear that while you are not required to fill out the survey, the study is quite beneficial, and he and the other "rank" officers appreciate any participation that you would be willing to give. He introduces to the group two persons dressed in semiformal attire, Dr. Smith and Dr. Wilson, both criminologists from the university located about 35 miles from your prison facility.

Once they are introduced, Dr. Smith explains, "Our study examines inmate sexual assault in prisons. Basically, there is a tendency for inmates around the nation to underreport incidents, and this naturally makes it impossible to truly know how prevalent it may be."

As the pair begin to pass out the surveys, Dr. Wilson adds, "So, we've decided to ask the custodial staff about their perceptions regarding this issue since you're the ones who spend the most time in proximity to the inmates who are involved in this behavior, when and if it occurs on your shift."

Once everyone has a survey, Dr. Wilson continues. "While on your shift and when you have time, we are asking that you complete this survey. Later we will follow up and interview some of you, as time permits throughout the night. We know that you're busy and that you may not wish to participate. If you wish to decline participation, we will naturally honor your wishes without comment."

After the doctor finishes speaking, the lieutenant lets the shift personnel know that he will not be keeping track of whoever participates. Rather, he and the sergeants will be staying out of sight as much as possible throughout the night to ensure that they do not obstruct the data collection efforts of the researchers. At that point, work details are assigned to all staff members and you are dismissed to relieve the prior shift and assume their duty posts.

As luck would have it, you are assigned to the escort crew, which means that your assignment is to patrol the prison on foot, going from one cell block to another to relieve officers for their mealtime or scheduled breaks, taking inmates to the infirmary or other areas of the prison as required, and conducting security checks throughout the night. You will also escort inmates to early morning breakfast later in your shift. This is considered a good detail because the mobility keeps the shift from becoming monotonous, and you get to talk and visit with numerous people throughout the night. There are about seven of you assigned to the escort crew. At one point in the evening, you go to relieve an officer on one of the cell blocks so that he can go have dinner.

While you are on the cell block, three other members of the escort crew join you. One of them says, "Hey, we're gonna hang out in the dayroom and fill out our surveys, watch some TV, and wait until the break relief is over. After that, Sarge says we gotta go to the laundry room and let some inmates get some supplies—we just wanted you to know."

You tell them okay and go to make your rounds. While making your rounds, you hear the other officers laughing. You approach the dayroom and hear one of them say, "Man, this survey is whacked—those brainiacs are crazy if they think that I am gonna tell them about what goes on with those convicts."

Another responds, "Yeah, man, like I give a damn if some convict gets his booty stolen—serves 'em right. The assholes will stay out of here if they don't like it."

All the officers laugh, and one says, "Look, I got an idea—let's just fill them out real fast and put that nothing ever happens. Maybe we could put that they are all lying and making false allegations against inmates that they owe money to! That would really mess up the study and give them something to scratch their heads about, huh?"

"Hell, yeah, great idea, we can make this fun! After all, the surveys are anonymous, right?"

You finish your security rounds while thinking about what you have just heard. You have not yet filled out your survey and have not even looked at it, but you begin to feel a little sorry for the researchers. You think to yourself, *Maybe this is why things never get better. No matter how much research they do, everybody is always lying to them. I wonder if they know how much of this probably takes place.*

You enter the dayroom just in time to see the last of the officers place his survey in the envelope that each officer was given and seal it. At this time, the officer you were relieving returns from dinner, and as you are walking off the cell block, you see the researchers walking toward the dayroom. They offer to collect the sealed envelopes from the other officers, who hand the doctors their envelopes and head off to the laundry room.

Dr. Smith looks at you and says, "If possible, we would like to interview you since you are next on our list. It will not affect your survey results, and we've told the sergeant that you might need to follow about 10 minutes behind everyone. Is that okay?"

Since the interview is really not optional (you know your supervisors would think it rude to decline the polite request of the researchers) and since you really are not in a hurry to go to the laundry room, you say, "Okay, sure." However, you know that they have no idea about the surveys that they just received, and you are a bit undecided as to what you think about all of this "research stuff," anyway. You contemplate what you may or may not say to them in the interview.

What would you do?

GLOSSARY

Absolute immunity: Protection for persons who work in positions that require unimpaired decision-making functions.

Administrative segregation: A nonpunitive classification that requires the separation of inmates from the general population for safety.

Administrator: An individual in an agency who operates in a managerial or leadership role beyond immediate supervisory capacity.

Aggravating circumstances: Magnify the offensive nature of a crime and tend to result in longer sentences.

Albert W. Florence v. Board of Chosen Freeholders of the County of Burlington, et al. (2012): Prison staff may strip search minor offenders and detainees within a jail or detention facility.

Alcatraz: A prison built on Alcatraz Island, California. First opened in 1934, it is considered to be the first U.S. supermax facility.

American Correctional Chaplains Association (ACCA): Provides representation and networking for chaplains who work in correctional environments.

Americans with Disabilities Act (ADA): Requires correctional agencies to make reasonable modifications to ensure accessibility for individuals with disabilities.

Armed disturbance control team: Deals with disturbances that have escalated to matters of life and death.

Assessment-evaluation cycle: Comparing assessment and evaluation data to determine the effectiveness of programs and to find areas for improvement.

Atkins v. Virginia (2002): Held that the execution of the mentally retarded is unconstitutional.

Attorney General's Review Committee on Capital Cases: Makes an independent recommendation to the attorney general regarding death penalty cases.

Auburn system: An alternative prison system located in New York.

Banishment: Exile from society.

Bastille: A fortification in Paris, France that was a symbol of tyranny and injustice for commoners and political prisoners.

Baxter v. Palmigiano (1976): Determined that inmates do not have the right to counsel for disciplinary hearings that are not part of a criminal prosecution.

Bell v. Wolfish (1979): Determined that body cavity searches of inmates after contact visits is permissible, as are searches of inmates' quarters in their absence. Double bunking does not deprive inmates of their liberty without due process of law.

Big House prisons: Typically large stone structures with brick walls, guard towers, and checkpoints throughout the facility.

Black Codes: Separate laws were required for slaves and free men who turned criminal.

Blind spots: In correctional facilities, these occur when areas of the prison are not easily viewed by security staff and/or surveillance equipment.

Blood in—blood out: The idea that for inmates to be accepted within a prison gang they must draw blood from an enemy of the gang.

Bounds v. Smith (1977): Determined that prison systems must provide inmates with law libraries or professional legal assistance.

Branding: Usually on thumb with a letter denoting the offense.

Bruscino v. Carlson (1988): Ruling that the high-security practices of USP Marion were constitutional.

Capital punishment: Putting the offender to death.

Centralized management: Tight forms of control in the communication process that ensure that decision-making power is reserved for only a small group of people.

Chivalry hypothesis: Contends that there is a bias in the criminal justice system against giving women harsh punishments.

Classical criminology: Emphasized that punishments must be useful, purposeful, and reasonable.

Code of Hammurabi: The earliest known written code of punishment.

Coker v. Georgia (1977): Ruling that the death penalty was unconstitutional for the rape of an adult woman if she had not been killed during the offense.

Collateral damage: Any damage incidental to an activity.

Compensatory damages: Payments for the actual losses suffered by a plaintiff.

Comprehensive Crime Control Act of 1984: Created a U.S. Sentencing Commission to establish federal sentencing guidelines favoring determinate sentences.

Conflict theory: Maintains that concepts of inequality and power are the central issues underlying crime and its control.

Consent decree: An injunction against both individual defendants and their agency.

Contract labor system: Utilized inmate labor through state-negotiated contracts with private manufacturers.

Convict: An inmate who is respected for being self-reliant and independent of other inmates or the system.

Convict boss/officer: A correctional officer with a keen understanding of convict logic and socialization.

Convict code: An inmate-driven set of beliefs that inmates aspire to live by.

Co-occurring disorders: When an offender has two or more disorders.

Cooper v. Pate (1964): Ruling that state prison inmates could sue state officials in federal courts.

Corrections: A process whereby practitioners engage in organized security and treatment functions to correct criminal tendencies among the offender population.

Crime control model: An approach to crime that increased the use of longer sentences, the death penalty, and intensive supervision probation.

Cruz v. Beto (1972): Ruling that inmates must be given reasonable opportunities to exercise their religious beliefs.

Custody level: Related to the degree of staff supervision that is needed for a given inmate.

Day reporting centers: Treatment facilities to which offenders are required to report, usually on a daily basis.

Decentralized management: The authority and responsibility of management are distributed among the supervisory chain, allowing decisions to correspond to the problems confronted by each level of management.

Declaratory judgment: A judicial determination of the legal rights of the person bringing suit.

Defamation: Some form of slander or libel that damages a person's reputation.

Detention: Secure confinement of juvenile offenders.

Determinate discretionary sentence: Type of sentence with a range of time to be served; the specific sentence to be served within that range is decided by the judge.

Determinate presumptive sentence: This type of sentence specifies the exact length of the sentence to be served by the inmate.

Determinate sentences: Consist of fixed periods of incarceration with no later flexibility in the term that is served.

Detoxification: The use of medical drugs to ease the process of overcoming the physical symptoms of dependence.

Diagnostic and Statistical Manual of Mental Disorders (DSM-V): A reference manual that sets forth the guidelines in applying a diagnosis of a mental disorder.

Direct supervision design: Cells are organized on the outside of the square space, with shower facilities and recreation cells interspersed among the typical inmate living quarters.

Discrimination: A differential response toward a group without providing any legally legitimate reasons for that response.

Disparity: Inconsistencies in sentencing and/or sanctions that result from the decision-making process.

Disturbance control teams: Specialized teams trained to respond to, contain, and neutralize inmate disturbances in prisons.

Domestic violence: Behaviors used by one person in a relationship to control the other.

Eastern State Penitentiary: Part of the Pennsylvania system located near Philadelphia.

Elderly first-time offenders: Those who commit their first offense later in life.

Electronic monitoring: The use of any mechanism worn by the offender for the means of tracking his or her whereabouts through electronic detection.

Elmira Reformatory: The first reformatory prison.

Emotional distress: Refers to acts that lead to emotional distress of the client.

English Penal Servitude Act: Established several rehabilitation programs for convicts.

Estelle v. Gamble (1976): Ruled that deliberate indifference to inmate medical needs constitutes cruel and unusual punishment and is unconstitutional.

Evaluative transparency: When an agency's evaluative process allows an outside person to have full view of the agency's operations, budgeting, policies, procedures, and outcomes.

Evidence-based practice (EBP): A significant trend throughout all human services that emphasizes outcomes.

Ex parte Hull (1941): Ruling that marked the beginning of the end for the hands-off doctrine.

Father of Parole: Alexander Maconochie, so named due to his creation of the mark system, a precursor to modern-day parole.

Federal Prison Industries Inc. (FPI): Organization for federal prison labor.

Fine: A monetary penalty imposed as a punishment for having committed an offense.

Ford v. Wainwright (1986): Ruling that defendants who could successfully invoke the insanity defense could avoid the death penalty.

Four standards of mental health care: Legal requirement for adequate health care also extends to mental health care.

Fulwood v. Clemmer (1962): Ruled that correctional officials must recognize the Muslim faith as a legitimate religion and allow inmates to hold services.

Furman v. Georgia (1972): Ruling that the death penalty was arbitrary and capricious and violated the prohibition against cruel and unusual punishment.

Gagnon v. Scarpelli (1971): Held that a parolee's sentence can only be revoked after a preliminary and final revocation hearing have been provided.

Gang cross-pollination: Occurs when a gang has developed such power and influence as to be equally effective regardless of whether its leadership is inside or outside of prison walls.

Gaol: A term used in England during the Middle Ages that was synonymous with today's jail.

General deterrence: Punishing an offender in public so other observers will refrain from criminal behavior.

General Educational Development (GED): The process of earning the equivalent of your high school diploma.

Glossip v. Gross (2015): Supreme Court case that determined that the Eighth Amendment does not require that a constitutional method of execution be free of any risk of pain.

Good faith defense: The person acted in the honest belief that the action taken was appropriate under the circumstances.

Great Law: Correctional thinking and reform in Pennsylvania that occurred due to the work of William Penn and the Quakers.

Gregg v. Georgia (1976): Held that death penalty statutes that contain sufficient safeguards against arbitrary and capricious imposition are constitutional.

Greyhounds: Older inmates who have acquired respect within the offender subculture due to their track record, criminal history, and criminogenic ideals.

Habitual elderly offenders: Have a long history of crime and also have a prior record of imprisonment throughout their lifetime.

Hall v. Florida (2014): Supreme Court case that held that executing an intellectually disabled person violates the Eighth Amendment's protection against cruel and unusual punishment.

Hands-off doctrine: The policy of the courts of avoiding intervention in prison operations.

Health Insurance Portability and Accountability Act (HIPAA): Guides professionals on matters regarding the confidentiality of medical information.

Hearing stage: Stage of a revocation proceeding that allows the probation agency to present evidence of the violation, which the offender is given the opportunity to refute.

Hedonistic calculus: A term describing how humans seem to weigh pleasure and pain outcomes when deciding to engage in criminal behavior.

HIV/AIDS: A chronic, potentially life-threatening condition caused by the human immunodeficiency virus.

Hogging: A term used to imply that a person is using others for some type of gain or benefit.

Holt v. Hobbs (2015): In cases of legitimate religious actions, the government must show that substantially burdening the religious exercise of an individual is "the least restrictive means of furthering that compelling governmental interest."

Holt v. Sarver I (1969): Ruled that prison farms in the state of Arkansas were operated in a manner that violated the prohibition against cruel and unusual punishments.

Home detention: The mandated action that forces an offender to stay within the confines of his or her home for a specified time.

Hudson v. Palmer (1984): Held that prison cells may be searched without the need of a warrant and without probable cause.

Hustle: Any action that is designed to deceive, manipulate, or take advantage of another person.

Hutto v. Finney (1978): Held that courts can set time limits on prison use of solitary confinement.

Implementation evaluation: Identifying problems and accomplishments during the early phases of program development for feedback to clinical and administrative staff.

Importation theory: Subculture within prisons is brought in from outside by offenders who have developed their beliefs and norms while on the streets.

Incapacitation: Deprives offenders of their liberty and removes them from society, ensuring that they cannot further victimize society for a time.

Indeterminate sentences: Sentences that include a range of years that will be potentially served by the offender.

Individual personality traits: Traits associated with criminal behavior.

Injunction: A court order that requires an agency to take some form of action(s) or to refrain from a particular action(s).

Institutional parole officer: Works to aid the offender in making the transition from prison life to community supervision.

Intensive supervision probation (ISP): The extensive supervision of offenders who are deemed the greatest risk to society or are in need of the greatest amount of governmental services.

Intentional tort: The actor, whether expressed or implied, was judged to have possessed intent or purpose to cause an injury.

Intermediate sanctions: A range of sentencing options that fall between incarceration and probation.

Isolation zone: Designed to prevent undetected access to the outer fencing of the prison facility.

Jail: A confinement facility, usually operated and controlled by county-level law enforcement, designed to hold persons who are awaiting adjudication or serving a short sentence of 1 year or less.

Jail reentry programs: Programs usually interlaced with probation and parole agencies as a means of integrating the supervisory functions of both the jail and community supervision agencies.

Johnson v. Avery (1969): Held that prison authorities cannot prohibit inmates from aiding other inmates in preparing legal documents.

Judicial Improvements Act of 1990: Extended the life of the Parole Commission until November 1, 1997.

Juvenile waiver: Occurs when the juvenile case is transferred to adult court.

Kennedy v. Louisiana (2008): A Supreme Court case where it was determined that it is a violation of the Eighth Amendment to impose the death penalty for the rape of a child when the crime did not result, and was not intended to result, in the death of the child.

Kingsley v. Hendrickson (2015): For claims of excessive force brought by pretrial detainees, it is only necessary to show that the force used was objectively unreasonable.

Labeling theory: Contends that individuals become stabilized in criminal roles when they are labeled as criminals.

Learning organizations: Have the inherent ability to adapt and change, improving performance through continual revision of goals, objectives, policies, and procedures.

Level of security: The type of physical barriers that are utilized to prevent inmates' escape and are related to public safety concerns.

Lex talionis: Refers to the Babylonian law of equal retaliation.

Libel: Written communication intended to lower the reputation of a person where such facts would actually be damaging to a reputation.

Little Hoover Commission: An internal state watchdog agency that was tasked with providing recommendations to the governor and legislature in California.

Madrid v. Gomez (1995): Case where the constitutionality of supermax facilities was questioned.

Major depressive disorder: Characterized by one or more major depressive episodes.

Malicious prosecution: Occurs when a criminal accusation is made without probable cause and for improper reasons.

Malingering: When inmates falsely claim symptoms of an illness.

Mandatory minimum: A minimum amount of time or a minimum percentage of a sentence must be served with no good time or early-release modifications.

Mark system: A system where the duration of the sentence was determined by the inmate's work habits and righteous conduct.

Martinson Report: An examination of a number of various prison treatment programs.

Maximum-security facilities: These high-security facilities use corrugated chain-link fence. These fences will be lit by floodlights at night and may even be electrified and eliminate "blind spots" in security where inmates can hide.

McCoy v. Louisiana (2018): Supreme Court case that ruled that it is a violation of an accused's Sixth Amendment rights if his or her defense counsel pleads guilty when the defendant objects and does not wish to concede guilt.

Media effects: The effect that the media has upon the public perception of prison or community supervision programs.

Medical model: An approach to correctional treatment that utilizes a type of mental health approach incorporating fields such as psychology and biology.

Medium-security facilities: Consist of dormitories that have bunk beds with lockers for inmates to store their possessions and communal showers and toilets. Dormitories are locked at night with one or more security officers holding watch.

Mental illness: Any diagnosed disorder contained within the DSM-5.

Minimal services view: Contends that inmates are entitled to no more than the bare minimum that is required by law.

Minimum-security facilities: These prisons usually consist of dormitory-style housing for offenders rather than cellblocks. Also, these prisons are designed to facilitate public works, such as farming or roadways, rather than being optimized for the offender's reform. A minimum-security facility generally has a single fence that is watched, but not patrolled, by armed guards. This is different from the typical prisons envisioned by the public.

Mitigating factors: Circumstances that make a crime more understandable and help to reduce the level of culpability that an offender might have.

Mood disorders: Disorders such as major depressive disorder, bipolar disorder, and dysthymic disorder.

Morris v. Travisono (1970): Resulted in a detailed set of procedures for classifying inmates that was overseen by the federal court system.

Mothers and Infants Nurturing Together (MINT): A Federal Bureau of Prisons program that promotes bonding and parenting skills for low-risk female inmates who are pregnant.

Mule: A person who smuggles drugs into prison for another inmate.

National Commission on Correctional Health Care (NCCHC): Sets the tone for standards of care in correctional settings.

National Institute of Corrections (NIC): An agency within the Federal Bureau of Prisons that is headed by a director appointed by the U.S. attorney general.

Negative punishment: The removal of a valued stimulus when the offender commits an undesired behavior.

Negative reinforcers: Unpleasant stimuli that are removed when a desired behavior occurs.

Negligence: Doing what a reasonably prudent person would not do in similar circumstances or failing to do what a reasonably prudent person would do in similar circumstances.

Offender Accountability Plan: Addresses needs for restitution, any particular arrangements that have been made with the victim, and provisions to ensure the offender's responsibility to the community at large.

Offender reentry: Includes all activities and programming conducted to prepare ex-convicts to return safely to the community and to live as law-abiding citizens.

Offender with special needs: A specialized offender who has some notable physical, mental, and/or emotional challenge.

Offenders-turned-elderly-in-prison: Inmates who have grown old in prison who have long histories in the system.

Office of Correctional Education (OCE): Created to provide national leadership on issues related to correctional education.

Old Newgate Prison: First prison structure in America.

O'Lone v. Estate of Shabazz (1987): Held that depriving an inmate of attending a religious service for "legitimate penological interests" was not a violation of the inmate's First Amendment rights.

One hand on, one hand off doctrine: More conservative rulings are being handed down from the Court, reflecting an eclipse of the hands-off doctrine.

Outcome evaluation: Involves quantitative research aimed at assessing the impact of the program on long-term treatment outcomes.

Pains of imprisonment: The various inconveniences and deprivations that occur as a result of incarceration.

Palmigiano v. Garrahy (1977): Ruling that attested to the importance of effective and appropriate classification systems.

Panopticon: Designed to allow security personnel to clearly observe all inmates without the inmates themselves being able to tell whether they are being watched.

Parole: The early release of an offender from a secure facility upon completion of a certain portion of his or her sentence.

Parole Commission and Reorganization Act: Established the U.S. Parole Commission as an independent agency within the Department of Justice.

Parole Commission Phaseout Act of 1996: Extended the life of the Parole Commission until November 1, 2002, but only in regard to supervising offenders who were still on parole from previous years.

Parole revocations officer: Primarily tasked with the routine holding of preliminary parole revocation hearings by reviewing allegations made by parole officers against parolees.

Passive agent: Views his or her job dispassionately as just a job and tends to do as little as possible.

Paternal officer: Uses a great degree of both control and assistance techniques in supervising offenders.

Patriarchy: A male-oriented and male-dominated social structure that defers to men and sees women in a subservient position to men.

Pell Grants: Need-based federal monies set aside for persons who pursue a college education.

Perimeter security system: A collection of components or elements that, when assembled in a carefully formulated plan, achieve the objective of confinement with a high degree of confidence.

Pods: Prefabricated sections in most modern prisons. Inmates will usually have individual cells with doors controlled from a secure remote control station.

Podular jail: Includes rounded architecture for living units and allows for direct supervision of inmates by security staff.

Positive punishment: Punishment where a stimulus is applied to the offender when the offender commits an undesired behavior.

Positive reinforcers: Rewards for a desired behavior.

Post-Release Employment Project (PREP): A study that demonstrated that UNICOR successfully prepared inmates for release and provided long-term benefits to society.

Preliminary hearing: Initial examination of the facts of the arrest to determine if probable cause does exist for a violation.

Presentence investigation report: A thorough file that includes a wide range of background information on the offender.

Prison Industry Enhancement (PIE) Certification Program: Partnership between the Texas Department of Criminal Justice and a private company that allows the company to employ offenders.

Prison Litigation Reform Act (PLRA): Limits an inmate's ability to file lawsuits and the compensation that he or she can receive.

Prison social media: Online dialogue that consists of information regarding visitation, rules of the correctional facility, or sending necessities to loved ones in prison.

Prisonization: The process of being socialized into the prison culture.

Private wrongs: Crimes against an individual that could include physical injury, damage to a person's property, or theft.

Probation: A control valve mechanism that mitigates the flow of inmates sent directly to the jailhouse.

Probationers: Criminal offenders who have been sentenced to a period of correctional supervision in the community in lieu of incarceration.

Process evaluation: Traditionally refers to assessment of the effects of the program on clients while they are in the program, making it possible to assess the institution's intermediary goals.

Procunier v. Martinez (1974): Prison officials may censor inmate mail only to the extent necessary to ensure security of the institution.

Progressive Era: A period of extraordinary urban and industrial growth and unprecedented social problems.

Protective custody: A security-level status given to inmates who are deemed to be at risk of serious violence if not afforded protection.

Public Safety Realignment (PSR): A California state policy designed to reduce the number of offenders in that state's prison system to 110,000.

Public wrongs: Crimes against society or a social group.

Punitive damages: Monetary awards reserved for the person harmed in a malicious or willful manner by the guilty party.

Punitive officer: Sees himself or herself as needing to use threats and punishment in order to get compliance from the offender.

Punk: A derogatory term for an inmate who engages in homosexual activity; implies that the inmate is feminine, weak, and subservient to masculine inmates.

Qualified immunity: Legal immunity that shields correctional officers from lawsuits, but first requires them to demonstrate the grounds for their possession of immunity.

Rational basis test: Sets guidelines for the rights of inmates that still allow correctional agencies to maintain security.

Recreation program administrator: Responsible for a number of duties, including surveying the recreational needs and interests of the offender population.

Rehabilitation: Offenders will be deterred from reoffending due to their having worthwhile stakes in legitimate society.

Reintegration model: Used to identify programs that looked to the external environment for causes of crime and the means to reduce criminality.

Release valve mechanism: When correctional systems use parole to reduce correctional populations.

Religious Land Use and Institutionalized Persons Act of 2000: Prohibits the government from substantially burdening an inmate's religious exercise.

Respect: An inmate's sense of standing within the prison culture.

Restorative justice: Interventions that focus on restoring the community and the victim with involvement from the offender.

Retribution: Offenders committing a crime should be punished in a way that is equal to the severity of the crime they committed.

Roadmap to Reentry: Identifies five evidence-based principles that guide efforts to improve correctional programs that are developed for those who will reenter society after being incarcerated.

Roper v. Simmons (2005): Ruled that the death penalty was unconstitutional when used with persons who were under 18 years of age at the time of their offense.

Ruiz v. Estelle (1980): Ruled that the Texas prison system was in violation of the prohibition against cruel and unusual punishments.

Rumor control: The active process of administrators to circumvent faulty information that is disseminated among staff or inmates and has the potential to cause unrest or disharmony throughout the facility.

Rural jail: Usually small jails in rural county jurisdictions that are often challenged by tight budgets and limited training for staff.

Sally port: Entry design that allows security staff to bring vehicles close to the admissions area in a secure fashion.

Sanctuary: A place of refuge or asylum.

Second Chance Pell Pilot Program: This program, announced by the Obama administration and stewarded by the Department of Education, selected 67 colleges and universities who partnered with more than 100 federal and state correctional institutions, enrolling approximately 12,000 incarcerated students in educational and training programs.

Security threat group (STG): A high-functioning, organized gang that uses an illegal industry to fund their objectives.

Selective incapacitation: Identifying inmates who are of particular concern to public safety and providing them with much longer sentences.

Sentencing stage: When a judge determines if the offender will be incarcerated or continue his or her probation sentence under more restrictive terms.

Shock incarceration: A short period of incarceration followed by a specified term of community supervision.

Short-term jail: A facility that holds sentenced inmates for no more than 1 year.

Slander: Verbal communication intended to lower the reputation of a person where such facts would actually be damaging to a reputation.

Smarter Sentencing Act of 2014: A bill that adjusts federal mandatory sentencing guidelines in an effort to reduce the size of the U.S. prison population.

Snitch: Term for an inmate who reveals the activity of another inmate to authorities.

Social learning theory: Contends that offenders learn to engage in crime through exposure to and adoption of definitions that are favorable to the commission of crime.

Special housing unit syndrome: The negative mental health effects of extended isolation.

Specific deterrence: The infliction of a punishment upon a specific offender in the hope that he or she will be discouraged from committing future crimes.

Stakeholders: Agency personnel, the surrounding community, and the offender population.

Strain theory/institutional anomie: Denotes that when individuals cannot obtain success goals, they will tend to experience a sense of pressure often called *strain*.

Strategic plan: A document that articulates agency goals and objectives and states how they might be realized.

Strategic planning: The planning of long-term goals and objectives, usually spanning the period of 1 year or more.

Supermax facility: A highly restrictive, high-custody housing unit within a secure facility, or an entire secure facility, that isolates inmates from the general population and from each other.

Supervised release: Post-release supervision.

Symbiotic prison relationship: When correctional staff and inmates develop negotiation behavior that is acceptable for institutional security and also meets inmates' basic human needs.

Systemwide administrators: Managers who are at the executive level and direct the entire system throughout the state.

Tactical planning: Ground-level planning that is narrow in focus and structured around the short-term resolution of particular issues.

Technical violations: Actions that do not comply with the conditions and requirements of a probationer's sentence.

Texas Correctional Industries (TCI): Provides offenders with marketable job skills to help reduce recidivism.

The Summary: The first inmate-operated prison newspaper in the world.

Therapeutic community: An environment that provides necessary behavior modifiers that allow offenders immediate feedback about their behavior and treatment progress.

Therapeutic recreation (TR): Programs designed to meet the needs of individuals with a variety of disabilities, impairments, or illnesses by providing specific services.

Thompson v. Oklahoma (1988): Held that the Eighth and Fourteenth Amendments prohibited the execution of a person who is under 16 years of age at the time of his or her offense.

Ticket of leave: A permit given to a convict in exchange for a certain period of good conduct.

Title IV of the Higher Education Act: Permitted inmates to apply for financial aid in the form of Pell Grants to attend college.

Tort: A legal injury in which a person causes injury as the result of a violation of one's duty as established by law.

Totality of the conditions: A standard used to determine if conditions in an institution are in violation of the Eighth Amendment.

Tower of London: One of the earliest examples of a jail used for confinement purposes.

Trial by ordeal: Very dangerous and/or impossible tests to prove the guilt or innocence of the accused.

Trop v. Dulles (1958): Developed a phrase that would be cited in future cases because it fit with many compelling arguments in favor of correctional reform.

Turner v. Safley (1987): A prison regulation that impinges on inmates' constitutional rights is valid if it is reasonably related to legitimate penological interests.

UNICOR Inc.: An organization for federal prison labor.

United States Parole Commission Extension Act of 2008: Provided for the continued performance of the U.S. Parole Commission.

United States v. Booker (2005): Determined judges no longer had to follow the sentencing guidelines that had been in place since 1987.

Universal design: Prison construction design that complies with ADA requirements and that accommodates all inmate needs in a universal fashion.

USP Florence ADMAX: A federal prison with a design that is nearly indestructible on the inside. Essentially, the offenders there have no contact with humans.

USP Marion: A special closed-custody unit designed to house the Federal Bureau of Prisons' worst inmates.

Victorian Era: Viewed women from a lens of inflexible femininity where women were to be considered pious and naïve of the evils of the world.

Vitek v. Jones (1980): Inmates are entitled to due process in involuntary transfers from prison to a mental hospital.

Walnut Street Jail: America's first attempt to incarcerate inmates with the purpose of reforming them.

Weekend confinement: Confinement that is restricted to the weekends or other times when the person in custody is off from work.

Welfare worker: Views the offender more as a client rather than a supervisee on his or her caseload.

Western State Penitentiary: Part of the Pennsylvania system located outside of Pittsburgh.

Wilson v. Seiter (1991): Deliberate indifference is required for liability to be attached for condition of confinement cases.

Windham School District: Secondary education program in Texas prisons.

Witherspoon v. Illinois (1968): Held that it was not constitutional to strike a potential juror from serving if the juror had doubts or reservations about the use of the death penalty.

Wolff v. McDonnell (1974): Inmates are entitled to due process in prison disciplinary proceedings that can result in the loss of good time credits or in punitive segregation.

Work Against Recidivism (WAR) program: Specifically targeted to successfully reintegrate offenders into society.

Work/education reform view: Claims that society is saved untold millions due to the lack of recidivism of inmates who have obtained employment or education.

Writ writer: An inmate who becomes skilled at generating legal complaints and grievances within the prison system.

Zebulon Brockway: The warden of Elmira Reformatory.

Zimmer Amendment: Restricted the purchase of several types of weight lifting equipment within the Federal Bureau of Prisons.

REFERENCES

Aday, R. H. (1994). Aging in prison: A case study of new elderly offenders. *International Journal of Offender Therapy and Comparative Criminology, 38*(1), 121.

American Correctional Association. (1998). *Causes, preventive measures, and methods of controlling riots and disturbances in correctional institutes*. Upper Marlboro, MD: Graphic Communications.

American Correctional Association. (2001). A short history of direct-supervision facility design. *Corrections Today*. Alexandria, VA: Author.

American Jail Association. (1993). *American Jail Association code of ethics*. Retrieved from http://www.corrections.com/aja/resolutions/index.html

American Probation and Parole Association. (1991). *Issue paper on caseload standards*. Washington, DC: Author.

American Probation and Parole Association. (2011). *Probation and parole FAQs*. Washington, DC: Author.

American Psychiatric Association. (2000). *Diagnostic and statistical manual of mental disorders* (4th ed., text rev.). Washington, DC: Author.

Anderson, J. C. (2008). Special needs offenders. In P. M. Carlson & J. S. Garrett (Eds.), *Prison and jail administration: Practice and theory* (2nd ed., pp. 361–372). Sudbury, MA: Jones & Bartlett.

Anderson, J. F., Mangels, N. J., & Dyson, L. (2010). *Significant prisoner rights cases*. Durham, NC: Carolina Academic Press.

Anno, B. J., Graham, C., Lawrence, J. E., & Shansky, R. (2004). *Correctional health care: Addressing the needs of elderly, chronically ill, and terminally ill inmates*. Washington, DC: National Institute of Corrections.

Appel, A. (1999). Accommodating inmates with disabilities. In P. M. Carlson & J. S. Garrett (Ed.), *Prison and jail administration* (pp. 346–352). Gaithersburg, MD: Aspen.

Applegate, B. K., & Sitren, A. H. (2008). The jail and the community: Comparing jails in rural and urban contexts. *The Prison Journal, 88*, 252–269.

Armstrong, J. J. (2008). Causes of institutional unrest. In P. M. Carlson & J. S. Garrett (Eds.), *Prison and jail administration: Practice and theory* (2nd ed., pp. 461–468). Sudbury, MA: Jones & Bartlett.

Ashford, J. B., Sales, B. D., & Reid, W. H. (2002). *Treating adult and juvenile offenders with special needs*. Washington, DC: American Psychological Association.

Associated Press. (2001). *Prison escape probe to focus on lax security*. Retrieved from http://www.clickonsa.com/ant/news/stories/news-20010108-085202.html

Atkins v. Virginia, 536 U.S. 304 (2002).

Augustus, J. (1972). *John Augustus' original report of his labors*. Montclair, NJ: Patterson Smith. (Original work published in 1852)

Barnes, H. E., & Teeters, N. K. (1959). *New horizons in criminology* (3rd ed.). Upper Saddle River, NJ: Prentice Hall.

Baxter v. Palmigiano, 425 U.S. 308 (1976).

Bell v. Wolfish, 441 U.S. 520 (1979).

Benestante, J. (1996, April 19). *Presentation before the Texas Board of Criminal Justice Special Committee on Prison Industries*. Austin: Texas Department of Criminal Justice.

Bernard, T. J., McCleary, R., & Wright, R. A. (1999). *Life without parole: Living in prison today* (2nd ed.). Los Angeles, CA: Roxbury.

Berzofsky, M., Maruschak, L. M., & Unangst, J. (2015). *Medical problems of state and federal prisoners and jail inmates, 2011–12*. Washington, DC: Bureau of Justice Statistics.

Bloom, B., Brown, M., & Chesney-Lind, M. (1996). Women on probation and parole. In A. J. Lurigio (Ed.), *Community corrections in America: New directions and sounder investments for persons with mental illness and co-disorders* (pp. 51–76). Washington, DC: National Institute of Corrections.

Bloom, B., Owen, B., & Covington, S. (2003). *Gender responsive strategies: Research, practice, and guiding principles for women offenders*. Washington, DC: National Institute of Corrections. Retrieved from http://www.nicic.org/Library/018017

Bounds v. Smith, 430 U.S. 817 (1977).

Bowers, W. J., & Pierce, G. L. (1980). Deterrence or brutalization: What is the effect of executions? *Crime & Delinquency, 26*(4), 453–484.

Braithwaite, J. (1989). *Crime, shame, and reintegration*. Cambridge, England: Cambridge University Press.

Branch-Johnson, W. (1957). *The English prison hulks*. London, England: C. Johnson.

Branham, L. S., & Hamden, M. S. (2009). *Cases and materials on the law and policy of sentencing and corrections* (8th ed.). St. Paul, MN: West.

Brennan, P. K., & Vandenberg, A. L. (2009). Depictions of female offenders in front-page newspaper stories: The importance of race/ethnicity. *International Journal of Social Inquiry, 2*(2), 141–175.

Brockway, Z. R. (1912). *Fifty years of prison service: An autobiography*. New York State Reformatory at Elmira. Annual report reprinted, Montclair, NJ: Patterson Smith, 1969.

Browne, J. (2010). Rooted in slavery: Prison labor exploitation. *Race, Poverty, & the Environment, 14*(2), 78–81.

Bruscino v. Carlson, 854 F. 2d 162 (7th Cir. 1988).

Bureau of Justice Assistance Center for Program Evaluation. (2007). *Reporting and using evaluation results*. Washington, DC: Author. Retrieved from http://www.ojp.usdoj.gov/BJA/evaluation/sitemap.htm

Bureau of Justice Statistics. (1998). *Substance abuse and treatment, state and federal prisoners, 1997*. Washington, DC: U.S Department of Justice.

Bureau of Justice Statistics. (2019). *Local jail inmates and jail facilities*. Washington DC: Author. Retrieved from https://www.bjs.gov/index.cfm?ty=tp&tid=12

Bureau of Labor Statistics. (2017a). *Employed persons by detailed occupation, sex, race, and Hispanic or Latino ethnicity*. Washington, DC: Author. Retrieved from: https://www.bls.gov/cps/cpsaat11.pdf

Bureau of Labor Statistics. (2017b). *Occupational employment statistics*. Washington, DC: Author.

Bureau of Labor Statistics. (2018). *Occupational outlook handbook May 2018: 21-1092 probation officers correctional treatment specialists*. Washington, DC: Author. Retrieved from https://www.bls.gov/ooh/community-and-social-service/probation-officers-and-correctional-treatment-specialists.htm

Burrell, B. (2006). *Caseload standards for probation and parole*. Washington, DC: National Institute of Corrections.

Carlson, P. M., & DiIulio, J. J. (2013). Organization and management of the prison. In P. M. Carlson (Ed.), *Prison and jail administration: Practice and theory* (3rd ed., pp. 269–286). Boston, MA: Jones & Bartlett.

Carlson, P. M., & Garrett, J. S. (Eds.). (2008). *Prison and jail administration: Practice and Theory* (2nd ed.). Sudbury, MA: Jones & Bartlett.

Carlson, P. M., Roth, T., & Travisono, A. P. (2008). History of corrections. In P. M. Carlson & J. S. Garrett (Eds.), *Prison and jail administration: Practice and theory* (2nd ed., pp. 3–18). Sudbury, MA: Jones & Bartlett.

Carp, R., & Stidham, R. (1990). *Judicial process in America*. Washington, DC: Congressional Quarterly Press.

Carroll, L. (1996). Lease system. In M. D. McShane & F. P. Williams (Eds.), *Encyclopedia of American prisons* (pp. 446–452). London, England: Taylor & Francis.

Carson, E. A. (2014). *Prisoners, 2013*. Washington, DC: Bureau of Justice Statistics.

Carson, E. A. (2015). *Prisoners, 2014*. Washington, DC: Bureau of Justice Statistics.

Carson, E. A. (2018). *Prisoners in 2016*. Washington, DC: Bureau of Justice Statistics.

Carson, E. A., & Sabol, W. J. (2016). *Aging of the state prison population, 1993–2013*. Washington, DC: Bureau of Justice Statistics.

Carter, R. (1996). Determinate sentences. In M. D. McShane & F. P. Williams (Eds.), *Encyclopedia of American prisons* (pp. 237–240). London, England: Taylor & Francis.

Center for Sex Offender Management. (2008). *An overview of sex offender treatment for a non-clinical audience*. Washington, DC: Office of Justice Programs, U.S. Department of Justice.

Center for Substance Abuse Treatment. (2005). *Substance abuse treatment for adults in the criminal justice system*. Treatment Improvement Protocol (TIP) Series 44. DHHS Publication No. (SMA) 05-4056. Rockville, MD: Substance Abuse and Mental Health Services Administration.

Center on Addiction and Substance Abuse. (2010). *Substance abuse and America's prison population*. New York, NY: Columbia University.

Centers for Disease Control and Prevention. (2003). *Prevention and control of infections with hepatitis viruses in correctional settings*. Atlanta, GA: U.S. Department of Health and Human Services.

Centers for Disease Control and Prevention. (2006). *Prevention and control of tuberculosis in correctional and detention facilities: Recommendations from CDC*. Washington, DC: Author. Retrieved from http://www.cdc.gov/mmwr/preview/mmwrhtml/rr5509a1.htm

Centers for Disease Control and Prevention. (2008). *Adverse childhood experiences study*. Atlanta, GA: Author.

Centers for Disease Control and Prevention. (2011). *Evaluation of large jail STD screening programs, 2008–2009*. Washington, DC: Author. Retrieved from https://npin.cdc.gov/publication/evaluation-large-jail-std-screening-programs-2008-2009

Chokprajakchat, S., & Kuanliang, A. (2015). *Research project to formulate the Offender Rehabilitation Act of Thailand*. Bangkok, Thailand.

Cochran, J. K., Chamlin, M. B., & Seth, M. (1994). Deterrence or brutalization: An impact assessment of Oklahoma's return to capital punishment. *Criminology, 32*, 107–133.

Coker v. Georgia, 433 U.S. 584 (1977).

Colorado Division of Criminal Justice. (2007). *Evidence based correctional practices*. Denver, CO: Office of Research and Statistics.

Cooper v. Pate, 378 U.S. 546 (1964).

Cox, S. M., Allen, J. M., Hanser, R. D., & Conrad, J. J. (2011). *Juvenile justice: A guide to theory, policy, and practice* (8th ed.). Thousand Oaks, CA: Sage.

Crawford, J. (2003). Alternative sentencing necessary for female inmates with children. *Corrections Today*. Retrieved from https://www.thefreelibrary.com/Alternative+sentencing+necessary+for+female+inmates+with+children.-a0123688020

Crist, D., & Spencer, D. (1991). *Perimeter security for Minnesota correctional facilities*. St. Paul: Minnesota Department of Corrections.

Cromwell, P., del Carmen, R., & Alarid, L. (2002). *Community-based corrections* (5th ed.). Belmont, CA: Wadsworth.

Cruz, T. (2015, February 12). *Senator Cruz: Smarter Sentencing Act is common sense: Joins bipartisan group in support of legislation to reduce mandatory minimums*. Retrieved from http://www.cruz.senate.gov/?p=press_release&id=2184

Cruz v. Beto, 405 U.S. 319 (1972).

Cullen, F. T., & Agnew, R. (2006). *Criminological theory: Past to present* (3rd ed.). Los Angeles, CA: Roxbury.

Dammer, H. R. (2002). Religion in corrections. In D. Levinson (Ed.), *The encyclopedia of crime and punishment* (Vol. 3, p. 1375). Thousand Oaks, CA: Sage. Retrieved from https://www.scranton.edu/faculty/dammer/ency-religion.shtml

Davis, S. F., Palladino, J. J., & Christopherson, K. (2012). *Psychology* (7th ed.). Upper Saddle River, NJ: Prentice Hall.

Death Penalty Information Center. (2004). *Constitutionality of the death penalty in America*. Retrieved from http://deathpenaltycurriculum.org/student/c/about/history/history-5.htm

Debro, J. (2008). The future of sentencing. In P. M. Carlson & J. S. Garrett (Eds.), *Prison and jail administration: Practice and theory* (2nd ed., pp. 503–510). Sudbury, MA: Jones & Bartlett.

del Carmen, R. V., Barnhill, M. B., Bonham, G., Hignite, L., & Jermstad, T. (2001). *Civil liabilities and other legal issues for probation/parole officers and supervisors*. Washington, DC: National Institute of Corrections.

del Carmen, R. V., Ritter, S. E., & Witt, B. A. (2005). *Briefs of leading cases in corrections* (4th ed.). Cincinnati, OH: Anderson.

DeWitt, C. B. (1988). *National directory of corrections construction*. Washington, DC: National Institute of Justice.

Dezhbakhsh, H., Rubin, P. H., & Shepherd, J. M. (2003). Does capital punishment have a deterrent effect? New evidence from post-moratorium panel data. *American Law & Economics Review, 5*(2), 344–376.

Dolan, L., Kolthoff, K., Schreck, M., Smilanch, P., & Todd, R. (2003). Gender-specific treatment for clients with co-occurring disorders. *Corrections Today, 65*(6), 100–107.

Donohue, J. J., & Wolfers, J. (2006). Uses and abuse of empirical evidence in the death penalty debate. *Stanford Law Review, 58*(1), 791–846.

Dorne, C., & Gewerth, K. (1998). *American juvenile justice: Cases, legislation, and comments*. San Francisco, CA: Austin & Winfield.

Dressler, D. (1962). *Practice and theory of probation and parole*. New York, NY: Columbia University Press.

Drowns, R., & Hess, K. M. (1990). *Juvenile justice*. St. Paul, MN: West.

Duncan, M. G. (1999). *Romantic outlaws, beloved prisons: The unconscious meanings of crime and punishment*. New York. NY: New York University Press.

Ehrlich, I. (1975). The deterrent effect of capital punishment: A question of life and death. *American Economic Review, 65*(3), 397–417.

Engelbert, P. (2001, July/August). Women in prison. *Agenda*. Retrieved from http://www-personal.umich.edu/~lormand/agenda/0107/womenprison.htm

Estelle v. Gamble, 429 U.S. 97 (1976).

Etter, S. (2005). *Technology redefined: 2005 in review*. Quincy, MA: Corrections.com. Retrieved from http://www.corrections.com/news/article/6274

Fader-Towe, H., & Osher, F. C. (2015). *Improving responses to people with mental illnesses at the pretrial stage*. New York, NY: Council of State Governments.

Federal Bureau of Investigation. (2013). *Crime in the United States: Table 74: Full-time law enforcement employees*. Washington, DC: Author. Retrieved from https://www.fbi.gov/about-us/cjis/ucr/crime-in-the-u.s/2012/crime-in-the-u.s.-2012/tables/74tabledatadecoverviewpdfs/table_74_full_time_law_enforcement_employees_by_population_group_percent_male_and_female_2012.xls

Federal Bureau of Prisons. (2001). *CFR Title 28, Part 524*. Washington, DC: Author. Retrieved from https://www.law.cornell.edu/cfr/text/28/part-524

Federal Bureau of Prisons. (2010). *Quick facts about the Bureau of Prisons*. Washington, DC: U.S. Department of Justice.

Feeley, M. M., & Little, D. L. (1991). The vanishing female: The decline of women in the criminal process, 1687–1912. *Law & Society Review, 24*, 719–757.

Fleisher, M. S. (2008). Gang management. In P. M. Carlson & J. S. Garrett (Eds.), *Prison and jail administration: Practice and theory* (2nd ed., pp. 355–360). Sudbury, MA: Jones & Bartlett.

Fleisher, M. S., & Rison, R. H. (1999). Inmate work and consensual management in the Federal Bureau of Prisons. In D. van Zyl Smit & F. Dunkel, (Eds.), *Prison labour—Salvation or slavery?* Aldershot, England: Ashgate.

Fletcher, M. A. (1999, July 22). Putting more people in prison can increase crime: Study says communities suffer when too many men gone. *Washington Post*. Retrieved from http://www.sfgate.com/cgi-bin/article.cgi?f=/c/a/1999/07/22/MN90235.DTL&ao=all

Ford v. Wainwright, 477 U.S. 399 (1986).

Freedman, E. B. (1981). *Their sisters' keepers: Women's prison reform in America, 1830–1930*. Ann Arbor: University of Michigan Press.

Furman v. Georgia, 408 U.S. 238 (1972).

Gagnon v. Scarpelli 411 U.S. 778 (1973).

George, R. S. (2008). Prison architecture. In P. M. Carlson & J. S. Garrett (Ed.), *Prison and jail administration: Practice and theory* (2nd ed., pp. 39–50). Sudbury, MA: Jones & Bartlett.

Giever, D. (2006). Jails. In J. M. Pollock (Ed.), *Prisons today and tomorrow* (2nd ed.). Sudbury, MA: Jones & Bartlett.

Gillespie, W. (2002). *Prisonization: Individual and institutional factors affecting inmate conduct*. El Paso, TX: LFB Scholarly.

Glaser, D. (1964). *The effectiveness of a prison and parole system*. New York, NY: Macmillan.

Glaze, L. E., & Bonczar, T. P. (2011). *Probation and parole in the United States, 2010*. Washington, DC: U.S. Department of Justice.

Gluck, S. (1997, June). Wayward youth, super predator: An evolutionary tale of juvenile delinquency from the 1950s to the present. *Corrections Today, 59*, 62–64.

Goldberg, R. (2003). *Drugs across the spectrum* (4th ed.). Belmont, CA: Wadsworth.

Golub, A. (1990). *The termination rate of adult criminal careers*. Pittsburgh, PA: Carnegie Mellon University Press.

Government Accounting Office. (1990). *Intermediate sanctions*. Washington, DC: Author.

GovTrack.us. (2010). *110th Congress 2007–2008*. Washington, DC: Author. Retrieved from http://www.govtrack.us/congress/bill.xpd?bill=s110-3294

GovTrack.us. (2015). *S. 1410 (3rd Congress): Smarter Sentencing Act of 2014*. Washington, DC: Author. Retrieved from https://www.govtrack.us/congress/bills/113/s1410

Gramsci, A. (1996). *Prison notebooks* (J. Buttigieg, trans.). New York, NY: Columbia University Press.

Greenblatt, A. (2014, July 18). *Drug sentencing guidelines reduced for current prisoners*. National Public Radio. Retrieved from http://www.npr.org/blogs/thetwo-way/2014/07/18/332619083/drug-sentencing-guidelines-reduced-for-current-prisoners

Gregg v. Georgia. 428 U.S. 153 (1976).

Hagstrom, A. (2018). MS-13 members nearly as dangerous in prison as they are on the street. *The Daily Caller*. Retrieved from http://dailycaller.com/2018/02/05/ms-13-prison/

Hanser, R. D. (2002). Inmate suicide in prisons: An analysis of legal liability under Section 1983. *The Prison Journal, 82*(4), 459–477.

Hanser, R. D. (2007). *Special needs offenders in the community*. Upper Saddle River, NJ: Prentice Hall.

Hanser, R. D. (2010a). Adrian Raine: Crime as a disorder. In F. T. Cullen & P. Wilcox (Eds.), *Encyclopedia of criminological theory*. Thousand Oaks, CA: Sage.

Hanser, R. D. (2010b). *Community corrections*. Thousand Oaks, CA: Sage.

Hanser, R. D. (2015). Using local law enforcement to enhance immigration law in the United States: A legal and social analysis. *Police Practice and Research: An International Journal, 16*(4), 303–315.

Hanser, R. D. (2018). Reentry and reintegration of adult special populations: Community involvement, police partnerships, and reentry councils. In K. Dodson (Ed.), *Routledge handbook on offenders with special needs* (pp. 485–499). London, England: Routledge.

Hanser, R. D. (2019). *Essentials of community corrections*. Thousand Oaks, CA: Sage.

Hanser, R. D., & Mire, S. (2010). *Correctional counseling*. Upper Saddle River, NJ: Pearson/Prentice Hall.

Hanson, G. R., Venturelli, P. J., & Fleckenstein, A. E. (2006). *Drugs and society* (9th ed.). Burlington, MA: Jones & Bartlett.

Hanson, G. R., Venturelli, P. J., & Fleckenstein, A. E. (2011). *Drugs and society* (11th ed.). Burlington, MA: Jones & Bartlett.

Harlow, C. W. (1999). *Prior abuse reported by inmates and probationers*. Washington, DC: Bureau of Justice Statistics.

Harrison, P. M., & Karberg, J. C. (2004). *Prison and jail inmates at midyear 2003*. Washington, DC: Bureau of Justice Statistics.

Hatton, J. (2006). *Betsy: The dramatic biography of prison reformer Elizabeth Fry*. Grand Rapids, MI: Kregel.

Hayes, L. M. (2010). *National study of jail suicide, 20 years later*. Washington, DC: National Institute of Corrections.

Herberman, E. J., & Bonczar, T. P. (2015). *Probation and parole in the United States, 2013*. Washington, DC: National Institute of Corrections.

Hercik, J. M. (2007). *Prisoner reentry, religion, and research*. Washington, DC: U.S. Department of Health and Human Services.

Hochstellar, A., & DeLisa, M. (2005). Importation, deprivation, and varieties of serving time: An integrated-lifestyle-exposure model of prison offending. *Journal of Criminal Justice, 33*(3), 257–266.

Hockenberry, S. (2013). *Juveniles in residential placement, 2010*. Washington, DC: National Center for Juvenile Justice.

Hockenberry, S. (2016). *Juveniles in residential placement, 2013*. Washington, DC: Office of Juvenile Delinquency and Prevention.

Hoffman, P. B. (2003). *History of the federal parole system*. Washington, DC: U.S. Parole Commission.

Hsia, H. M. (2004). *Disproportionate minority confinement 2002 update*. Washington, DC: Office of Juvenile Justice and Delinquency Prevention.

Hudson v. Palmer, 468 U.S. 517 (1984).

Huizinga, D., Thornberry, T., Knight, K., & Lovegrove, P. (2007). *Disproportionate minority contact in the juvenile justice system: A study of differential minority arrest/referral to court in three cities*. Washington, DC: U.S. Department of Justice.

Hutto v. Finney, 437 U.S. 678 (1978).

Immigration and Customs Enforcement. (2011). *2009 Immigration detention reforms*. Washington, DC: Department of Homeland Security. Retrieved from http://www.ice.gov/factsheets/2009detention-reform

Inciardi, J. A., Rivers, J. E., & McBride, D. C. (2008). *Drug treatment*. In P. M. Carslon & J. S. Garrett (Eds.), *Prison and jail administration: Practice and theory* (2nd ed., pp. 403–412). Sudbury, MA: Jones & Bartlett.

Ingram, G. L., & Carlson, P. M. (2008). Sex offenders. In P. M. Carlson & J. S. Garrett (Eds.), *Prison and jail administration* (2nd ed., pp. 373–382). Gaithersburg, MD: Aspen.

International Association for Chiefs of Police. (2008). *Tracking sex offenders with modern technology: Implications and practical uses with law enforcement*. Alexandria, VA: Author.

James, D. J., & Glaze, L. E. (2001). *Mental health problems of prison and jail inmates*. Washington, DC: Bureau of Justice Statistics.

James, D. J., & Glaze, L. E. (2006). *Mental health problems of prison and jail inmates*. Washington, DC: U.S. Department of Justice.

Johnson v. Avery, 393 U.S. 483 (1969).

Johnson, B. R. (2004). Religious programs and recidivism among former inmates in prison fellowship programs: A long-term follow-up study. *Justice Quarterly, 21*(2), 329–354.

Johnson, B. R. (2012, January). Can a faith-based prison reduce recidivism? *Corrections Today*, 60–62.

Johnson, B. R., & Larson, D. B. (2003). *The InnerChange Freedom Initiative: A preliminary evaluation of America's first faith-based prison*. Retrieved from http://www.baylor.edu/content/services/document.php/25903.pdf

Johnson, H. A., Wolfe, N., & Jones, M. (2008). *History of criminal justice* (4th ed.). Southington, CT: Anderson.

Johnson, L. B. (2008). Food service. In P. M. Carlson & J. S. Garrett (Eds.), *Prison and jail administration: Practice and theory* (2nd ed., pp. 149–158). Sudbury, MA: Jones & Bartlett.

Johnson, R., & Dobrzanska, A. (2005). *Life with the possibility of life: Mature coping among life-sentence prisoners*. Paper presented at the annual meeting of the American Society of Criminology, Royal York, Toronto, Canada. Retrieved from http://citation.allacademic.com/meta/p31858_index.html

Johnson, S. C. (1999). Mental health services in a correctional setting. In P. M. Carlson & J. S. Garrett (Eds.), *Prison and jail administration: Practice and theory* (2nd ed., pp. 107–116). Sudbury, MA: Jones & Bartlett.

Johnston, N. (2009). *Prison reform in Pennsylvania*. Philadelphia: Pennsylvania Prison Society.

Jones, J. (2007). *Pre-release planning and re-entry process: Addendum 02*. Tulsa: Oklahoma Department of Corrections.

Justice Policy Institute. (2008). *Substance abuse treatment and public safety*. Washington, DC: Author.

Kaeble, D. (2018). *Probation and parole in the United States, 2016*. Washington, DC: Bureau of Justice Statistics.

Kaeble, D., & Bonczar, T. P. (2017). *Probation and parole in the United States, 2015*. Washington, DC: Bureau of Justice Statistics.

Karberg, J. C., & James, D. J. (2005). *Substance dependence, abuse, and treatment of jail inmates, 2002*. Washington, DC: U.S. Department of Justice.

Karpowitz, D., & Kenner, M. (2001). *Education as crime prevention: The case for reinstating Pell Grant eligibility for the incarcerated*. Annandale-on-Hudson, NY: Bard College.

Katz, L., Levitt, S. D., & Shustorovich, E. (2003). Prison condition, capital punishment, and deterrence. *American Law & Economics Review, 5*(2), 318–343.

Kauffman, K. (1988). *Prison officers and their world*. Cambridge, MA: Harvard University Press.

Kerle, K. (1982). Rural jail: Its people, problems and solutions. In D. Shanler (Ed.), *Criminal justice in rural America* (pp. 189–204). Washington, DC: National Institute of Justice.

Kerle, K. (1999). Short term institutions at the local level. In P. M. Carlson & J. S. Garrett (Eds.), *Prison and jail administration: Practice and theory* (pp. 59–65). Gaithersburg, MD: Aspen.

King, E., & Baker, M. (2014). *Respectful classification practices with LGBTI inmates: Trainer's manual*. New York: New York State Department of Corrections and Community Supervision.

Kohen, A., & Jolly, S. K. (2006). *Deterrence reconsidered: A theoretical and empirical case against the death penalty*. Paper presented at the annual meeting of the Midwest Political Science Association, Chicago, IL. Retrieved from http://citeseerx.ist.psu.edu/viewdoc/download?doi=10.1.1.585.590&rep=rep1&type=pdf

Kuhles, B. (2013). *SHSU tests prison education programs*. Huntsville, TX: Sam Houston State University. Retrieved from http://www.shsu.edu/~pin_www/T@S/2013/prisoned.html

Kurshan, N. (1991). *Women and imprisonment in the United States: History and current reality*. Baltimore: Monkeywrench Press.

Kurshan, N. (1996). *Women and imprisonment in the United States: History and current reality*. Retrieved from http://www.freedomarchives.org/Documents/Finder/DOC3_scans/3.kurshan.women.imprisonment.pdf

Latessa, E. J., & Allen, H. E. (1999). *Corrections in the community* (2nd ed.). Cincinnati, OH: Anderson.

Levinson, D. (Ed.). (2002). *The encyclopedia of crime and punishment* (Vol. 3). Thousand Oaks, CA: Sage.

Lewis, K. R., Lewis, L. S., & Garby, T. M. (2012). Surviving the trenches: The personal impact of the job on probation officers. *American Journal of Criminal Justice, 38*, 67–84.

Lilly, J. R., Cullen, F. T., & Ball, R. A. (2014). *Criminological theory: Context and consequences* (6th ed.). Thousand Oaks, CA: Sage.

Lindemuth, A. L. (2007). Designing therapeutic environments for inmates and prison staff in the United States: Precedents and contemporary applications. *Journal of Mediterranean Ecology, 8*, 87–97.

Linder, D. (2005). *A history of witchcraft persecutions before Salem*. Retrieved from http://law2.umkc.edu/faculty/projects/Ftrials/salem/witchhistory.html

Lindner, C. (2006). John Augustus, father of probation, and the anonymous letter. *Federal Probation, 70*(1), 150–165.

Liptak, A. (2011, May 31). Justices, 5–4, tell California to cut prison population. *New York Times*. Retrieved from http://www.nytimes.com/2011/05/24/us/24scotus.html

Little Hoover Commission. (2004). *Breaking the barriers for women on parole*. Retrieved from https://lhc.ca.gov/sites/lhc.ca.gov/files/Reports/177/Report177.pdf

Lofstrom, M., & Martin, B. (2017). *California's county jails*. San Francisco: Public Policy Institute of California.

Luan, L. (2018). *Profiting from enforcement: The role of private prisons in U.S. immigration detention*. Washington, DC: Migration Policy Institute. Retrieved from https://www.migrationpolicy.org/article/profiting-enforcement-role-private-prisons-us-immigration-detention

MacCormick, A. (1931). *The education of adult prisoners: A survey and a program* [Reprint 1976]. New York, NY: AMS Press.

Madrid v. Gomez, 889 F. Supp. 1146 (1995).

Mann, M. (2016). *How RFID technology can help corrections facilities. Corrections one*. Retrieved from https://www.correctionsone

.com/careers/articles/188622187-How-RFID-technology-can-help-corrections-facilities/

Mann, S. C. (2017). The death penalty: A longitudinal study of national violent crime rates from a rational choice perspective. *American Journal of Social Sciences, 2,* A29–A59.

Martin, C. (2016). Determining the deterrent impact of the death penalty. *Susquehanna University Political Review, 7*(2).

Martinson, R. (1974). What works? Questions and answers about prison reform. *Public Interest, 35.*

Maruschak, L. M. (2006). *Medical problems of jail inmates.* Washington, DC: U.S. Department of Justice, Office of Justice Programs.

McKeiver v. Pennsylvania, 403 U.S. 528 (1971).

McNeece, C. A., Springer, D. W., & Arnold, E. M. (2002). Treating substance abuse disorders. In J. B. Ashford, B. D. Sales, & W. H. Reid (Eds.), *Treating adult and juvenile offenders with special needs* (pp. 131–170). Washington, DC: American Psychological Association.

McNeil, D. E., Binder, R. L., & Robinson, J. C. (2005). Incarceration associated with homelessness, mental disorder, and co-occurring substance abuse. *Psychiatric Services, 56,* 840–846.

McShane, M. D. (1996a). Chain gangs. In M. D. McShane & F. P. Williams (Eds.), *Encyclopedia of American prisons* (pp. 144–117). London, England: Taylor & Francis.

McShane, M. D. (1996b). Historical background. In M. D. McShane & F. P. Williams (Eds.), *Encyclopedia of American prisons* (pp. 455–457). London, England: Taylor & Francis.

Meko, J. A. (2008). A day in the life of a warden. In P. M. Carlson & J. S. Garrett (Eds.), *Prison and jail administration: Practice and theory* (2nd ed., pp. 235–242). Sudbury, MA: Jones & Bartlett.

Merton, R. K. (1938). Social structure and anomie. *American Sociological Review, 3,* 672–682.

Messina, N., & Grella, C. (2006). Childhood trauma and women's health: A California prison population. *American Journal of Public Health, 96*(10), 1842–1848.

Messner, S. F., & Rosenfeld, R. (2001). *Crime and the American dream* (3rd ed.). Belmont, CA: Wadsworth.

Michigan Department of Corrections. (2003). *Michigan presentence investigation.* Lansing, MI: Author.

Miller v. Alabama 132 S. Ct. 2455 (2012).

Minton, T. D., & Golinelli, D. (2014). *Jail inmates at midyear 2013—statistical tables.* Washington, DC: Bureau of Justice Statistics.

Minton, T. D., & Zeng, Z. (2015). *Jail inmates at midyear 2014—statistical tables.* Washington, DC: Bureau of Justice Statistics.

Mitchell, M. (2011). Texas prison boom going bust. *Star-Telegram.*

Mobley, A. (2011). Garbage in, garbage out? Convict criminology, the convict code, and participatory prison reform. In M. Maguire & D. Okada (Eds.), *Critical issues in crime and justice: Thought, policy, and practice* (pp. 333–349). Thousand Oaks, CA: Sage.

Mocan, H. N., & Gittings, R. K. (2003). Getting off death row: Commuted sentences and the deterrent effect of capital punishment. *Journal of Law and Economics, 46*(2), 283–322.

Montana Department of Corrections. (2010). *Recreation programs DOC 5.5.3.* Helena, MT: Author.

Moore, E. O. (1981). A prison environment's effect on health care service demands. *Environmental Systems, 11,* 17–34.

Morris v. Travisono, 310 F. Supp. 857 (1970).

Morrissey v. Brewer 408 U.S. 471 (1972).

Morton, J. B. (1992). *An administrative overview of the older inmate.* Washington, DC: National Institute of Corrections.

Morton, J. B. (2005, October 1). *ACA and women working in corrections.* Washington, DC: Corrections Today.

Myers, P. L., & Salt, N. R. (2000). *Becoming an addictions counselor: A comprehensive text.* Burlington, MA: Jones & Bartlett.

National Advisory Commission on Criminal Justice Standards and Goals. (1973). *Corrections.* Washington, DC: Government Printing Office.

National Center for Women and Policing. (2003). *Hiring and retaining more women: The advantages to law enforcement agencies.* Beverly Hills, CA: Author.

National Commission on Correctional Health Care. (2002). *The health status of soon-to-be-released inmates: A report to Congress.* Chicago, IL: Author.

National Gang Center. (2013). *National youth gang survey analysis.* Washington, DC: Author. Retrieved from http://www.nationalgangcenter.gov/Survey-Analysis

National Gang Intelligence Center. (2009). *National gang threat assessment 2009: Prison gangs.* Washington, DC: Author.

National Institute of Corrections. (1986). *Protective custody: Data update and intervention considerations.* Washington, DC: Author.

National Institute of Corrections. (1993). *The intermediate sanctions handbook: Experiences and tools for policymakers.* Washington, DC: Author.

National Institute of Corrections. (2001). *NIC research on small jail issues: Summary findings.* Washington, DC: U.S. Department of Justice.

National Institute of Justice. (1998). *Restorative justice: An interview with visiting fellow Thomas Quinn.* Washington, DC: U.S. Department of Justice.

National Institute of Justice. (2005). *Implementing evidence based practice in corrections.* Washington, DC: U.S. Department of Justice.

National Institute of Justice. (2012). *Challenges of conducting research in prisons.* Washington, DC: Author. Retrieved from http://www.nij.gov/journals/269/pages/research-in-prisons.aspx

National Institute of Justice. (2013a). *Racial profiling.* Washington, DC: U.S. Department of Justice.

National Institute of Justice (2013b). *Racial profiling and traffic stops.* Washington, DC: U.S. Department of Justice.

National Institute of Justice. (2016). *Five things about deterrence.* Washington, DC: U.S. Department of Justice. Retrieved from https://nij.gov/five-things/Pages/deterrence.aspx

National PREA Resource Center. (2015). *Prison Rape Elimination Act.* Washington, DC: Bureau of Justice Assistance.

Neubauer, D. W. (2019). *Courts and the criminal justice system* (11th ed.). Belmont, CA: Wadsworth.

New York Correction History Society. (2008). *The nation's first reformatory: Elmira.* Retrieved from http://www.correctionhistory.org/index.html

Noonan, M. E., & Ginder, S. (2014). *Mortality in local jails and state prisons, 2000–2012.* Washington, DC: Bureau of Justice Statistics.

Nunez-Neto, B. (2008). *Offender reentry: Correctional statistics, reintegration into the community, and recidivism.* Washington, DC: Congressional Research Service.

Office of Juvenile Justice and Delinquency Prevention. (2016). *Juvenile residential facility census 2016* [machine-readable data files]. Washington, DC: Author.

Office of Juvenile Justice and Delinquency Prevention. (2017, June 1). *Statistical briefing book.* Retrieved from https://www.ojjdp.gov/ojstatbb/corrections/qa08211.asp?qaDate=2015

Office of National Drug Control Policy. (2001). *Fact sheet: Drug treatment in the criminal justice system.* Washington, DC: Author.

Office of Program Policy Analysis and Governmental Accountability. (2010). *Intermediate sanctions for non-violent offenders could produce savings*. Tallahassee, FL: Author.

O'Lone v. Estate of Shabazz, 482 U.S. 342 (1987).

Olson, L. (2004). An exploration of therapeutic recreation in adult federal medical centers and Wisconsin correctional facilities. *UW-L Journal of Undergraduate Research, VII*, 1–3.

Osher, F., D'Amora, D. A., Plotkin, M., Jarrett, N., & Eggleston, A. (2012). *Adults with behavioral health needs under correctional supervision: A shared framework for reducing recidivism and promoting recovery*. Washington, DC: National Institute of Corrections.

Owen, B., Pollock, J., Wells, J., & Leahy, J. (2015). *Critical issues impacting women in the justice system: A literature review*. Washington, DC: National Institute of Corrections.

Palmigiano v. Garrahy, 443 F. Supp. 956 (1977).

Philadelphia Prison System. (2010). *About PPS: History of the Philadelphia Prison System*. Retrieved from http://legacy.phila .gov/prisons/inmatelocator/index.htm

Philips, D. E. (2001). *Legendary Connecticut*. Willimantic, CT: Curbstone Press.

Pollak, O. (1950). *The criminality of women*. Philadelphia: University of Pennsylvania Press.

Pollock, J. (1986). *Sex and supervision: Guarding male and female inmates*. New York, NY: Greenwood Press.

Pollock, J. M. (2014). *Women's crimes, criminology, and corrections*. Long Grove, IL: Waveland Press.

Procunier v. Martinez, 416 U.S. 396 (1974).

Rafter, N. H. (1985). *Partial justice: Women in state prisons 1800–1935*. Boston, MA: New England University Press.

Redding, H. (2004). *The components of prison security*. Naples, FL: International Foundation for Protection Officers. Retrieved from https://www.ifpo.org/resource-links/articles-and-reports/ protection-of-specific-environments/the-components-of- prison-security/

Riveland, C. (1999). *Supermax prisons: Overview and general considerations*. Washington, DC: National Institute of Corrections.

Robbins, S. P. (2005). *Organizational behavior* (12th ed.). Upper Saddle River, NJ: Prentice Hall.

Roper v. Simmons. 543 U.S. 551 (2005).

Ross, P. H., & Lawrence, J. E. (2002). Healthcare for women offenders: Challenge for the new century. In R. L. Gido & T. Alleman (Eds.), *Turnstile justice: Issues in American corrections* (pp. 73–88). Englewood Cliffs, NJ: Prentice Hall.

Roth, M. P. (2006). Chain gangs. In M. P. Roth, *Prisons and prison systems: A global encyclopedia* (pp. 56–57). Westport, CT: Greenwood Press.

Roth, M. P. (2011). *Crime and punishment: A history of the criminal justice system*. Belmont, CA: Cengage Learning.

Ruddell, R., & Mays, G. L. (2007). Rural jails: Problematic inmates, overcrowded cells, and cash-strapped counties. *Journal of Criminal Justice, 35*, 251–260.

Ruiz v. Estelle, 503 F. Supp. 1265 (S.D. Tex. 1980).

Sabol, W. J., & Minton, T. D. (2008). *Jail inmates at midyear 2007*. Washington, DC: Bureau of Justice Statistics.

Samuel, B. (2001). *Elizabeth Gurney Fry (1780–1845): Quaker prison reformer*. Retrieved from http://www.quakerinfo.com/fry.shtml

Schriro, D. (2009). Is good time a good idea? A practitioner's perspective. *The Federal Sentencing Reporter, 21*(3), 181.

Schlossman, S., & Spillane, J. (1995). *Bright hopes, dim realities: Vocational innovation in American correctional education*. Berkeley, CA: National Center for Research in Vocational Education.

Schuster, T. (2015). *PREA and LGBTI rights*. Hagerstown, MD: American Jail Association.

Sedlak, A. J., & McPherson, K. S. (2010). *Youth's needs and services: Findings from the survey of youth in residential placement*. Washington, DC: Office of Juvenile Justice and Delinquency Prevention.

Sellin, T. (1959). *The death penalty: A report for the model penal code project of the American Law Institute*. Philadelphia, PA: American Law Institute.

Sellin, T. (1970). The origin of the Pennsylvania system of prison discipline. *Prison Journal, 50*(Spring/Summer), 15–17.

Shepherd, J. M. (2004). Murders of passion, execution delays, and the deterrence of capital punishment. *Journal of Legal Studies, 33*(2), 283–322.

Shepherd, J. M. (2005). Deterrence versus brutalization: Capital punishment's differing impacts among States. *Michigan Law Review, 104*(2), 203–256.

Sickmund, M., & Puzzanchera, C. (2014). *Juvenile offenders and victims, 2014 national report*. Washington, DC: National Center for Juvenile Justice.

Sieh, E. W. (2006). *Community corrections and human dignity*. Sudbury, MA: Jones & Bartlett.

Silverman, I. J. (2001). *Corrections: A comprehensive view* (2nd ed.). Belmont, CA: Wadsworth.

Solomon, L., & Baird, S. C. (1982). Classification: Past failures, future potential. In L. Fowler (Ed.), *Classification as a management tool: Theories and model for decision-makers*. College Park, MD: American Correctional Association.

Stanko, S., Gillespie, W., & Crews, G. (2004). *Living in prison: A history of the correctional system with an insider's view*. Westport, CT: Greenwood Press.

Stepp, E. A. (2008). Emergency management. In P. M. Carlson & J. S. Garrett (Eds.), *Prison and jail administration: Practice and theory* (2nd ed., pp. 469–478). Sudbury, MA: Jones & Bartlett.

Stohr, M., & Walsh, A. (2011). *Corrections: The essentials*. Thousand Oaks, CA: Sage.

Stohr, M., Walsh, A., & Hemmens, C. (Eds.). (2009). *Corrections: A text/ reader*. Thousand Oaks, CA: Sage.

Stohr, M., Walsh, A., & Hemmens, C. (Eds.). (2013). *Corrections: A text/ reader* (2nd ed.). Thousand Oaks, CA: Sage.

Sykes, G. M. (1958). *The pain of imprisonment*. Princeton, NJ: Princeton University Press.

Texas Department of Criminal Justice. (2010). *Organizational charts: Correctional Institutions Division*.

Thompson v. Oklahoma, 487 U.S. 815 (1988).

Torres, S. (2005). Parole. In R. A. Wright & J. M. Mitchell (Eds.), *Encyclopedia of criminology*. New York, NY: Routledge.

Treatment Advocacy Center. (2014). *The treatment of persons with mental illness in prisons and jails: A state survey*. Arlington, VA: Author.

Trop v. Dulles, 356 U.S. 86 (1958).

Turner v. Safley, 482 U.S. 78 (1987).

United States v. Booker, 543 U.S. 220 (2005).

U.S. Census Bureau. (2017). *Quick facts*. Washington, DC: U.S. Department of Commerce. Retrieved from https://www .census.gov/quickfacts/fact/map/US/INC110216

U.S. Commission on Civil Rights. (2015). *With liberty and justice for all: The state of civil rights at immigration detention facilities*. Washington, DC: Author.

U.S. Congress. (2008). *United States Parole Commission Extension Act of 2008*. Washington, DC: Author. Retrieved from https:// www.govtrack.us/congress/bills/110/s3294/text

U.S. Department of Justice. (2006). *Commonly asked questions about the Americans with Disabilities Act and law enforcement.* Washington, DC: Disability Rights Section. Retrieved from http://www.ada.gov/q&a_law.htm

U.S. Department of Justice. (2012). *Prison Rape Elimination Act: Juvenile facility standards* (28 C.F.R Part 115). Washington, DC: Author.

U.S. Department of Justice. (2016). *Federal interagency reentry council.* Washington, DC: Author. Retrieved from https://www.justice.gov/reentry/federal-interagency-reentry-council

U.S. Department of Justice, Civil Rights Division, Disability Rights Section. (2010). *Justice Department's 2010 ADA standards for accessible design go into effect.* Washington, DC: Author.

U.S. Sentencing Commission. (2007). *2007 federal guidelines manual.* Washington, DC: Author. Retrieved from https://www.ussc.gov/guidelines/archive/2007-federal-sentencing-guidelines-manual

U.S. Sentencing Commission. (2014a, April 12). *News release: U.S. Sentencing Commission votes to reduce drug trafficking sentences.* Washington, DC: Author.

U.S. Sentencing Commission. (2014b). *Quick facts: Women in the federal offender population.* Washington, DC: Author.

Van Baalen, S. M. (2008). Religious programming. In P. M. Carlson & J. S. Garrett (Eds.), *Prison and jail administration: Practice and theory* (2nd ed., pp. 127–138). Sudbury, MA: Jones & Bartlett.

Van Keulen, C. (1988). *Colorado alternative sentencing programs: Program guidelines.* Washington, DC: National Institute of Corrections. Retrieved from http://www.nicic.org/pubs/pre/007064.pdf

Vesely, R. (2004). *California rebuked on female inmates.* Women's E-News. Retrieved from https://womensenews.org/2004/12/california-rebuked-female-inmates/

Vitek v. Jones, 445 U.S. 480 (1980).

Volkow, N. D. (2006, August 19). Treat the addict, cut the crime rate [Editorial]. *Washington Post*, p. A17.

Wallenstein, A. (2014). American jails: Dramatic changes in public policy. In P. M. Carlson (Ed.), *Prison and jail administration: Practice and theory* (3rd ed., pp. 11–26). Burlington, MA: Jones & Bartlett.

Wallenstein, A., & Kerle, K. (2008). American jails. In P. M. Carlson & J. S. Garrett (Eds.), *Prison and jail administration: Practice and Theory* (2nd ed., pp. 19–38). Sudbury, MA: Jones & Bartlett.

Ward, D. A. (1994). Alcatraz and Marion: Confinement in super-maximum custody. In J. W. Roberts (Ed.), *Escaping prison myths: Selected topics in the history of federal corrections* (pp. 81–94). Washington, DC: American University Press.

White House. (2016). *Fact sheet: President Obama announces new actions to reduce recidivism and promote reintegration of formerly incarcerated individuals.* Washington, DC: White House Office the Press Secretary. Retrieved from https://obamawhitehouse.archives.gov/the-press-office/2016/06/24/fact-sheet-president-obama-announces-new-actions-reduce-recidivism-and

Whitten, L. (2013). *Probation officers' stress and burnout associated with caseload events.* Washington, DC: National Institute of Corrections. Retrieved from https://nicic.gov/probation-officers-stress-and-burnout-associated-caseload-events-2013

Wiliszowski, C. H., Fell, J. C., McKnight, A. S., Tippetts, A. S., & Ciccel, J. D. (2010). An evaluation of three intensive supervision programs for serious DWI offenders. *Association for the Advancement of Automotive Medicine, 54*(1), 375–387.

Wilson v. Seiter, 501 U.S. 294 (1991).

Wintersteen, M. B., Diamond, G. S., & Fein, J. A. (2007). Screening for suicide risk in the pediatric emergency and acute care setting. *Current Opinion in Pediatrics, 19*(4), 398–404.

Witherspoon v. Illinois, 391 U.S. 510 (1968).

Wolf, S. (2012). Mara Salvatrucha: The most dangerous street gang in the Americas? *Latin American Politics and Society, 54*(1), 65–99.

Wolff v. McDonnell, 418 U.S. 539 (1974).

Women's Prison Association. (2003). *A portrait of women in prison.* New York, NY: Author.

Wooldredge, J. (1996). American Correctional Association. In M. D. McShane & F. P. Williams (Eds.), *Encyclopedia of American prisons* (pp. 45–52). London, England: Taylor & Francis.

Wright, R. A. (1994). *In defense of prisons.* Westport, CT: Greenwood Press.

Zeng, Z. (2018). *Jail inmates in 2016.* Washington, DC: U.S. Department of Justice.

Zimmerman, P. R. (2004). State executions, deterrence and the incidence of murder. *Journal of Applied Economics, 7*(2), 163–193.

INDEX

ABOUT THE AUTHOR

Robert D. Hanser is a full professor and chair of the criminal justice program at the University of Louisiana at Monroe. Rob has also administered a regional training academy in northeastern Louisiana (North Delta Regional Training Academy) that provides training to correctional officers, jailers, and law enforcement throughout a 12-parish region in Louisiana. Rob was also the first male president for the Board of Directors of the Louisiana Coalition Against Domestic Violence (LCADV), which demonstrates an understanding of victim needs and services as well as offender rehabilitation. Rob is the program director of the Blue Walters Substance Abuse Treatment Program at Richwood Correctional Center, a prison-based substance abuse treatment program in Louisiana, and he is the director of Offender Programming for LaSalle Corrections. Further, Rob is the director and lead facilitator for the Fourth Judicial District's Batterer Intervention Program. He serves as the board president for Freedmen Inc., a faith-based organization that provides reentry services for offenders in Louisiana. Lastly, Rob currently holds a gubernatorial appointment on the statewide Louisiana Reentry Council, which ensures that regional efforts are aligned with statewide initiatives. He has dual licensure as a professional counselor in Texas and Louisiana, is a certified anger resolution therapist, and has a specialty license in addictions counseling.